Polydiacetylenes

NATO ASI Series

Advanced Science Institutes Series

A Series presenting the results of activities sponsored by the NATO Science Committee, which aims at the dissemination of advanced scientific and technological knowledge, with a view to strengthening links between scientific communities.

The Series is published by an international board of publishers in conjunction with the NATO Scientific Affairs Division

A	**Life Sciences**	Plenum Publishing Corporation
B	**Physics**	London and New York
C	**Mathematical and Physical Sciences**	D. Reidel Publishing Company Dordrecht and Boston
D	**Behavioural and Social Sciences**	Martinus Nijhoff Publishers Dordrecht/Boston/Lancaster
E	**Applied Sciences**	
F	**Computer and Systems Sciences**	Springer-Verlag Berlin/Heidelberg/New York
G	**Ecological Sciences**	

Series E: Applied Sciences – No. 102

Polydiacetylenes

Synthesis, Structure and Electronic Properties

Edited by

D. Bloor

Queen Mary College
London, UK

R.R. Chance

Allied Corporation
Morristown, USA

1985 **Martinus Nijhoff Publishers**
Dordrecht / Boston / Lancaster
Published in cooperation with NATO Scientific Affairs Division

Proceedings of the NATO Advanced Research Workshop on Polydiacetylenes, Stratford-upon-Avon, UK, September 12-17, 1984

Library of Congress Cataloging in Publication Data

ISBN 90-247-3251-4 (this volume)
ISBN 90-247-2689-1 (series)

Distributors for the United States and Canada: Kluwer Boston, Inc., 190 Old Derby Street, Hingham, MA 02043, USA

Distributors for the UK and Ireland: Kluwer Academic Publishers, MTP Press Ltd, Falcon House, Queen Square, Lancaster LA1 1RN, UK

Distributors for all other countries: Kluwer Academic Publishers Group, Distribution Center, P.O. Box 322, 3300 AH Dordrecht, The Netherlands

Printed in The Netherlands

PREFACE

The polydiacetylenes are a class of polymers that are attracting increasing attention worldwide. There are many reasons for this interest one of the most important being the availability of many polydiacetylenes as macroscopic, high quality, single crystals. This fact was first reported in the pioneering work of Prof. G. Wegner on the solid-state polymerization of disubstituted diacetylenes in the late 1960s. Since then studies of the polymerization process and the properties of monomers and polymers have advanced understanding of solid-state reactivity and the physics of quasi-one-dimensional materials. More recently work on soluble polydiacetylenes, gels and films has been of interest for both academic and technological reasons.

Progress in this area has required a combination of research disciplines ranging from synthetic organic chemistry to solid-state physics. The interdisciplinary effort required for successful research in polydiacetylenes was reflected in the mix of chemists, physicists and materials scientists who attended the Workshop. The emerging potential for commercial applications of polydiacetylenes was also evident in the nearly equal participation of academic and industrial/government scientists.

The Workshop was the first major international meeting to focus solely on polydiacetylenes. It provided a forum in which problems of mutual interest could be discussed by scientists with diverse backgrounds and interests. It also satisfied the need for a review of the science of these materials at a time when this basic understanding is leading to technological applications.

Recent rapid advances, and consequent lively controversies, brought out in the main presentations and the working group sessions. The first focus of interest was the polymerization process. Work at Stuttgart and Bayreuth has provided details of the process at the molecular level but only for a limited number of monomers. More data is required for the formation of a proper overview and as input to macroscopic theories. Several such models exist but none is currently compatible with all the available kinetic data.

A vigorous debate centred on the behaviour of soluble polymers, in particular whether the solvato- and thermo-chromic effects observed in solutions are due to a conformational change in isolated chains or an aggregation phenomenon. The possibility of the occurrence of cis-conformations in random coils in good solvent also attracted attention. Both topics will be the subject of active debate for some time.

Electrical conductivity is a further subject under discussion. The view tht the drift velocity saturates at the lowest applied field, implying a very high initial mobility is gaining favour but it was felt that there was a need for further experimental corroboration of this hypothesis. The interpretation of optical and Raman data in terms of an exciton model appears to be well established. The observation of photo-induced absorption at Bell Labs. suggests the presence of polaronic states but alternatively the effect may be due to defects. The preparation of Langmuir-Blodgett multilayer thin films, mechanical properties and applications to electronic, optical and time-temperature indicating devices were less controversial topics. The commercialisation of time-temperature indicators by the Allied Corp. is exciting and will bring polydiacetylenes to the attention of a wider community.

In addition to the main talks there were twenty contributions highlighting recent progress. It was generally agreed that the Workshop has provided a timely opportunity to review this rapidly expanding area at a critical stage of its development. It is to be hoped that synergistic interactions will have been fostered among the participants in the Workshop.

This volume brings together the bulk of the invited and contributed papers and the summaries of the working group sessions.

The abbreviation PDA is widely used to indicate polydiacetylenes in general. There is, however, less agreement on abbreviations to describe particular monomers and polymers. These are usually identified by their pendant groups. While many of the PDAs mentioned in this volume are specifically identified by chemical formula and abbreviation some, more commonly studied PDAs, are identified by an abbreviation only. The table below gives formulae and abbreviations for these PDAs, note symmetric substitution is assumed unless specified.

Chemical Structure	Abbreviations
$-\ CH_2OSO_2C_6H_4CH_3$	TS, PTS, TSHD, TS-6
$-\ CH_2OSO_2C_6H_4F$	FBS, PFBS
$-\ (CH_2)_4OSO_2C_6H_4CH_3$	TS-12
$-\ CH_2NC_{12}H_8$	DCH, DCHD
$-\ CH_2OCONHC_2H_5$	ETU, EUHD
$-\ CH_2OCONHC_6H_5$	PU, PUHD
$-\ (CH_2)_4OCONHC_2H_5$	ETCD
$-\ (CH_2)_4OCONHC_6H_5$	TCDU
$-\ (CH_2)_4OCONHCHCH_3$	IPUDO
$-\ (CH_2)_nOCONHCH_2COOC_4H_9$	nBCMU

$- C_6H_4NHCOCH_3$	DADD
$- (CH_2)_2\ OH$	OD
$- CH_2CO_2(CH_2)_8CH_3$	CAP
$- CH_3;\ - CH_2OH$	1OH
$- (CH_2)_mCH_3;\ -(CH_2)_nCOOH$	m/n AFA
$- C_6H_4OCO(CH_2)_3OCOC_6H_4$-(Cyclic)	BPG

We wish to thank the members of the organising committee (Gerhard Wegner, Michel Schott and Alan Heeger) for their assistance. The success of the Workshop was due to the efforts of all the participants both in their individual contributions and in the discussions of the working groups. The Workshop was made possible by generous support of the NATO Scientific Affairs Division and contributions from the Allied Corporation, GTE Laboratories Inc., ICI plc., GEC Research Laboratories (Hirst Research Centre) and BDH Chemicals Ltd..

D. Bloor, R.R. Chance

CONTENTS

LIST OF PARTICIPANTS

Dr. J-P. Aime,
Group de Physique des Solides,
L'Ecole Normale Superieure,
Tour 23,
Place Jussieu,
75251 Paris Cedex 05,
France.

Dr. P.A. Albouy,
Laboratoire de Physique des Solides,
Universite de Paris-Sud,
Batiment 510,
91405 Orsay,
France

Dr. S. Allen,
ICI plc,
Electronics Group,
P.O. Box 11,
Runcorn,
Cheshire WA7 4QE,
UK.

Dr. N. Balcioglu,
Hacettepe University,
Department of Chemistry,
Organic Chemistry Section,
Ankara,
Turkey.

Prof. Dr. H. Bassler,
Phillips-Universitat Marburg,
FB 14,
Hans-Meerwein-Strasse,
3550 Marburg,
W.Germany.

Dr. D.N. Batchelder,
Dept. of Physics,
Queen Mary College,
Mile End Road,
London E1 4NS
UK

Prof. R.H. Baughman,
Allied Corporation,
P.O. Box 1021R,
Morristown,
New Jersey 07960,
USA.

Prof. D. Bloor,
Dept. of Physics,
Queen Mary College,
Mile End Road,
London E1 4NS
UK.

Dr. V.P. Bodart,
Facultes Universite Notre-Dame de la Paix-Namur,
Dept. de Chimie,
rue de Bruxelles, 61,
B-5000 Namur,
Belgium.

Dr. R.R. Chance,
Allied Corporation,
P.O. Box 1021R,
Morristown,
New Jersey 07960,
USA.

Prof. D. Chapman,
Royal Free Hospital,
School of Medicine,
Dept. of Biochemistry,
Roland Hill Street,
London NW3 2PF
UK.

Dr. K. Donovan,
Dept. of Physics,
Queen Mary College,
Mile End Road,
London E1 4NS
UK.

Miss M. Drury,
B.P. Research Centre,
Chertsey Road,
Sunbury-on-Thames,
Middlessex TW16 7LN,
UK.

Dr. M. Dudley,
Dept. of Pure and Applied
Chemistry,
Unversity of Strathclyde,
295 Cathedral Street,
Glasgow G1 1X
UK.

Dr. H. Eckhardt,
Allied Corporation,
P.O. Box 1021R,
Morristown,
New Jersey,
USA.

Dr. V. Enkelmann,
Albert-Ludwigs-Universitat,
Institut fur Makromoleculare
Chemie,
Hermann-Staudinger-Haus,
Stefan-Meier-Strasse 31,
7800 Freiburg i Br.,
W.Germany.

Dr. M. Goodyear,
BDH Chemicals Ltd.,
Broom Road,
Poole,
Dorset BN12 4NN
UK.

Dr. W. Guttler,
Universitat Bayreuth,
Lehrstuhl fur Experimental
Physik II,
Postfach 3008,
8580 Bayreuth,
W.Germany

Dr. F. Kajzar,
DEIN/LERA,
CEN Saclay,
91191 Gif sur Yvette Cedex,
France.

Dr. A. Karpfen,
Institut fur Theoretische
Chemie,
Universitat Wien,
Wahringer Strasse 17,
A-1090 Wien,
Austria.

Dr. H. Kiess,
Laboratories RCA Ltd.,
Badenerstrasse 569,
8048 Zurich,
Switzerland.

Prof. J.B. Lando,
Dept. of Macromolecular
Science,
Case Western Reserve University,
Cleveland,
Ohio 44106,
USA.

Dr. I. Ledoux,
CNET,
Paris B,
196, rue de Paris,
92220 Bagneaux,
France.

Dr. S. Mann,
Marconi Research Centre,
West Hanningfield Road,
Great Baddow,
Chelmsford,
Essex CM2 8HN,
UK.

Dr. J. Messier,
DEIN/LERA,
CEN Saclay,
91191 Gif sur Yvette Cedex,
France.

Dr. P.C. Morgan,
British Aerospace plc,
Dynamics Group,
P.O. Box 5,
Filton,
Bristol BS12 7QW,
UK.

Dr. B. Movaghar,
GEC Research Laboratories,
Hirst Research Centre,
East Lane,
Wembley,
Middlesex HA9 7PP,
UK.

Prof. E. Mulazzi,
Universita Degli Studi di Milano,
Departimento di Fisica,
via Celoria 16,
20133 Milano,
Italy.

Dr. J. Orenstein,
Bell Laboratories,
Murray Hill,
New Jersey 07974,
USA.

Dr. I.R. Peterson,
GEC Research Laboratories,
Hirst Research Centre,
East Lane,
Wembley,
Middlesex HA9 7PP,
U.K.

Dr. R.G. Priest,
Naval Research Laboratories,
Washington,
D.C. 20375,
USA.

Dr. T. Prusik,
Allied Corporation,
P.O. Box 1021R,
Morristown,
New Jersey 07960,
USA.

Dr. D.J. Sandman,
GTE Laboratories,
Waltham,
Massachusetts,
USA

Prof. M. Schott,
Groupe de Physique des Solides,
L'Ecole Normale Superieure,
Tour 23,
Place Jussieu,
75251 Paris Cedex 05,
France.

Dr. R. Sethi,
Plessey Research (Caswell) Ltd.,
Allen Clarke Research Centre,
Caswell, Towcester,
Northants NN12 8EQ,
UK.

Prof. R. Silbey,
Department of Chemistry,
MIT,
Cambridge,
Massachusetts 02139,
USA.

Prof. Dr. H. Sixl,
Universitat Stuttgart,
Physicalisches Institut,
Teil 3,
Pfaffenwaldring 57,
D-7000 Stuttgart 80,
W.Germany.

Dr. S. Suhai,
German Cancer Research
Centre,
Im Neunheimer Feld 280,
6900 Heidelberg,
W.Germany.

Dr. S.K. Tripathy,
GTE Laboratories,
Waltham,
Massachusetts 02254,
USA.

Prof. Dr. G. Wegner,
Albert-Ludwigs-Universitat,
Institute fur Makromoleculare
Chemie,
Hermann-Staudinger-Haus,
Stefan-Meier-Strasse 31,
7800 Freiburg i Br.,
W.Germany.

Prof. Dr. G. Weiser,
Phillips Universitat Marburg,
FB Physik,
Renthof 5,
D-3550 Marburg,
W.Germany.

Dr. E.G. Wilson,
Department of Physics,
Queen Mary College,
Mile End Road,
London El 4NS,
U.K.

Prof. F. Wudl,
Institute for Polymers and
Organic Solids,
University of California-SB,
Santa Barbara,
California 93106,
USA.

Dr. J. Yardley,
Allied Corporation,
P.O. Box 1021R,
Morristown,
New Jersey 07960,
USA.

Dr. R.J. Young,
Dept. of Materials,
Queen Mary College,
Mile End Road,
London El 4NS,
UK.

DIACETYLENE POLYMERIZATION KINETICS

D. Bloor

Department of Physics, Queen Mary College, London E1 4NS, U.K.

ABSTRACT

Theoretical and experimental studies of the polymerization kinetics of disubstituted diacetylenes are reviewed. The problem of relating kinetics measured by macroscopic methods to the microscopic processes occuring in the solid-state is discussed. Areas of controversy and those deserving more detailed study are identified.

1. INTRODUCTION

Polydiacetylenes are unusual since the most interesting samples, those with single crystal or highly aligned fibrous morphologies, are obtained by solid-state rather than solution polymerization. This contrasts with most commercial polymers for which the reverse is true. Furthermore diacetylene polymerization is a topochemical reaction which places severe constraints on the lattice packing of the monomers and their intermolecular interactions, factors which determine monomer reactivity and the morphology of the product polymer. These strictures are the reason for the relative importance of solution and solid-state rections in synthetic chemistry. In solution reacting molecules sample a vast range of reaction pathways and rules for selecting the most likely product are well established. In the solid-state one, or a few, molecular conformations occur with well defined molecular contact, which for large molecules cannot be predicted accurately. Thus solid-state chemistry has been a hit-or-miss subject. The advent of rapid crystal structure determination with automated four-circle x-ray diffractometers is allowing progress to be made in the prediction of solid-state reactions. Studies of diacetylenes are making a valuable contribution to progress in this area (1-3).

We are not concerned, here, however, with details of the crystal structures, it suffices to say that if the separation of the reacting carbon atoms, C(1) and C(4'), of Fig. 1, is less than 4A a solid state reaction is probable, in agreement with the criterion proposed by Schmidt. Nor do we want to discuss the restrictions that the requirement of an unique polymerization direction places on the symmetry of monomer and polymer crystals. These topics are in the literature (5-7). Here we concentrate on the reaction kinetics of reactive monomers.

In view of the comments above it is not surprising that solution polymerization dominates most discussions of polymerization kinetics, e.g. in ref. 8 the extensive discussion includes four sentences on solid-state polymerization, which is dismissed as 'only of academic interest'. The application of diacetylenes as time-temperature indicators seems likely to change this. The basic ideas of conventional discussions are, however, relevant even in the solid state. The polymerization is viewed as three distinct processes, (a) initiation, (b) propagation and (c) termination of the polymer chain. The details of these processes depend on the type of polymerization reaction involved. The polymerization kinetics are then deduced by solving the rate equations for the three processes. In order to do this it is usual to assume steady-state conditions, when the population of propagating polymer chains remains constant as a result of an exact balance between the rates of initiation and termination (8).

Under steady state conditions the reaction rate at constant temperature is given by:

$$R = -\frac{\partial[M]}{\partial t} = f([I],[M]) \qquad (1)$$

where [I] and [M] are concentrations of initiator and monomer respectively. For a unimolecular reaction, such as a radical polymerization, R is proportional to [M] and the reaction is termed first order. Then integrating equation 1 gives:

$$[M] = [M_o] \exp(-kt) \qquad (2)$$

where $[M_o]$ is the initial concentration and k is the rate constant. The effect of temperature is to vary the rate constant according to an Arrhenius expression

$$k = A \exp(-E_a/kT). \qquad (3)$$

This assumes that the transition state is reached by activation over a barrier of height E_a, the barrier crossing being attempted at a frequency A. In general A and E_a will be some functions of the process occurring during polymerization. For radical polymerization the total activation energy is

$$E_a = E_p + E_i/2 - E_t/2 \qquad (4)$$

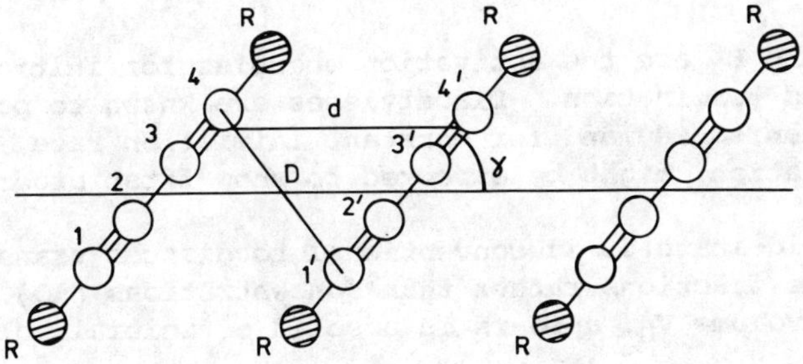

Figure 1. Packing of diacetylene monomers in a reactive array with D < 4Å.

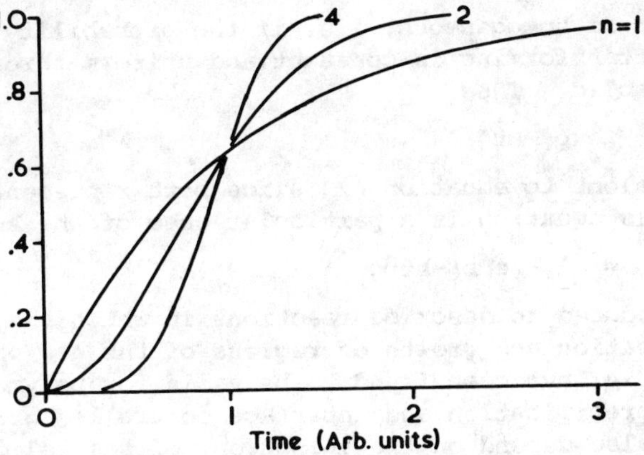

Figure 2. Kinetic curves calculated from the general Avrami equation for different values of the exponent n.

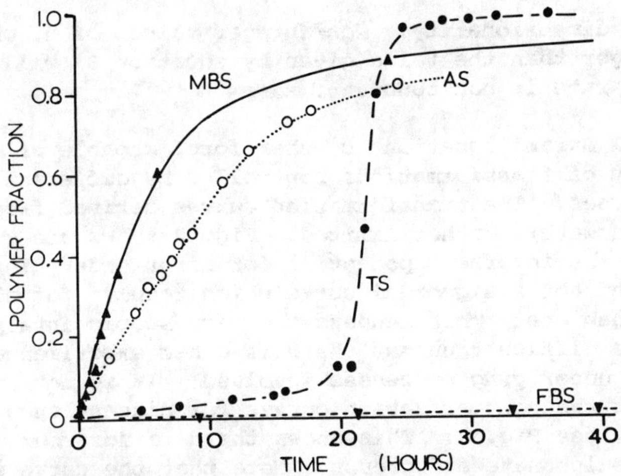

Figure 3. Data for the thermal polymerization of four diacetylenes with end groups $R = -CH_2OSO_2-C_6H_4-X$ where X = -F (FBS), $-CH_3$(TS), $-OCH_3$(MBS) and $-N{=}N-C_6H_5$(ATS). Data for TS, MBS from (11), FBS from (12) and ATS from (13).

where E_i, E_p and E_t are the activation energies for initiation, propagation and termination. Diacetylenes are known to polymerize by a radical mechanism (9) and for constant initiation rate, e.g. radiation polymerization, might be expected to show first order kinetics.

In the solid-state it is conventional to discuss transformations in terms of volume fractions rather than concentrations (10). Thus a new phase, with a volume V_n, appears in a solid of initial volume V at a rate:

$$\frac{dV_n}{dt} = k\,(V-V_n) \qquad (5)$$

if the reaction is homogeneous, i.e. if the probability of an elementary volume transforming is constant and uniform throughout the untransformed solid. Thus

$$V_n/V = 1 - \exp(-kt) \qquad (6)$$

which is equivalent to equation (2) since both represent first order processes. This equation is a particular case of the Avrami equation

$$X = V_n/V = 1 - \exp(-kt^n) \qquad (7)$$

which was introduced to describe reactions in which the transformation occurs by nucleation and growth of regions of the new phase. Values of n from 1 to >4 have been found to be valid for polymorphic changes, discontinuous precipitation and interface controlled growth in solids (10). n will also depend on the anisotropy of the solid, for constant nucleation rate of the new phase:

$$n = d + 1 \qquad (8)$$

where d is the dimensionality. Non-integer values of n, which may be smaller or larger than the value given by equation 8, will occur if the nucleation rate is not constant.

The general Avrami equation is, therefore, capable of describing a great variety of transformations controlled by different underlying physical processes. The transformation curves derived from this equation are, however, rather limited. Fig. 2 shows the two forms generated, (a) the inverse exponential for first order kinetics (equation 6) and (b) a sigmoidal curve which results for all values of n greater than one. This renders the microscopic interpretation of kinetic data difficult unless there is other experimental evidence concerning the underlying processes involved. It is not surprising, therefore, that the polymerization curves for diacetylenes display the same forms, see Fig. 3. This shows the data for four monomers with modified sulphonate sidegroups. Note that the curve for the fluorine terminated end group is sigmoidal but that the induction period is approximately 80 hours at 60°C.

Fig. 3 illustrates another interesting fact that though the packing of the monomer units is very similar in the four compounds, the reaction kinetics are markedly different in both form and rate. It is interesting to note that the reaction from diacetylene monomer to polydiacetylene (single–triple–single–double bond alternation) is photochemically allowed but thermally forbidden if the Woodward-Hoffmann (WH) rules are rigidly applied (14). Such apparent contradictions are known elsewhere (15) and strict application of the WH rules must be treated with caution, particularly for concerted reactions such as diacetylene polymerization. Before returning to consider the experimental data in more detail, we will first outline the models of diacetylene polymerization which have been presented in the literature.

2. KINETIC MODELS

Simple thermodynamic and structural arguments lead to the conclusion that a topotactic solid-state polymerization can proceed either in a single phase or lead to phase separation (6,16,17). In the single phase case the ideal situation is a homogeneous reaction with the monomer randomly nucleated in the polymer. This will be possible if the interfacial energy between polymer and monomer is low (18). If the interfacial energy is high nucleation is favoured in the vicinity of preformed chains. At a high enough energy phase separation must occur but there will be intermediate cases where either a single phase or a phase segregated reaction can occur, depending on the relative kinetics. Alternatively if lattice strain increases during polymerization but there is no phase transition, the reaction can terminate at an intermediate conversion. The following discussion will be restricted to the cases of an ideal, single phase polymerization and polymerization with phase segregation.

For a homogeneously polymerizing material the polymer chains will, in general, be subject to elastic strain because of mismatch between the length of the polymer repeat unit and the separation of monomer units in the polymerization direction. An exact match of monomer and polymer dimensions as originally assumed for diacetylenes by Hirschfeld and Schmidt (19) is a rare occurrence and is now known not to be essential for the production of single crystals of polymer. The model of polymerization kinetics controlled by lattice strain has been developed by Baughman (20) and Baughman and Chance (21). They assumed (a) the presence of a reservoir of active precursor monomers capable of initiating polymerization, (b) thermal generation of precursors at a constant rate at all stages of the polymerization and (c) deactivation, without initiating polymerization, at a constant rate. This latter rate was assumed to be much greater than the competing rate for polymer chain initiation. Finally it was assumed that the propagating chains have a conversion independent lifetime, if propagation is not terminated by a defect or a previously polymerized monomer.

The polymerization rate, if the reaction is far from equilibrium, is

$$\frac{dX}{dt} = B\,Y(1-X)\exp(-F_I/kT) \tag{9}$$

where X is the fraction of polymer present, B is a constant for a given monomer and reaction conditions, F_I is the activation energy for initiation and Y the mean polymer chain length. Both F_I and Y are dependent on X. Y will be reduced by the presence of previously formed polymer but is likely to increase as the propagation rate changes during polymerization. The propagation rate will be given by:-

$$k_p = (Z_o/\tau)\exp(-F_p/kT) \tag{10}$$

where Z_o is the number of monomer units polymerized at X = 0, τ is the propagation life-time and the activation energy for propagation, F_p is also a function of X. At any conversion X the propagating chains add Z ($=k_p\tau$) monomers unless they encounter defects or pre-polymerized sites. Thus if the probability of the site next to a propagating chain end being an unreacted monomer is S then

$$Y = 2\sum_{n=0}^{Z} S^n \quad \text{and} \quad S = (1-X)/W+1-X \tag{11}$$

where W is the fraction of lattice sites occupied by defects or polymer. Assuming some initial concentration of defect sites W_o equations 9-11 can be solved iteratively for any specified X dependence of F_I and F_p.

The elastic behaviour of the partially polymerized lattice is model dependent. The two simplest cases are when (a) the two components are subjected to equal strain and (b) the two components are subjected to equal stress (22). Case (a), the Voight limit, corresponds to components coupled in parallel, i.e. long polymer chains, and (b), the Reuss limit, corresponds to components coupled in series, i.e. short polymer chains. Baughmann (20) used case (a) to calculate the lattice strain.

The local strain approximation is used to calculate the conversion dependence of F_I and F_p as shown in Fig. 4. Monomer units are subjected to a force f(X) due to lattice strain and initiation and propagation produce identical local lattice distortions. The crystal is assumed to relax to its new average strain after the reaction barrier has been crossed locally. Then:

$$F_I(X) = F_p(X) = (d_m-d_p)^2\,E_m\left[\left[1+\frac{X}{1-X}\frac{E_p}{E_m}\right]^{-2} -1\right] \tag{12}$$

where d_m and d_p are the initial monomer separation and the final polymer unit size and E_m and E_p are the elastic modulii of monomer and polymer in the polymerization direction.

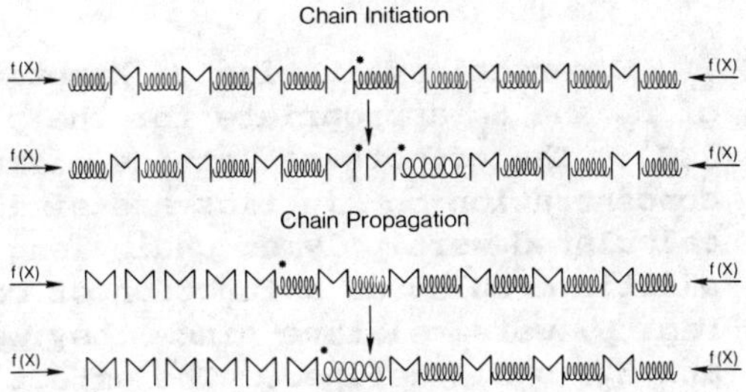

Figure 4.
Schematic representention of chain initiation and propagation showing the local deformation at the reaction site. (After 20).

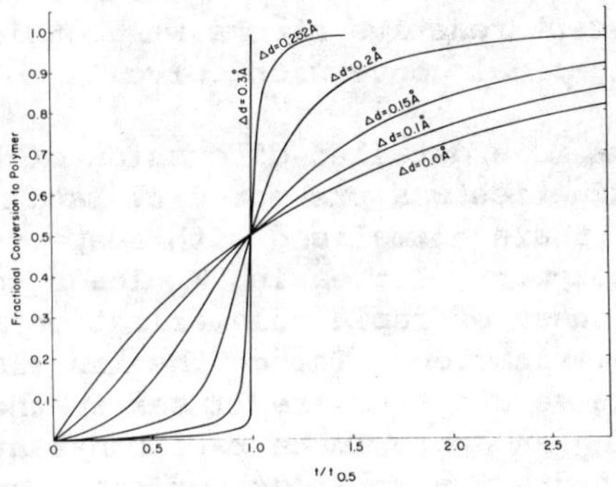

Figure 5.
Effect of changes in lattice strain on reaction kinetics, elastic constant appropriate for TS. (After 20).

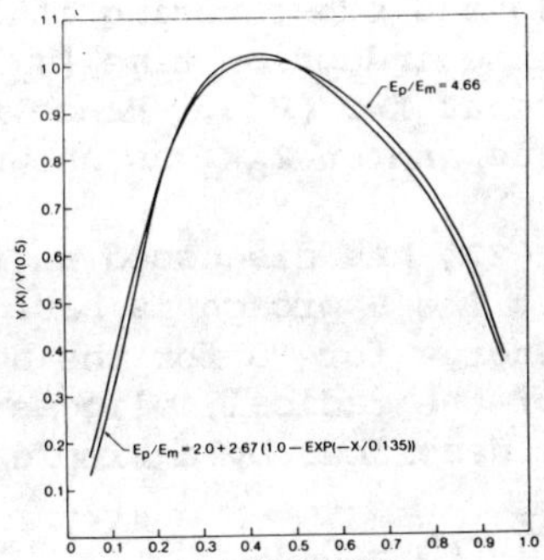

Figure 6.
Chain length variation predicted for TS by the crystal strain model, note values of Y(X) are relative to the value at 50% conversion. (After 20).

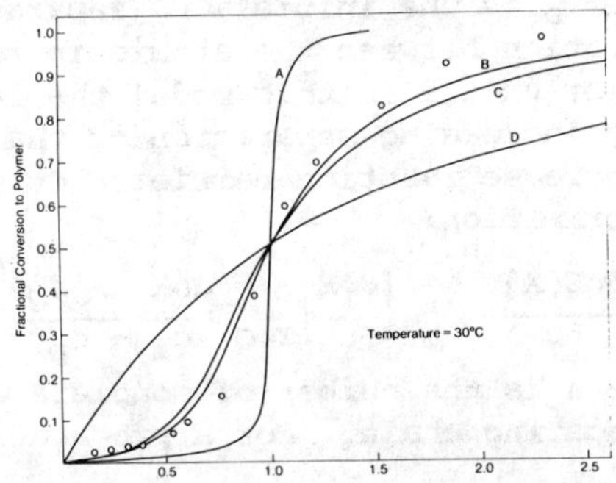

Figure 7.
Calculated kinetics for A, F_I and F_P dependent on X; B, F_P independent of X; C, F_I independent of X and D, F_P and F_I independent of X. The circles are experimental data for the γ-radiation polymerization of TS. (After 20).

Numerical evaluation of kinetic curves was performed using values of E_m and E_p appropriate for the para-toluene sulphonate monomer (TS) (20). The effects of lattice mismatch, Fig. 5, and blocking defect concentration on kinetics are similar, Fig. 5. Other parameters calculated were polymer chain length and effective polymer chain elastic modulus as a function of conversion. The values of Y(X), Fig. 6, were relative since they were calculated in the limit $Z_oW_o \ll 1$ and Z_o is ill defined. The effect of making either F_I or F_p or both independent of X is shown in Fig. 7. F_I constant, i.e. a constant initiation rate, should be appropriate for radiation polymerization. For either F_I or F_p constant the kinestics become first order as expected. The case d_m=d_p with interactions between monomers in adjacent reactive stacks was considered, strong interactions lead to a sigmoidal conversion curve.

A more detailed discussion of the effect of chain blocking defects on kinetics was presented by Baughman and Chance (21). The curves of Fig. 5 are normalised with respect to the time to reach 50% conversion to polymer. Increasing W_o leads to a slowing of the reaction, delaying the onset of rapid polymerization and reducing the maximum rate of polymerization. Though the kinetics are not first order plots of log (1-X) versus time are linear in the region of rapid polymerization giving an approximate rate constant, K. The induction time can be defined a the intercept of the linear section of such plots with the time axis, see Fig. 11. By developing a functional form for Y (equation 11) and substituting it in equation 9 it was shown that K and the inverse induction time (τ_i^{-1}) are both proportional to the reaction rate at X=O (R_o). Hence the product $K\tau_i$ is independent of R_o but does depend on Z_oW_o as shown in Fig. 8.

Bassler (23) has discussed an alternative model for polymer chain propagation. The approach is based on the experimental fact that the activation energy for TS for the addition of a monomer increases as the length of the radical, oligomer intermediates increases (24). This data is described by a simple linear expression:

$$E = \frac{\delta \Delta E}{\delta n} (D - D_o) \tag{13}$$

where D_o is the initial C-C separation, cf. Fig. 1, and D is the separation between the atomic in reacting chain end and the next monomer unit. In this model the lattice mismatch leads to a gradually increasing separation of chain end and adjacent monomer and an increase reaction barrier. At conversion X using the Voight approximation:

$$\frac{\delta \Delta E(X)}{\delta n} = \left[\frac{\delta \Delta E}{\delta n}\right]_{X=O} \frac{d(X) - d_p}{d_m - d_p} \tag{14}$$

where n is the number of monomers already incorporated into the propagating chain. For a propagation time τ n can be evaluated from

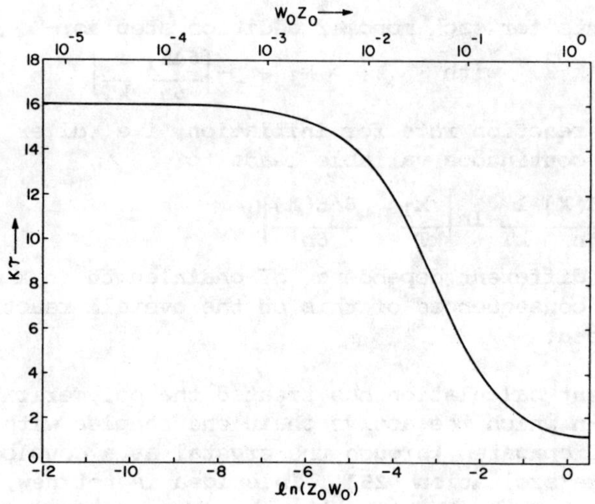

Figure 8. Variation of $K\tau_i$ with defect concentration for the case of TS. (After 21).

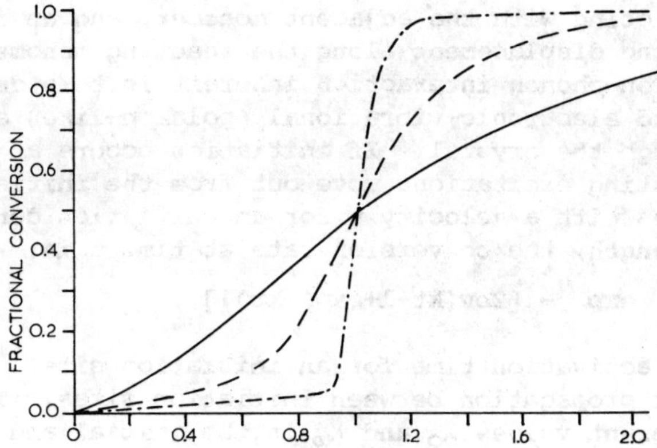

Figure 9. Kinetic curves calculated for a bi-phase polymerization for rate constant ratios of 3, 30 and 250 (solid, dash and chain curves). (Ref. 2).

the rate constants for each monomer addition step as:-

$$n = \tau \sum_{i=1}^{n} k_i \quad \text{with} \quad k_i \triangleq k_1 \exp-\left[\frac{\delta\Delta E_i}{\delta n}\frac{1}{kT}\right] \tag{15}$$

where k_1 is the reaction rate for initiation, i.e. dimer formation. Treating i as a continuous variable leads to:

$$n\ (X) = \frac{\delta E(X)}{\delta n}\frac{1}{kT} \ln\left[\frac{\tau k_1}{kT} \cdot \frac{\delta\Delta E(X)}{\delta n}\right] \tag{16}$$

This leads to a different dependence of chainlength on X than that of Fig. 6. The consequences of this on the overall reaction kinetics were not discussed.

Another recent calculation has treated the polymerization as a chain reaction in which the active chain end couples with lattice vibrations and propagates through the crystal as a coupled polaron (or soliton) like excitation (25). This idea is not new, the concept of a coupled exciton-vibration has been discussed previously (26,27) but not fully developed. With current interest in the application of non-linear wave equations to polymers the re-emergence of this idea is not surprising. The polymerization rate is determined from the transition from an initial to a final electronic state of the crystal, the energies being dependent on the nuclear configurations, through a transition state in which the energies are equal. Departure from the equilibrium states is expressed in terms of vibrational oscillators. Two modes are considered, one a bending of the triple bond, which at large amplitude produces the non-bonding p-orbital capable of interacting with the adjacent monomer, and an intermolecular mode involving displacement along the reacting monomer stack. The strong electron-phonon interaction inherent in this description leads to a coupled electronic-vibrational (polaron-like) excitation propagating through the crystal. If initiation occurs at random and the chain propagating excitations move out from the initiation site in both directions with a velocity v for an initiation density of ρ per unit chain length, the conversion rate at time t is:

$$1 - \left[\frac{\partial X}{\partial t}\right]_t = \exp - [2\rho v\{kt-1+\exp(-kt)\}] \tag{17}$$

where k^{-1} is the activation time for an initiation site. $(\rho v)^{-1}$ is then the time for propagation between initiation sites. It is assumed that ρv has different values λ_o and λ_∞ in the initial and fast polymerization regimes of a sigmoidal conversion curve, and that the switch from one regime to the other occurs at time t_c. Then:

$$1 - \left[\frac{\partial X}{\partial t}\right]_t = \exp\ (-\lambda_o t_c - \lambda_\infty k(t-t_c)^2\ H(t-t_c)) \tag{18}$$

where $H(t-t_c)$ is the Heaviside step function. As we will see below all three models have been shown to reproduce the experimental data for the monomer TS.

Little attention has been devoted to polymerization which either leads to phase segregation or induces a phase transition in the monomer during polymerization. Kinetic curves can be constructed by assuming that two phases occur and that polymerization proceeds by a first order process in both phases but with different rate constants (2). The form of the kinetic curve depends on the nature of the transition between the phases. In general the initial phase will exist up to some conversion X_1, both phases will exist between X_1 and X_2 and the final phase exists above X_2. Special cases can be modelled by appropriate choice of X_1, X_2 and the rate constants. Conversion curves calculated in this way are shown in Fig. 9. The sigmoidal forms are similar to those deduced from the models described above reinforcing the requirement for additional data to confirm the correct choice of model.

3. EXPERIMENTAL DATA

3.1 Toluene sulphonate

The monomer which has been most extensively studied is TS. Numerous studies of thermal, photo and γ-ray polymerization kinetics have been undertaken, typical data appears in Fig. 3 for thermal and in Fig. 7 for for γ-ray polymerization. Further description of the measurements can be found in refs. (28-38). Parallel studies of other physical properties have shown that the polymerization of TS is best described as a homogeneous single phase reaction (2, 39-44).

Baughman (20) used the Voight model to derive the ratio of the elastic modulii for monomer and polymer and obtained a good fit for the thermal and γ-irradiation-polymerization kinetics using the crystal strain theory, see Fig. 7 and 10. The polymerization has been effected by uv radiation (32) and chlorine (33). Uv polymerization is restricted by the penetration depth of the actinic radiation; an analysis using the crystal strain model has been reported (32). Polymerization with chlorine is complicated by the occurrence of chemical degradation of the polymer and the kinetic data has not been analysed in detail.

Independent measurements of the elastic modulii give conflicting results. Brillouin scattering measurements gave an elastic constant ratio $E_p/E_m \sim 4.5$ (42,43) close to the value, 4.66, deduced by Baughman (20). Recently, however, ultrasonic studies (45) have given a much larger value of 6.5. This value is in reasonable agreement with the observed variation in thermal expansion coefficient during polymerization (2). If the larger value is correct then the validity of the crystal strain model for TS must be questioned.

Drastic changes in reaction kinetics are observed to occur for (a) polymerization of TS crystals under hydrostatic pressure (37,38),

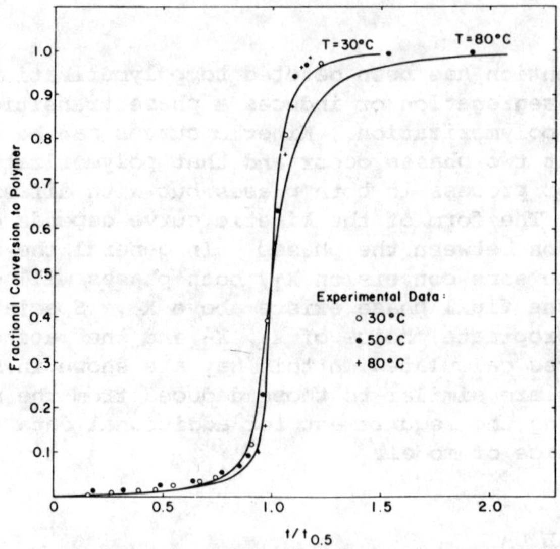

Figure 10. Kinetic data for TS with predicted curves deduced from the crystal strain model. (After 20).

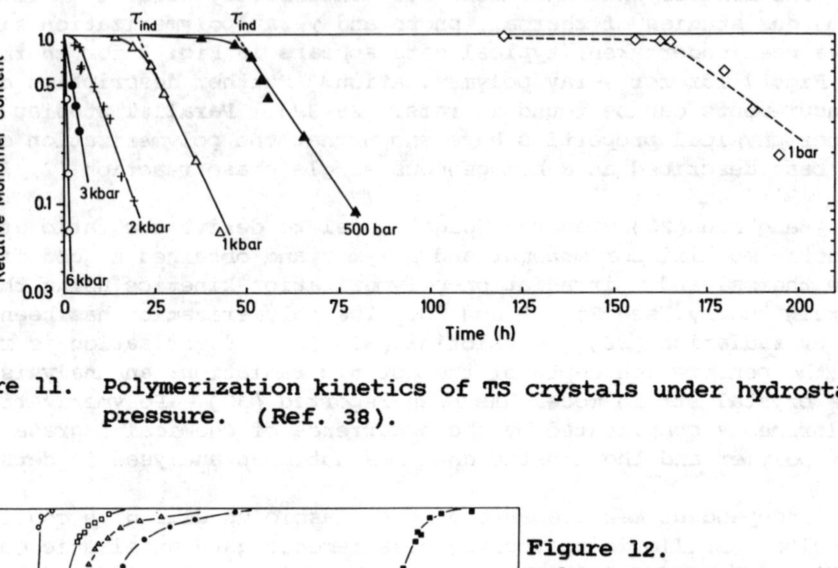

Figure 11. Polymerization kinetics of TS crystals under hydrostatic pressure. (Ref. 38).

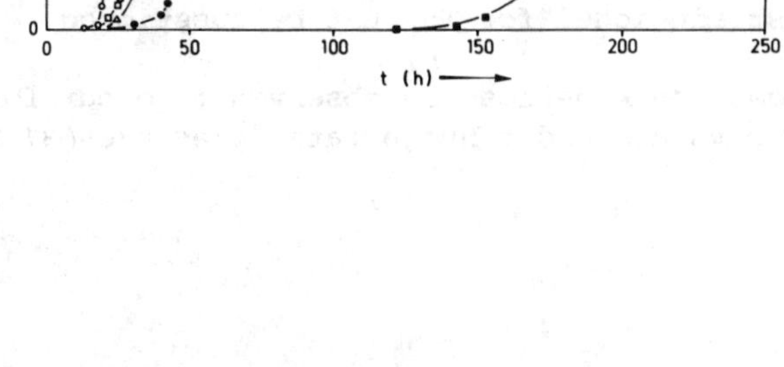

Figure 12. Polymerization kinetics for mixed TS/FBS crystals at 60°C for concentrations of CBS of 0.01 to 0.62 increasing from left to right. (Ref. 48).

see Fig. 11, (b) an isotopic substitution (46), (c) for mixed crystals of TS and closely related monomers where the terminal methyl moiety of the end groups are replaced by a methoxy or a halide (47-49), see Fig. 12 and, (d) pure crystals of related monomers with either isomorphous or similar crystal structures, Fig. 3.

Baughman and Chance (21) attributed these effects to either variations in defect densities or the increased importance of defects as external influences, e.g. pressure, produce a significant increase in the propagation length of the growing chains. The dependence of kinetics on defect concentration has been demonstrated for TS samples of different purity (35) so that this proposition is not unreasonable. For pure TS crystals the observed suppression of the induction period by pressure can be explained by this approach, in the other cases the arguments appear less convincing. The data for isotopically substituted material requires a specific change of defect density relative to pure TS and the subtle effects of deuteration of different sites are difficult to account for. For mixed crystals the defect concentration is likely to increase leading to a reduction in polymerization rate but experimentally both reduction and acceleration have been observed. The kinetics illustrated in Fig. 3 differ dramatically despite very similar lattice packing in the three cases. Furthermore the least perfect crystals are those of MBS and yet the reaction rate is highest. The crystal strain model does not, of course, address the problem of initiation rate changes which could produce all of the effects catalogued above. The initiation process must involve channelling of energy into the excited precursor molecule. Little attention has been paid to studying this complex process.

The crystal strain model predicts a particular variation in chain length during polymerization, Fig. 3. Since TS polymer is insoluble at all except the lowest conversion levels (50) it is not possible to measure chain lengths directly. Even at low conversion accurate molecular weights were not deduced (50) and the solutions are metastable (51) rendering reliable measurements difficult. Attempts to determine chain lengths directly have involved electron microscopy of crystals photopolymerized by closely contacted masks (52,53) and production of holographic gratings (54,55). Both address photopolymerization rather than thermal polymerization. The former method is insensitive indicating a maximum chain length of 4 μm, 8000 repeat units, but is rendered unreliable by scattering and possible secondary nucleation. An initial chain length of 0.4 μm was indicated. A similar value of 0.6 μm was deduced from the holographic studies, and of 0.2 μm by electron beam irradiation (60). The ratio of initial to maximum length is similar to that predicted by the crystal strain model.

Diffuse x-ray scattering has been used to determine the variation in chain length at low conversion (40,56). Above 10% conversion the

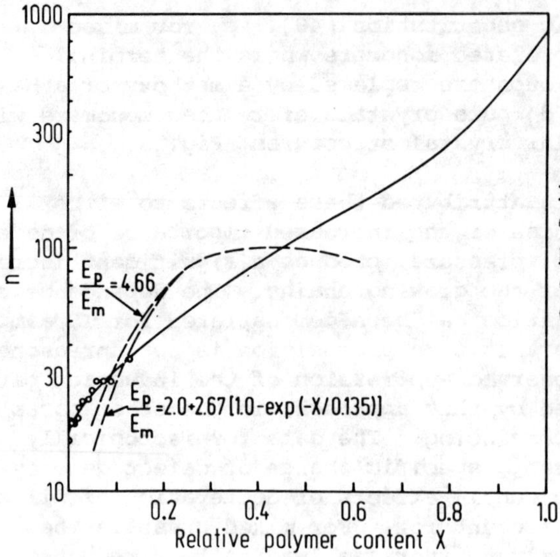

Figure 13.
Comparison of the theoretical predictions of refs. (20) and (23) with the data of ref. 40. (Ref. 23).

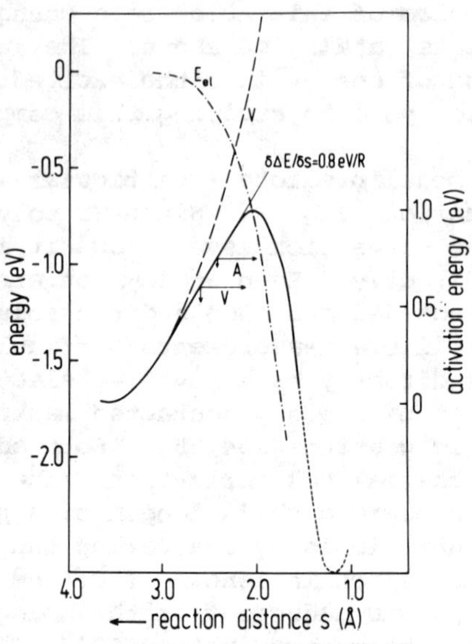

Figure 14. Potential barrier for polymerization deduced for TS, D is the separation of reacting carbons. (Ref. 23).

distribution is narrower than the instrumental resolution but broadens again at high conversion. This is surprising in view of the low strain implied by high resolution x-ray studies (57). The length variation is similar to that deduced by Brillouin scattering (42) but the overall chain lengths are greater and similar to the values indicated by solution studies (50,51). This variation is distinctly different from the prediction of the crystal strain model, but is in good agreement with the mis-match model of Bassler, Fig. 13. Unfortunately the range of data available for comparison is limited so this is not a crucial test of the models. The potential curve used in the latter calculation was determined using the results of the studies of the pressure dependence of TS kinetics (37,38) and is shown in Fig. 14. This curve can be used to deduce the dependence of activation energy on reaction distance. The result of this calculation is shown in Fig. 15 for values of D greater than the van der Waals contact of adjacent diacetylene units. The dependence is strong and the effect on kinetics assuming thermal activation with a constant prefactor is marked, see Fig. 16. These predictions do not appear to agree with the available rather fragmentary data. For the compounds of Fig. 3 the reaction rates vary widely but the activation energies are similar, 22.8 kcal/mol for TS (32), 24.1 kcal/mol for MBS (11) and 26 kcal/mol for FBS (12). It would appear that differences in kinetics have more to do with changes in prefactor rather than barrier height.

The model of Flytzanis (25) also fits the thermal polymerization characteristic of TS. Some support for the importance of particular vibrational modes in the polymerization process is given by the observation of a constant prefactor in the kinetics for reactive, oligomeric intermediates (24,58). These, and early flash photochemical experiments (59) are the only studies of the microscopic kinetics of diacetylenes reported to date. They are, however, an important contribution. The development of the intermediates after uv flash initiation at 270 K is shown in Fig. 17. From time resolved spectra the species A-E were identified as dimer through to hexamer diradical intermediates. The rate constants and activation energies were determined leading to a constant value of 10^{11} s^{-1} for the exponential prefactor and an increasing activation energy as the intermediates became longer. There were no significant differences in the time constants observed for perdeuterated samples, indicating that isotopic substitution affects chain initiation rather than propagation. At room temperature the diradicals have lifetimes of the order 1 μsec. Subsequently longer lived species occur, these were identified as long dicarbene intermediates, which eventually, on a time scale of millisec, convert into stable polymer, in agreement wih the earlier work which followed only the growth of the polymer (59). These elegant studies show in great detail the kinetics at the microscopic scale, avoiding the problems of interpretation inherent in macroscopic measurements of conversion. Further measurements of this kind are required for other diacetylenes.

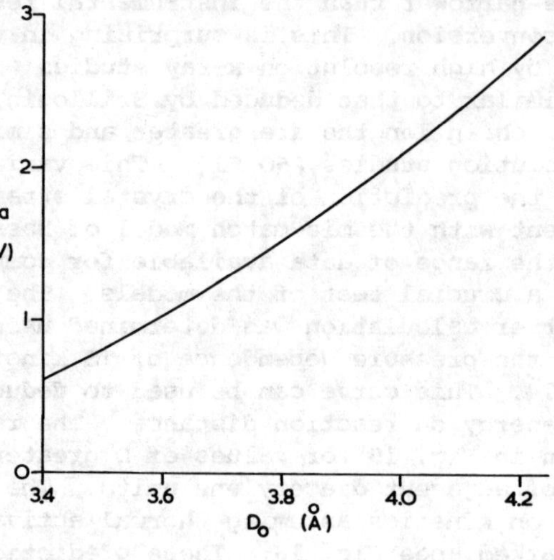

Figure 15. Calculated variation of reaction barrier height using the reaction potential of Figure 19.

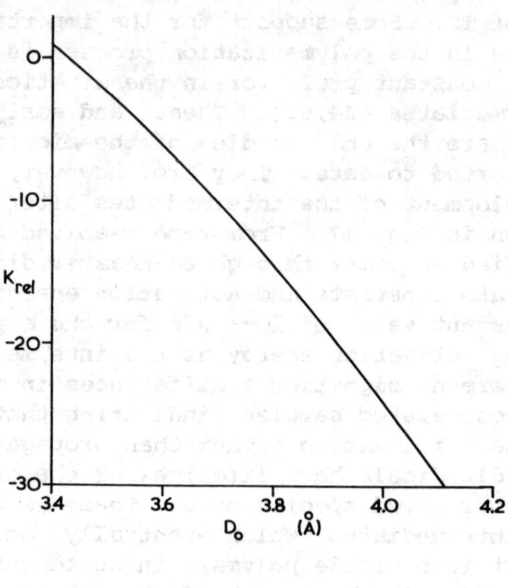

Figure 16. Thermal reaction rate as a function of reaction distance deduced from the data of Figure 20.

The influence of defects on kinetics has been mentioned above. Early studies (61) indicated the occurrence of faster polymerization in the vicinity of dislocations. Raman studies (39) indicate that imperfect samples are inhomogeneous supporting this finding. While dislocations do not play any role in the radiation polymerization of TS (62) this appears not to be so for thermal polymerization (63). Little attention has been paid to this problem and it deserves further study. Thus, though the kinetics of TS have been extensively investigated there are still a number of unresolved questions.

3.2 Other monomers

The polymerization of monomers with substituent groups similar to those of TS has been mentioned above (11-13). In many cases minor changes in substituent lead to significant changes in crystal structure so that many of these monomers are unreactive (48,64). The differences in reactivity of similar reactive monomers are large (see Fig. 3). In the case of MBS this was attributed to large lateral dimensional changes during polymerization (21). Measurements of unit cell dimensions have shown that this is not so, the dimensional changes being similar for MBS and TS (65). Resolution of this problem requires further studies of microscopic kinetics and effects of dislocations.

Thorough studies of kinetics have been made for only a few other monomers. These are TS-12 (3,66-69), the BCMU's and to a lesser degree DCH (75-77) and the urethanes (16,78-79). TS-12 and the BCMUs are monomers giving soluble polymers and TS-12, DCH and the ethyl urethane are systems where polymerization does not occur in a single phase.

Studies at low temperatures established that the intermediate species occurring in TS-12 are the same as those for TS, but unlike TS the monomer to polymer transformation involves a phase change (67). This phase change occurs homogeneously and hence the polymerization is not drastically affected. The monomer contains two asymmetric units in parallel reactive stacks. The final polymer contains a single symmetric repeat unit which requires different displacements in alternate sheets of monomer molecules during polymerization. As the polymer is soluble absolute molecular weights could be determined by GPC analysis (68). The molecular weight distribution was found to be sensitive to the polymerization method (3,67-69). Typical results are shown in Fig. 18. For γ-ray polymerization the number average for initially formed chains is 60 above $0^{o}C$ and 400 below $0^{o}C$. As conversion is increased at the higher temperature the distribution shifts and is dominated by a higher molecular weight component with a number average of the order 1000. This information has been used to determine the kinetic chain length as a function of conversion (Fig. 19). A significant finding is that the initial lower molecular weight component remains even at higher conversions

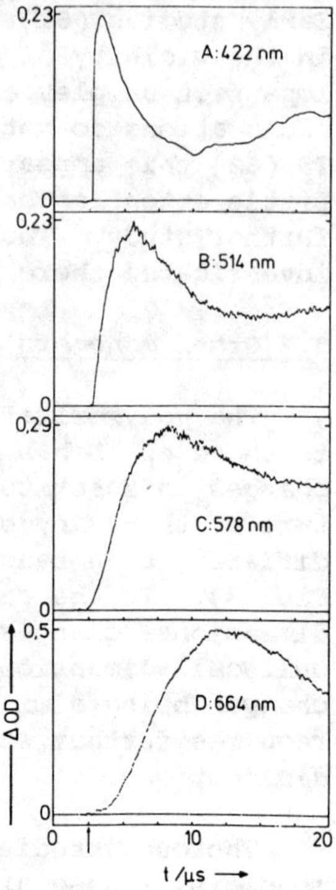

Figure 17. Kinetics of reaction intermediates at 270 K after uv flash initiation. (Ref. 24).

Figure 18. Molecular weight distributions for γ-ray polymerized TS-12 ($R=R'=-(CH_2)_4-OSO_2C_6H_4CH_3$) at various conversions between 0.5% and 65%. (Ref. 68).

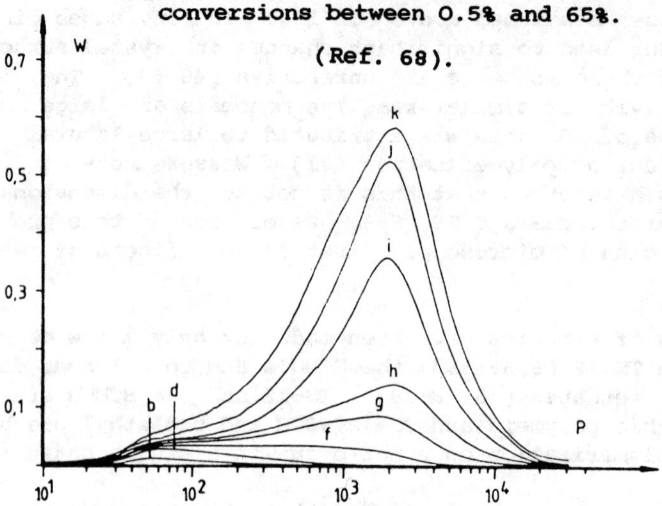

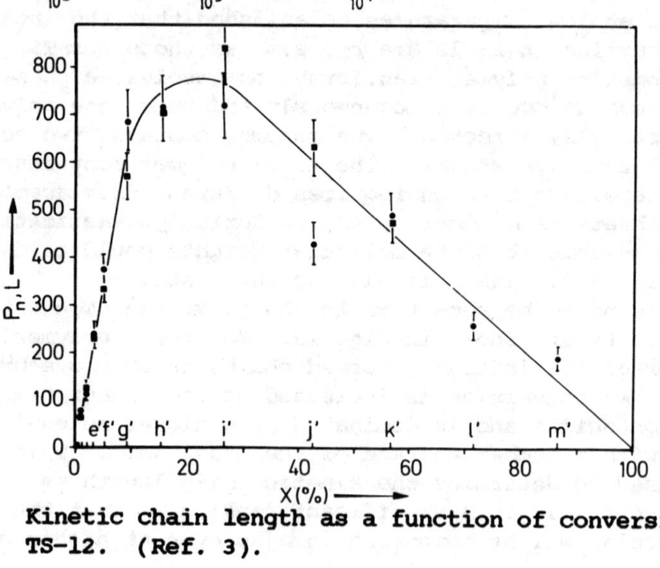

Figure 19. Kinetic chain length as a function of conversion for TS-12. (Ref. 3).

so that the ends of the initially formed chains are inactive. TS-12 is the only diacetylene for which detailed studies of the molecular weight distribution as a function of conversion have been reported.

The BCMUs are readily polymerized photochemically but show little thermal polymerization. They have been studied in detail by photoacoustic calorimetry and DSC (70,73,74), by actinometry (71) and by conventional measurements of polymer content (72). The quantum yield for 4 BCMU (n=4) was shown to be significantly larger than that of TS the initial quantum yield being in excess of 50 compared with 0.06 for TS (71). This was attributed to the nearly exact lattice match of BCMU monomers and polymers (3) leading to a large kinetic chain length (~ 2000 monomer units). Conversion studies of thin films gave a value of 70 for the initial quantum yield (72), much higher than estimated by photoacoustic measurements (70). It was also concluded that quenching of the monomer excited states by previously formed polymer was important. The apparent discrepancy with the photoacoustic data was subsequently shown to be due to problems of interpretation (73) and that the initial quantum yield was greater than 100. An even higher value of 125 was deduced by DSC measurements (74). It was suggested that the high efficiency for photo-reaction and the low thermal reactivity could be understood in terms of the reaction diagram since a high barrier for thermal polymerization would favour decay of the excited state to a dimer rather than a monomer. X-ray studies of 3BCMU have shown that adjacent sheets of monomer must rotate in opposite directions during polymerization which could effect the reactivty (3).

There is less data on DCH, this shows typical sigmoidal conversion curves (75,76) but x-ray studies indicate that a coherent first order phase transition occurs at intermediate conversion during γ-ray polymerization. Thermal polymerization of DCH is inhomogeneous leading to break up of the crystals. This is a case in which the nature of the polymerization is determined by competition between single and biphase routes. Strong quenching of monomer fluorescence by polymer has been observed (77). The related dianthracene and asymmetric anthracene-carbazole monomers are unreactive and co-crystalisation of DCH and the asymmetric monomer leads to a drastic reduction in rate of γ-ray polymerization.

The urethanes are further monomers for which small changes in substituent lead to large changes in kinetics. The ethyl urethane (ETU) has a sigmoidal curve for thermal polymerization (2,79) while the phenyl urethane does not have an induction period (cf. TS, MBS) (82). The data for ETU are apparently described by the crystal strain model, however, x-ray diffraction and Raman studies (79,80) show that phase segregation occurs during thermal polymerization. The results are well described by second order kinetics suggesting that the ph transformation is important in determining the kinetics. In con' γ-ray polymerization occurs in a single phase but Raman spectra and mechanical properties show that the polymer chains are much shorter than those in thermally polymerized crystals (80,81).

The photochemistry of Langmuir-Blodgett multilayers has been discussed (83,84) as has that of phenazine-diacetylene carbonic acid complexes (85). γ-ray polymerization of the complexes and pure acids have also been studied. The complexes are unusual in that the initiation can occur via a photochemical reaction between the phenazine and acid. Thermal polymerization has been reported for bis(1,3-pentadiynl mercury (88) and a number of nitrophenoxy substituted monomers (89-91). The thermal polymerization of hexadiynediol has also been reported (92). With the exception of the LB multilayers the kinetics were not analysed in detail.

4. CONCLUSIONS

The principal problem in any study of reaction kinetics is that a wide variety of processes lead to a small number of kinetic behaviours. Diacetylenes illustrate this, giving either sigmoidal or approximately first order kinetic data for reactions occurring either homogeneously in a single phase or involving co-operative phase transitions or phase segregation. Such a variety of microscopic behaviours cannot be described by a single model. Even for the best behaved material (TS) there are three models which can describe some, but not all, of the available data.

Further progress requires more detailed studies principally of the microscopic kinetics and crystal structures. The wide variation in rates of polymerization for monomers which are able to react appears to derive from two causes. First is the nature of the molecular rearrangement during polymerization. The examples of TS-12 and 3BCMU were mentioned above, other examples exist which involve complex rearrangements of the molecules during polymerization (93-95). In such cases thermal polymerization is slow and γ-ray polymerization often does not go to completion. Secondly, the initiation process can be identified as an important factor. The studies of microscopic kinetics is perprotonated and perdeuterated TS show that differences in macroscopic kinetics do not derive from differences in chain propagation. How energy is transported to precursor molecules remains an important question but one that will not be easy to tackle. Also at the microscopic level the influence of defects on kinetics requires further attention.

To date mixed crystals have received little attention but significant changes in kinetics have been observed for a wide range of monomers (47-49,77,96-97). Further work is required on this topic.

Despite the extensive literature on kinetics there are numerous open questions. One reason for this is that very few monomers have been studied in great detail, the literature being littered with much fragmentary data. This probably reflects the relative youth of studies of polydiacetylenes and that as it matures more attention will be given to such 'academic' subjects as kinetics. It must always be borne in mind that though diacetylenes are regarded as a single class of compounds the diversity of substituent groups means that generalisations concerning the properties of both monomers and polymers should be treated with caution.

(1) D.Bloor, Developments in Crystalline Polymers - I, Ed. D.C. Bassett, Applied Sci.Publ., London, 1982 p.151.
(2) D. Bloor, Mol.Cryst.Liq.Cryst., 93, 183 (1983).
(3) V. Enkelmann, G. Wenz, H.A. Muller, M. Schmidt and G. Wegner, Mol.Cryst.Liq.Cryst., 105, 11 (1984).
(4) G.M.J. Schmidt, Solid State Photochemistry, Verlag Chemie, Weinheim, 1976.
(5) R.H. Baughman, J.Polym.Sci., Polym.Phys.Ed. 1, 1511 (1974).
(6) R.H. Baughman, Contemporary Topics in Polymer Science, Vol.2, Eds. E.M. Pearce and J.R. Schaefgen, Plenum Publishing Corp., N.Y., 1977, p.205.
(7) R.H. Baughman and K.C. Yee, J.Polym.Sci.Macromol.Revs. 13, 219 (1978).
(8) M.J. Bowden, Macromolecules - An Introduction to Polymer Science, Eds. F.A. Bovey and F.H. Winslow, Academic Press, N.Y., 1979, P.23.
(9) H. Sixl and W. Neumann, Mol.Cryst.Liq.Cryst., 105, 41 (1984) and references therein.
(10) J.W. Christian, The Theory of Transformations in Metals and Alloys, Pt. I. Equilibrium and General Kinetic Theory, Pergamon Press, Oxford (1975).
(11) D. Bloor, D.J. Ando, C.L. Hubble and R.L. Williams, J.Polym. Sci., Poly.Phys.Ed., 18, 779 (1980).
(12) K.C. Yee, J.org.Chem., 44, 2571 (1979).
(13) G.N. Patel, Macromol., 14, 1170 (1981).
(14) J.K. Burdett, J.Am.Chem.Soc. 102, 5458 (1980).
(15) N.T. Anh, Quantum Theory of Chemical Reactions, Vol. 1 D. Reidel, Dordrecht, 1979, p.177.
(16) J. Kaiser and G. Wegner, Reactivity of Solids, Chapman and Hall, London, 1972, p.506.
(17) J. Kaiser, G. Wegner and E.W.Fischer, Israel J.Chem. 10, 157 (1972).
(18) I.M. Papissov and V.A. Kabanov, J.Polym.Sci.C. 16, 911 (1967).
(19) F.L. Hirschfeld nd G.M.J. Schmidt, J.Polym.Sci.A, 2, 2181 (1964).
(20) R.H. Baughman, J.Chem.Phys. 68, 3110 (1978).

(21) R.H. Baughman and R.R. Chance, J.Chem.Phys. 73, 4113 (1980).
(22) R.E. Robertson and R.P. Kambour, in Polymer Science, Ed. A. Jenkins, North-Holland, London 1972.
(23) H. Bassler, Adv.in Polym.Sci. 63, 1 (1984).
(24) H. Niederwald and H. Schwoerer, Z.Naturforsch 38a, 749 (1983).
(25) C. Flytzanis, Non-linear optical properties of Organic and Polymeric Materials, Ed. D.J. Williams, ACS Symposium Series 233, 167 (1983).
(26) N.N. Semenov, J.Polym.Sci. 55, 563 (1961).
(27) E.I. Adirovich, Dokl.Akad.Nauk.SSR, 136, 119 (1961).
(28) G.W. Wegner, Makromol.Chem. 145, 85 (1981).
(29) D. Bloor, L. Koski, G.C. Stevens, F.H. Preston and D.J. Ando, J.Mater.Sci. 10, 1678 (1975).
(30) R.R. Chance and J.M. Sowa, J.Amer.Chem.Soc. 99, 6703 (1977).
(31) G.N. Patel, R.R. Chance, E.A. Turi and Y.P. Kanna, J.Amer.Chem. Soc. 100, 6644 (1978).
(32) R.R. Chance and G.N. Patel, J.Polym.Sci.Polym.Phys.Ed. 16, 859 (1978).
(33) G.N. Patel, H.R. Battacharjee and A.F. Preziosi, J.Polym.Sci. Polym.Letts.Ed. 19, 511 (1981).
(34) A.R. McGhie, P.S. Kalyanaraman and A.F. Garito, J.Polym.Sci. Polym.Letts. Ed. 16, 335 (1978).
(35) A.R. McGhie, P.S. Kalyanaraman and A.F. Garito, Mol.Cryst. Liq.Cryst. 50, 287 (1979).
(36) E.M. Barrall II, T.C. Clarke and A.R. Gregges, J.Polym.Sci. Polym.Phys.Ed. 16, 1355 (1978).
(37) K. Lochner, Th. Hinrichsen, W. Hofberger and H. Bassler, Phys.Stat.Solidi (a) 50, 95 (1978).
(38) K. Lochner, H. Bassler, Th. Hinrichsen, Ber.Bunsenges,Phys. Chem. 83, 899 (1979).
(39) D. Bloor, R.J. Kennedy and D.B. Batchelder, J.Polym.Sci. Polym.Phys.Ed. 17, 1355 (1979).
(40) P.A. Albouy, P. Keller and J.P. Pouget, J.Amer.Chem.Soc. 104, 6556 (1982).
(41) R.J. Leyrer, G. Wegner and W. Wettling, Ber.Bunsenges,Phys. Chem. 82, 6917 (1978).
(42) V. Enkelmann, R.H. Leyrer and G. Wegner, Makromol.Chem. 180, 1787 (1979).
(43) K. Lochner and H. Bassler, Ber.Bunsenges.Phys.Chem. 84, 880 (1980).
(44) W. Hoptner, J.U. von Schutz and H.C. Wolf, J.Polym.Sci.Polym. Phys. Ed. 18, 469 (1980).
(45) W. Rehwald, A. Vonlanthen and W. Meyer, Phys.Stat.Solidi (a) 75, 219 (1983).
(46) C. Krohnke, V. Enkelmann and G. Wegner, Chem.Phys.Lett. 71, 38 (1980).
(47) V. Enkelmann, Makromol.Chem. 179, 2811 (1978).
(48) V. Enkelmann, Makromol.Chem. 184, 1945 (1983).

(49) D.J. Ando, D. Bloor and B. Tieke, Makromol.Chem.Rapid Commun. 1, 385 (1980).
(50) G.N. Patel, J.Polym.Sci.Polym.Phys.Ed. 17, 1591 (1979).
(51) D. Bloor, D.B. Batchelder, D.J. Ando, R.T. Read and R.J. Young, J.Polym.Sci.Polym.Phys.Ed. 19, 321 (1981).
(52) R. Mondong and H. Bassler, Chem.Phys.Lett. 78, 371 (1981).
(53) F. Bruanshweig and H. Bassler, Chem.Phys.Lett. 90, 41 (1982).
(54) K.H. Richter, W. Guttler and M. Schwoerer, Chem.Phys.Lett. 92, 4 (1982).
(55) K.H. Richter, W. Guttler and M. Schwoerer, Appl.Phys.A. 32, 1 (1983).
(56) P.A. Albouy, J.N. Patillon and J.P. Pouget, Mol.Cryst.Liq. Cryst. 93, 239 (1983).
(57) D.N. Batchelder, J.Polym.Sci.Polym.Phys.Ed. 14, 1235 (1976).
(58) H. Niederwald, H. Eichele and M. Schwoerer, Chem.Phys.Lett. 72, 242 (1980).
(59) R.J. Leyrer and G. Wegner, Ber.Bunsenges.Phys.Chem. 83, 470 (1979).
(60) H. Niederwald, H. Siedel, W. Guttler and M. Schwoerer, J.Phys.Chem. 88, 1933 (1984).
(61) W. Shermann, G. Wegner, J.O. Williams and J.M. Thomas, J.Polym.Sci.Polym.Phys.Ed. 13, 753 (1975).
(62) M. Dudley, J.N. Sherwood, D.J. Ando and D. Bloor, Mol.Cryst. Liq.Cryst. 93, 223 (1983).
(63) J.N. Sherwood, private communication.
(64) D.J. Ando, D. Bloor, C.L. Hubble and R.L. Williams, Makromol. Chem. 181, 453 (1980).
(65) R.L. Williams, Thesis, Univ. of London (1982).
(66) G. Wenz and G. Wegner, Makromol.Chem.Rapid Commun. 3, 231 (1982).
(67) D. Siegel, H. Sixl, V. Enkelmann and G. Wenz, Chem.Phys. 72, 201 (1982).
(68) G. Wenz and G. Wegner, Mol.Cryst.Liq.Cryst. 96, 99 (1983).
(69) G. Wenz, M.A. Muller, H. Schmidt and G. Wegner, Macromol. 17, 837 (1984).
(70) R.R. Chance and M.L. Shand, J.Chem.Phys. 72, 948 (1980).
(71) H.R. Battacharjee and G.N. Patel, J.Photochem. 16, 85 (1981).
(72) A. Prock, M.L. Shand and R.R. Chance, Macromol. 15, 238 (1982).
(73) A. Prock, M.L. Shand and R.R. Chance, J.Chem.Phys. 76, 5834 (1982).
(74) H. Eckhardt, T. Prusik and R.R. Chance, Macromol. 16, 732 (1983).
(75) K.C. Yee and R.R. Chance, J.Polym.Sci., Polym.Phys.Ed. 16, 431 (1978).
(76) V. Enkelmann, R.J. Leyrer, G. Schleier and G. Wegner, J.Mat.Sci. 15, 168 (1980).
(77) V. Enkelmann, G. Schleier and H. Eichele, J.Mat.Sci. 17, 533 (1982).
(78) G. Wegner, Makromol.Chem. 154, 35 (1972).

(79) C. Galiotis, R.J. Young, D.J. Ando and D. Bloor, Makromol.Chem. 184, 1083 (1983).
(80) C. Galiotis, R.J. Young and D.N. Batchelder, J.Polym.Sci.Polym. Phys.Ed. 21, 2483 (1983).
(81) C. Galiotis and R.J. Young, Polymer, 24, 1023 (1983).
(82) G. Wegner, Z. Naturforschung 24b, 824 (1969).
(83) B. Tieke and G. Wegner, Makromol.Chem. 179, 1639 (1978).
(84) J.P. Fouassier, B. Tieke and G. Wegner, Israel J.Chem. 18, 277, (1979).
(85) C. Bubeck, T.H. Nguyen Xuan, H. Sixl, B. Tieke and D. Bloor Ber.Bunsenges.Phys.Chem. 87, 1149 (1983).
(86) B. Tieke and D. Bloor, Makromol.Chem. 182, 133 (1981).
(87) B. Tieke, D. Bloor and R.J. Young, J.Mat.Sci. 17, 1156 (1982).
(88) M. Steinbach and G. Wegner, Makromol.Chem. 178, 1671 (1977).
(89) P.S. Kalyanaraman, A.F. Garito, A.R. McGhie and K.N. Desai, Makromol.Chem. 180, 1393 (1979).
(90) A.F. Garito, C.J. Horner, P.S. Kalyanaraman and K.N. Desai, Makromol.Chem. 181, 1605 (1980).
(91) C.J. Horner, A.F. Garito and K.N. Desai, Makromol.Chem. 182, 1363 (1981).
(92) D. Bloor and G.C. Stevens, J.Polym.Sci.Polym.Phys.Ed. 15, 703 (1977).
(93) V. Enkelmann, J.Chem.Res. (M) p.3901 (1981).
(94) G.N. Patel, E.N. Duesler, D.Y. Curtin and I.C. Paul, J.Amer. Chem.Soc. 102, 461 (1980).
(95) R.B. Wilson, E.N. Duesler, D.Y. Curtin, I.C. Paul, R.H. Baughman and A.F. Preziosi, J.Amer.Chem.Soc. 104, 509 (1982).
(96) G.N. Patel, U.S. Patent 4189399 (1980).
(97) G.G. Miller and G.N. Patel, Polym.J. 13, 1075 (1981).

SOLID-STATE PHOTOPOLYMERIZATION OF DIACETYLENES

H. Eckhardt, T. Prusik, and R.R. Chance

Allied Corporation, Corporate Technology
Morristown, N.J. 07960

A detailed calorimetric and spectroscopic study of the UV photopolymerization of highly radiation-sensitive diacetylenes is presented. The polymerization energetics and kinetics, quantum yield, and variation of optical properties on polymerization are discussed. The quantum yield of the diacetylene 4BCMU is very high at low conversion, nearly 100 polymer repeat units per absorbed photon. The decrease of the polymerization rate with increasing conversion is explained through quenching of the excited monomer by the polymer chain. This process leads to near complete quenching of the photoreaction at about 40% in the 4BCMU case. The calorimetric results suggest, that the energetics of photopolymerization is strongly influenced by the hydrogen bonding network in 4BCMU.

I. INTRODUCTION

Over the last decade the solid-state polymerization of diacetylenes has been the subject of much chemical and physical research worldwide. This interest was triggered largely by the work of Wegner(1) who identified the polymerization process as a 1,4-addition reaction. The C1 and C4 carbon atoms of adjacent diacetylene monomers link up to form fully-conjugated polymer chains. In many cases the reaction product is a single crystal of macroscopic dimensions with preservation of the structural order of the monomer single crystal. Initiation of the polymerization requires the formation of a biradical dimer; this process is uphill in energy by roughly 1 eV.(2,3) Also, bond formation, for both initiation and propagation of the chains, involves a molecular rotation of the monomer molecules which has to overcome the repulsive forces of the lattice, thus providing an additional energy barrier associated with the initiation process. The energy to overcome this barrier can be supplied by heat, UV-photons, or high energy radiation.

Much of the work on polydiacetylenes has centered on the kinetics and mechanism of the polymerization process. Due to the low degree of disorder in the crystals, experimental techniques such as high resolution optical and magnetic spectroscopy have yielded a great deal of information about the individual reaction steps and the intermediates involved in the polymerization process.(4) This has led to a better understanding of the diacetylene polymerization process and topochemical reactions in general.

Most of the studies in the literature have focussed on 2,4-hexadiyne-1,6 diol bis(p-toluenesulfonate) (TS). This compound polymerizes thermally and is also moderately reactive to UV-light. The conversion to polymer is "autocatalytic", i.e. the conversion curve displays a S-shape with the polymerization rate increasing abruptly by a factor of 200 at about 10% conversion to polymer.(5) This autocatalytic effect is not a general feature of diacetylenes. The thermal polymerization process in TS and many other diacetylenes is temperature activated by about 1eV. It is now well established that the first intermediate to chain propagation is the biradical dimer.(4) The activation energy in the thermal process is therefore roughly the energy difference between the monomer and the dimer. Each addition of a monomer unit to the growing chain in TS releases ~1.6 eV as shown by Differential Scanning Calorimetry.(2,3)

In the photopolymerization process, the activation barrier is overcome by the photon energy. At high temperatures (at least room temperature and above) the monomer excited state reacts with a monomer ground state to form the biradical dimer, which can then initiate the chain propagation as in the thermal case. At low temperatures the process is bimolecular requiring two excited states.(6) Since the activation energy (E_a) for UV polymerization of TS is small compared to E_a for the thermal process, the latter energy is primarily associated with chain initiation and not with the chain propagation event.

This paper focuses on the photopolymerization of another diacetylene, 4BCMU. BCMU stands for butoxycarbonylmethylurethane with the sidegroup $R=(CH_2)_4OCONHCH_2COOC_4H_9$. The behavior of this monomer differs from TS in several important aspects. There is virtually no interference from substituent absorption in the UV spectrum of 4BCMU; in TS the toluene moiety shows significant absorption in the same region as the diacetylene group. 4BCMU is not thermally reactive and the conversion to polymer shows no autocatalytic behavior. The latter indicates the absence of a mismatch of the monomer and polymer lattice constant; a large mismatch of these lattice constants is the origin of the autocatalytic effect in TS. 4BCMU also shows a much higher sensitivity to UV-photons (at room temperature) than TS and all other diacetylenes that have been reported. Study of the 4BCMU system is facilitated by the fact that poly-4BCMU dissolves readily in common organic solvents. The interesting solution properties of this class of polydiacetylenes is summarized elsewhere in these proceedings.(16)

We discuss the energetics of the photopolymerization in Section II and the kinetics of the reaction and the change in the optical properties of 4BCMU during photopolymerization in Section III. The structure and sprectroscopy of the intermediates will be addressed only briefly since these questions are discussed by Sixl.(17) For a complete review of the photopolymerization of diacetylenes the reader is referred to the recent review article by Baessler.(7)

II. ENERGETICS OF DIACETYLENE PHOTOPOLYMERIZATION

The energy barrier for initiation of the thermal polymerization in TS and many other diacetylenes is about 1 eV. Patel et al.(3) have argued that this energy should represent roughly the energy difference between two diacetylene monomers and the biradical dimer. Baessler(7), on the other hand, argues that there are significant contributions to this energetic barrier by the energy of the librational motion required to shorten the C1-C4 reaction distance of the reacting molecular pair. Whatever the nature of this barrier, once this thermal energy has been supplied, the chain propagation is an exothermic process. This has been demonstrated by Chance et al.(2,3) using Differential Scanning Calorimetry. The DSC experiments yield the reaction enthalpy per monomer addition event. Each addition of a monomer to a running chain releases about 1.6 eV. The energetics of the thermal polymerization of TS is summarized in Figure 1 (solid curve).

In the photopolymerization process, the activation energy is overcome by the photon energy with the monomer excited state reacting with an adjacent monomer in its ground electronic state to form the biradical dimer. This dimer can then undergo fission back to the two monomers or initiate a chain. Irradiation with UV-light of wavelength less than 300 nm polymerizes 4BCMU. As can be seen from Figure 2 this is the region of the first monomer absorption band (first singlet state), which shows a progression of about 2300 cm^{-1} consistent with values for the triple-bond stretching frequencies observed in Raman spectra. However the active precursor for the biradical dimer is most likely a triplet state and not the singlet state excited by the UV-photon.(4,7) This triplet state is populated via intersystem crossing from the singlet state.

Calorimetric techniques have also proved to be useful tools for determining energetic parameters in the photopolymerization process. For instance, photoacoustic spectroscopy (PA) has been used by Chance et al.(8,9) to study the reaction kinetics of 4BCMU. In a PA experiment the heat evolved as a result of a photochemical reaction as well as that due to photon absorption is detected by a microphone located in the gas surrounding the sample or by a piezoelectric transducer attached to the sample. The information extracted from this experiment is the reaction enthalpy for photopolymerization (per chain initiation event). The experiments demonstrated indeed that photopolymerization is a strongly exothermic process. However quantitative analysis of the data is complicated by the fact that heat diffusion inside the sample and the acoustic coupling between solid and gas have to be taken into account, which requires detailed knowledge of thermal parameters and application of a fairly complicated one-dimensional diffusion model.(9)

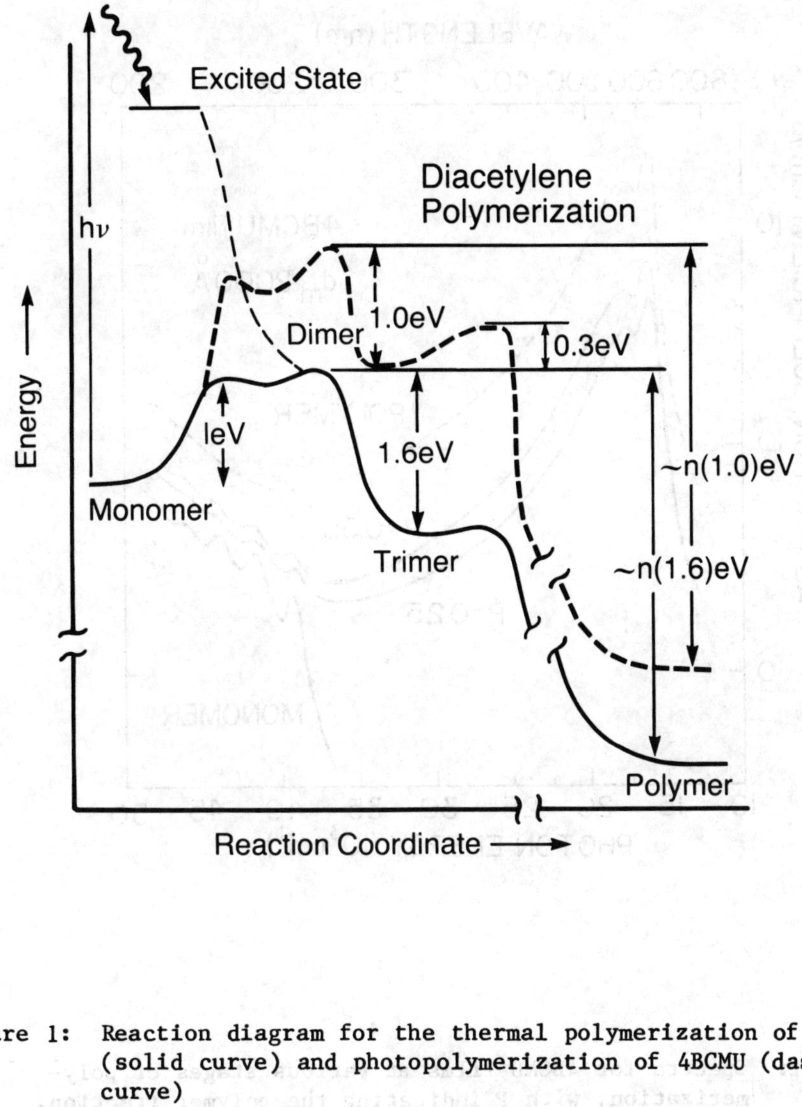

Figure 1: Reaction diagram for the thermal polymerization of TS (solid curve) and photopolymerization of 4BCMU (dashed curve)

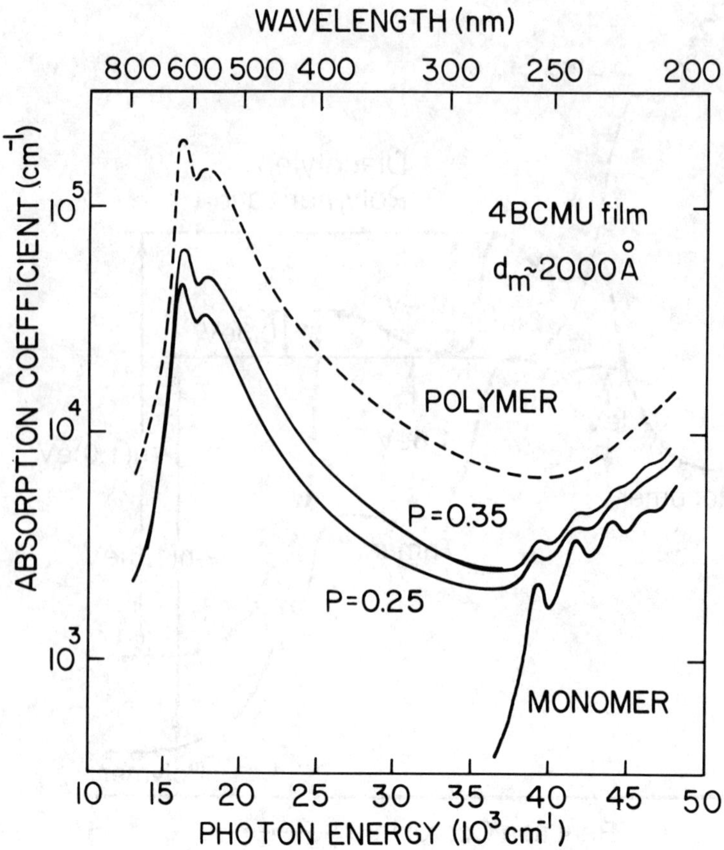

Figure 2: Spectra for 4BCMU films at various stages of polymerization, with P indicating the polymer fraction.

A more direct measurement of the heat of photopolymerization has been obtained with a Differential Scanning Calorimeter modified to allow UV-initiation of the reaction.(10) In this experiment the diacetylene monomer is held at a constant temperature in the DSC chamber and irradiated at time t=0 with UV light. The heat evolved is then measured as a function of time until the light is turned off and the polymerization process is terminated. Figure 3 shows a typical DSC-trace. The heat due to light absorption is small compared to that due to the exothermic diacetylene polymerization. Therefore, the heat of reaction per monomer addition can be calculated with good accuracy from the area under the DSC-curve and the number of moles of polymer produced during the experiment. The latter is determined spectroscopically by measuring the absorbance of the polymer dissolved in chloroform. Table I gives the results for 4BCMU and another urethane substituted diacetylene, ETCD [$R=(CH_2)_4OCONHC_2H_5$] as well as the thermal value for TS.

Table I: Summary of thermal and photopolymerization results for 4BCMU, ETCD, and TS.

	4BCMU	ETCD	TS
Heat of polymerization ΔH_p for thermal reaction (kcal/mole)	――	――	36
Heat of polymerization ΔH_p for photoreaction (kcal/mole)	21.8	23.0	――
Thermal activation energy E_a (kcal/mole)	――	――	23
Activation energy for photopolymerization E_a (kcal/mole)	~7	~7	~3

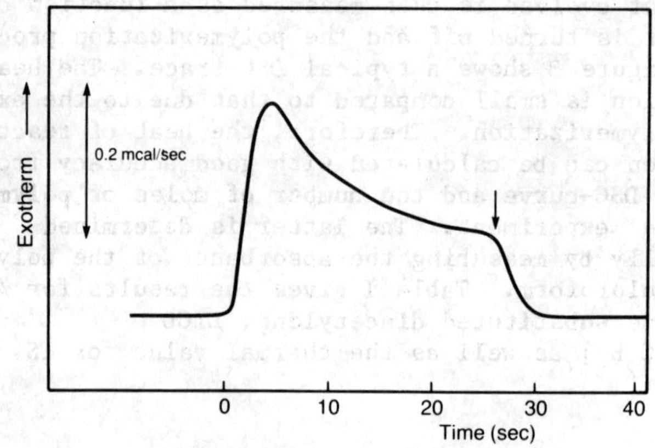

Figure 3: Typical DSC trace for photopolymerization of 4BCMU; t=0 denotes beginning of irradiation; the arrow denotes the end of the irradiation.

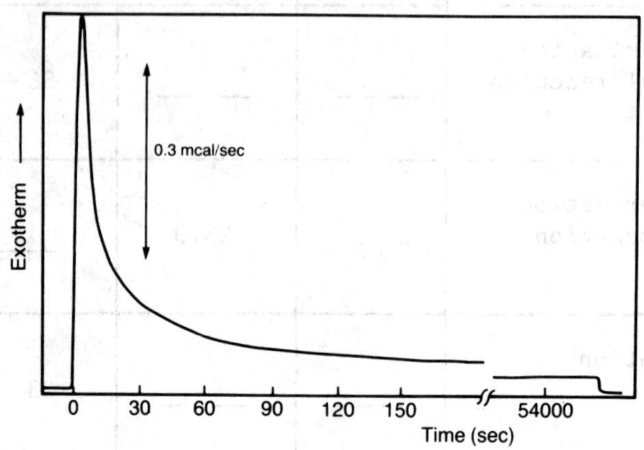

Figure 4: DSC signal for photopolymerization. The polymerization rate is maximum at t=0 and decays to a constant level for $t \to \infty$.

If the biradical dimer is the intermediate in both the thermal polymerization and photopolymerization, then the energy released on addition of a monomer unit to a growing chain should be the same for both processes. Unfortunately, both 4BCMU and ETCD are thermally inactive and a direct comparison is not possible. Just the opposite is true for TS. The high thermal reactivity of TS allows easy measurement of the thermal heat of polymerization, but the low photosensitivity, compared to 4BCMU, prohibits any quantitative determination of the enthalpy of the photoreaction. Therefore a direct proof that the energetics of chain propagation is independent of the mode of initiation is not possible with these materials.

A possible explanation for the unexpectedly small ΔH_p value in 4BCMU and ETCD lies in the fact that the urethane substituted diacetylenes show strong intramolecular hydrogen bonding between the C=O and N-H groups on adjacent substituents.(11) The hydrogen bonding in the monomer is between reacting monomer units, so that in the polymer the hydrogen bonding is intramolecular. A considerable distortion in the four CH_2 groups linking the sidechain to the backbone in the polymer is required to enable intramolecular hydrogen bonding in the polymer. Therefore polymerization is probably accompanied by an energetically unfavorable reorganization in the hydrogen network at the expense of the total reaction enthalpy. This also provides a straightforward explanation why 4BCMU is thermally inactive, in that formation of the dimer requires similar rearrangement of the hydrogen bonding network; this increases the energy barrier for chain initiation by perhaps as much as 0.6 eV, thus making 4BCMU (and ETCD) thermally unreactive materials. The low ΔH_p value measured for the 4BCMU photopolymerization suggests the dashed curve in Figure 1 as a description of the reaction diagram for 4BCMU.(10) The dimer state is now closer in energy to the excited state (than is the case for TS) leading to the possibility of stronger nonradiative coupling between the two states. If this is true a "compensation" effect could exist where high photosensitivity is accompanied by low thermal sensitivity and vice versa. It would be interesting as a test of this hypothesis to find a diacetylene system in which the calorimetric techniques can measure the thermal and photo energy parameters.

III. KINETICS OF PHOTOPOLYMERIZATION

As we have discussed in the previous section the photopolymerization of diacetylenes is a multistep process. In the first step an excited monomer state is created through photon absorption. The excited state has a probability q of initiating a polymer chain and a probability 1-q of returning to the ground state nonradiatively. (Diacetylenes do not fluoresce to any significant extent.) Once this first energy barrier is overcome the chains grow to a kinetic chain lenght n (in polymer repeat units). In the early stage biradical oligomers are formed up to 6 repeat units. Beyond 6 units the bicarbene structure is energetically more favorable for the oligomers.(4) One parameter, which is important for practical applications, is the quantum yield, defined as the number of polydiacetylene repeat units (r.u.) produced per absorbed photon and written as the product nq; it is this quantity that is measured directly in polymer conversion studies. In order to determine the primary quantum yield q for optical chain initiation the kinetic chain length n has to be measured independently. In TS the chain length is conversion dependent with $n \sim 25$ r.u. at low conversion and much longer at higher conversions. Measurement of nq in TS is only practicle in the low conversion limit. The factors determining the chain length are the strain caused by the mismatch between monomer and polymer repeat distance in the reaction direction and the collision of propagating chains at high conversion. Due to the insolubility of p-TS the measurement of the actual chain length is very difficult. Chain lengths have been measured for 4BCMU,(12) since poly-4BCMU is highly soluble. The results of gamma-ray polymerization studies(12) indicate that strain is not important and n does not vary dramatically with conversion. Consequently the measured average chain length in poly-4BCMU is much higher than in poly-TS ($n \sim 2400$ r.u.)(12).

The DSC experiment described in section I provides a convenient and accurate way for determination of the quantum yield nq. The heat A(t) measured in the photoinitiated polymerization process is composed of two terms, the energy released in the photochemical reaction, $\Delta H_p dP(t)/dt$, and the energy deposited in the sample by the absorbed light $h\nu I_o$, where hv is the photon energy and I_o is the light intensity. From the ratio of the signal at t=0 and $t=\infty$ (taken from DSC curves similar to the one in figure 4) the product nq can be deduced provided the heat of polymerization ΔH_p is known. For 4BCMU nq is about 60 at room temperature in good agreement with optical measurements (see below).(13) This gives a primary quantum yield of q~0.025. For TS nq~0.1 represents a reasonable estimate for the overall quantum yield in the low conversion limit.(7) If n is ~25 at low conversions, q~0.004 is implied. It follows that for 4BCMU both the larger kinetic chain length n and the higher primary quantum yield q contribute to the much higher overall yield.

Thin film absorption spectroscopy is a second reliable way for determination of the quantum yield.(13) In a typical experiment spectra of the polymerized film are taken after various irradiation dosages; the polymer film is then dissolved and the polymer content determined from the absorbance of the polymer solution. Figure 4 shows the evolution of the polymer absorption spectra with increasing degree of polymerization. The measured polymer absorption band in these films agrees well with that of a single crystal. This suggests that the crystallite size in the film is large enough to have no effect on the spectra. Figure 5 gives the polymer fraction P versus irradiation time t or dosage. Only for low conversion is P linear with time, i.e. dP/dt is constant. For P>3% the conversion rate decreases with increasing conversion. Surprisingly the maximum polymer fraction reached in the UV photopolymerization is only about 40% even after very long irradiation times. This is in sharp contrast to gamma polymerization of 4BCMU where almost complete conversion is achieved. From the linear part of the conversion curve a quantum yield nq of ~70 can be calculated in good agreement with the DSC results.

The surprising decrease of the polymerization rate with increasing conversion can be explained through quenching of the excited monomer by the polymer chain produced during the reaction.(13) This process has been shown to play an important role with diacetylenes which have fluorescent substituent groups. However, the urethane substituted diacetylenes do not fluoresce and a direct probe of the quenching mechanism is therefore not available. In order to take quenching into account, a redefinition of the chain initiation probability q is required. Usually, q is defined as(5)

$$q = \xi_i/(\xi_i + \xi_d) \simeq \xi_i/\xi_d$$

where ξ_i is the rate constant for chain initiation and ξ_d the sum of all other rate constants for deactivation of the excited state. By including the second-order rate constant for quenching, ξ_q, we have

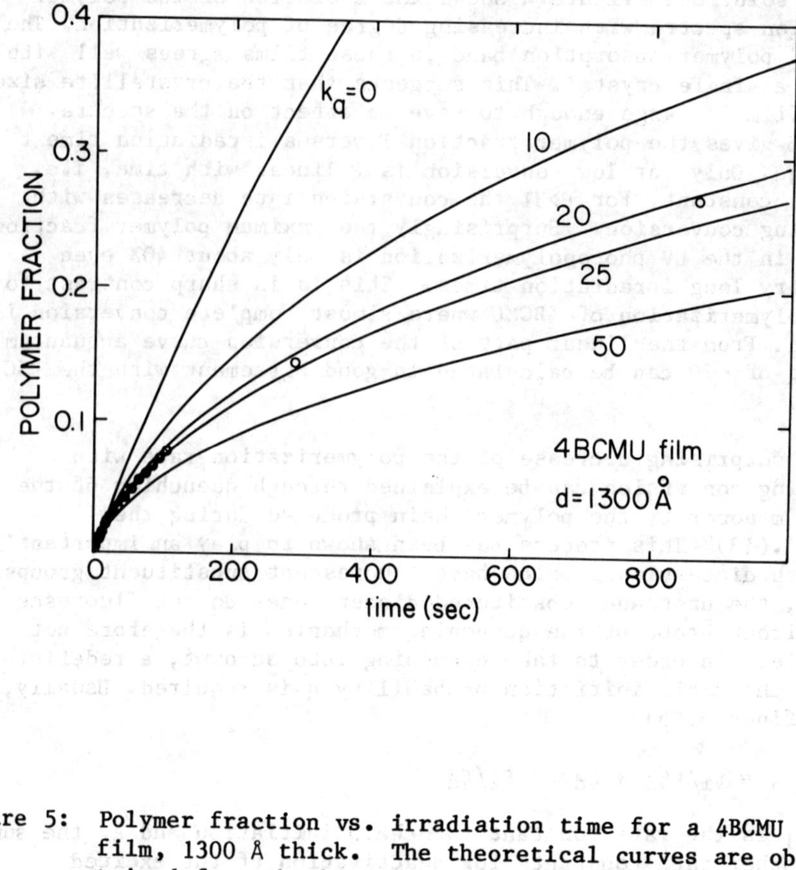

Figure 5: Polymer fraction vs. irradiation time for a 4BCMU film, 1300 Å thick. The theoretical curves are obtained from the kinetic model (see text) with indicated values of quenching parameters, k_q.

$$q' = q/(1 + k_q P(x,t)/\rho)$$

where $k_q=\xi_q/\xi_d$, ρ is the density, and P(t) is the polymer conversion at time t and distance x from the crystal surface.(15) This model yields good fits to the experimental conversion curve (see solid line, Figure 5). An important conclusion from the discussion so far is that meaningful values of the quantum yield for photopolymerization can only be obtained if quenching of the excited monomer state is taken into account or the experiments are done in the low conversion limit.

A comment about the temperature dependence of the quantum yield is appropiate at this point. Both experiments, DSC and optical spectroscopy, give an activation energy E_a of ~0.3 eV for the temperature range -60C to +60C. In the photoinitiated polymerization process it is chain propagation (and/or possibly chain termination) which probably determines the temperature dependence of the yield and not the initiation step which dominates the E_a measured for the thermal reaction. The activation energy for the photopolymerization is included as an energy barrier for monomer addition in our reaction diagram, FOgure 1. However, it is doubtful, that the temperature dependence can be described by a single activation energy. First, nothing is known about the temperature dependence of q. In addition, low temperature time resolved spectroscopy on TS reveals that the activation energy per addition reaction increases with increasing chain length from 0.25 eV for the dimer-trimer addition to about 0.4 eV for n~20.(14) This can be attributed to the increasing distance of the oligomer to the adjacent monomer with increasing chain length. Since the activation energy is associated with the excitation of the librational motion of the diacetylene molecule, an increase in reaction distance means an increase in activation energy. Because lattice mismatch is not important in 4BCMU the measured activation energy of 0.3 eV could represent the activation energy for each individual step. Clearly, a detailed crystal structure analysis and determination of chain length as a function of temperature are required to fully understand the photopolymerization in 4BCMU.

IV. SUMMARY

We have described the energetics and kinetics of the UV-initiated polymerization of 4BCMU. The quantum yield for photopolymerization of 4BCMU is nearly 100 polymer repeat units per absorbed photon, easily the highest value reported for diacetylenes. This large value is probably attributable for the most part to the long propagation lengths in 4BCMU. The measured quantum yield is only valid at low polymer conversions due to a quenching of the reaction process observable at polymer concentrations as low as 3%. This quenching is adequately described by a model based on decrease in the singlet state lifetime of the diacetylene due to nonradiative energy transfer to the polymer in the lattice. This process eventually leads to near complete quenching of the photoreaction so that the maximum polymer concentration attainable in the 4BCMU case is about 40%. The energetics of the 4BCMU reaction have been monitored with DSC techniques. The results suggest that the hydrogen bonding network in 4BCMU may play an important role in the overall energetics of polymerization.

ACKNOWLEDGEMENTS

We acknowledge the collaboration of A. Prock and M.Shand in the optical studies.

REFERENCES

1. G. Wegner, Z. Naturforschung 24B, 824(1969).
2. R.R. Chance, G.N. Patel, E.A. Turi, and Y.P. Khanna, J. Am. Chem. Soc. 100, 1307(1978).
3. G.N. Patel, R.R. Chance, E.A. Turi, and Y.P. Khanna, J. Am. Chem. Soc. 100, 6644(1978).
4. H. Gross, Advances in Polym. Sci. 63, 49(1984), and references therein.
5. R.R. Chance and G.N. Patel, J. Polym. Sci: Polym. Phys. Ed. 16, 859(1978).
6. H. Gross, W. Neumann, and H. Sixl, Chem. Phys. Lett 95, 584(1983).
7. H. Baessler, Advances in Polym. Sci. 63, 49(1984).
8. R.R. Chance and M.L. Shand, J. Chem. Phys. 72, 948(1980).
9. A. Prock, M.L.Shand, and R.R. Chance, J. Chem. Phys. 76, 5834(1982).

10. H. Eckhardt, T. Prusik, and R.R. Chance, Macromolecules 16, 732(1983).
11. G.N. Patel, R.R. Chance, and J.D. Witt, J. Chem. Phys. 70, 4387(1979).
12. G.N. Patel and E.K. Walsh, J. Polym. Sci.: Polym. Lett. Ed. 17, 203(1979).
13. A. Prock, M.L. Shand, and R.R. Chance, Macromolecules 15, 238(1982).
14. H. Niederwald and M. Schwoerer, Z. Naturforschung 38A, 749(1983).
15. R.R. Chance and R.H. Baughman, Eighth Molecular Crystal Symposium (Santa Barbara, 1977) p.81.
16. R.R. Chance et al, these proceedings.
17. H. Sixl et al, these proceedings.

PHOTOPOLYMERIZATION REACTION INTERMEDIATES IN DIACETYLENE CRYSTALS

HANS SIXL

Physikalisches Institut, Teil 3, Universität Stuttgart,
Pfaffenwaldring 57, D-7000 Stuttgart 80, West-Germany.

ABSTRACT

Recent spectroscopic investigations and model calculations concerning the reaction intermediates of the solid state polymerization in diacetylene crystals are reviewed. It will be shown, that diacetylenes are a unique model system for the study of reaction mechanisms. The reaction intermediates and final reaction products are best described by diradical (DR), asymmetric carbene (AC), dicarbene (DC) and stable oligomer (SO) molecules of different chain lengths. The respective singlet (S=0), triplet (S=1) or quintet (S=2) states and their roles in the polymerization process are investigated in detail by ESR and optical absorption spectroscopy. The optical and thermal reaction kinetics are characterized by the photoinitiation, the chain propagation and chain termination reactions. A one-dimensional electron gas model and an exciton model is applied to the different absorption series of the intermediates. The structural transition from diradicals to dicarbenes observed at a critical chain length of six repeat units is quantitatively explained by the generation of "up" and "down" radical electrons at the chain ends of the reaction intermediates.

INTRODUCTION

In a solid state polymerization reaction monomer diacetylene crystals are transformed to polymer crystals in successive reaction steps. Nearly perfect polymer single crystals are obtained thermally (kT) or by UV- or X-ray irradiation (hν) of the monomer crystals [1-3]. Within the class of diacetylene molecules (R-C≡C-C≡C-R) which show this unusual chemical reaction, the TSHD (with side groups R = $-CH_2SO_3$-Ø-CH_3) is the best known representative , which is characterized by a variety of reaction intermediates [4-19]. The unconventional reactivity and the unusual properties of the polymer crystal have attracted the interest of both, physicists and chemists. The general feature of the low temperature photopolymerization reaction is shown schematically in Fig. 1 by example of the diradical DR-intermediates.

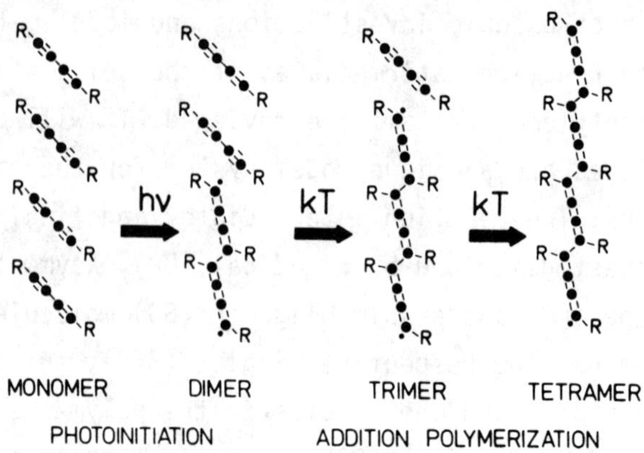

FIGURE 1 Low-temperature polymerization reaction of diacetylene crystals.

The dimer photoinitiation reaction requires two UV-photons (e.g. two monomer molecule excitations), which give rise to a nonlinear intensity dependence of the dimer quantum yield [12]. In successive reaction steps further monomer molecules are added thermally (upon annealing at about 100 K) or optically (upon direct photoexcitation of the adjacent monomer molecules). In the chain termination reactions (not shown in the Figure) the reactive chain ends are transformed into unreactive configurations by removal of the radical electrons.

The reaction intermediates of this extraordinary solid state reaction have been correlated to dimer, trimer, tetramer molecules etc. with different chemical structures. The physical properties of the reactive short chain intermediates and final reaction products are best characterized by the approximate structures represented in Figure 2. These structures give the principal contributions to the real structures of the DR_n, DC_n and AC_n intermediates and stable oligomer SO_n molecules. The length of the oligomer molecules is determined by the number n of monomer units. The diradical (DR_n) with $2 \leq n \leq 5$ and the dicarbenes (DC_n) with $n \geq 8$ show a symmetric configuration. At short chain lengths only diradicals are observed. At long chain length only dicarbenes are present. A bond defect structure is present at the ends of the reaction

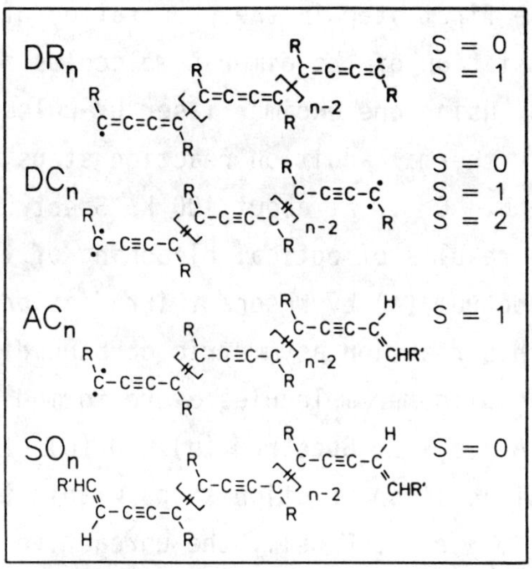

FIGURE 2 Principal structures of the reaction intermediates and final reaction products of the solid state polymerization in diactylene crystals. The real electronic structure is given by additional admixtures of mesomeric configurations.

intermediates, which is correlated to a carbene $-\ddot{C}-$ with two free valence electrons and a radical carbon atom $-\dot{C}=$ with one free valence electron. Both cases represent reactive chain ends which allow reactions of the chain with the neighbouring monomer molecules. The unreactive chain ends of the SO_n final photoproducts are thermally and optically stable up to room temperature.

In order to obtain a deeper insight into the reaction mechanism and to elucidate the real electronic structures we have performed a careful study of the optical and ESR-absorptions of the reaction products. Model calculations will be presented to explain the different chain length dependent optical and ESR absorption series.

EXPERIMENTAL RESULTS

Optical absorption spectra

The polarized optical absorption spectra of the DR, AC and SO low-temperature photoproducts of Figure 2 are shown in the spectra of Figures 3-5. The first step in the preparation of all spectra is the UV-photoinitiation of the dimer A-molecules shown in spectrum (a) of Fig. 3, using one excimer laser UV-pulse. Spectrum (b) then is obtained by thermal addition reaction steps within the diradical series A → B → C ... at about 100 K. Spectrum (a) of Figure 4 shows the results of optical bleaching of the dimer A-absorption of spectrum (a) by resonant irradiation into the A-absorption. In this reaction asymmetric carbene dimer intermediates a and stable oligomer molecules α are formed according to the photoreaction A → a → α. Spectrum (b) and (c) of Figure 4 is obtained by optical addition reaction steps within the asymmetric carbene series a → b → c ... Finally the unreactive stable oligomer absorptions of Figure 5 are obtained by optical bleaching of the whole DR series of Figure 3 via the reaction $DR_n \rightarrow AC_n \rightarrow SO_n$.

The correlation of the A, B, C,... photoproduct absorption series to diradical DR dimer, trimer, tetramer,... intermediates

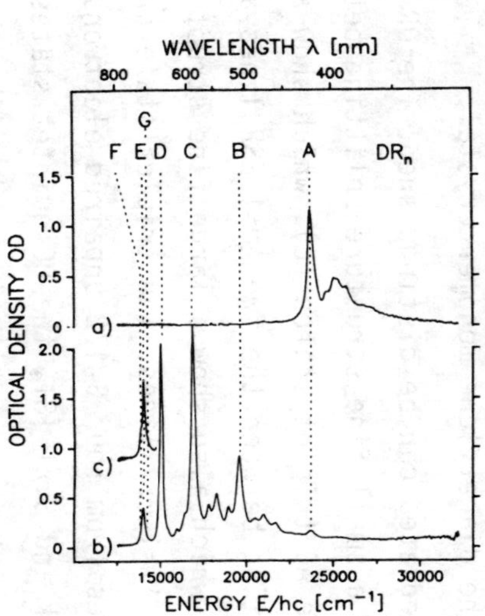

FIGURE 3

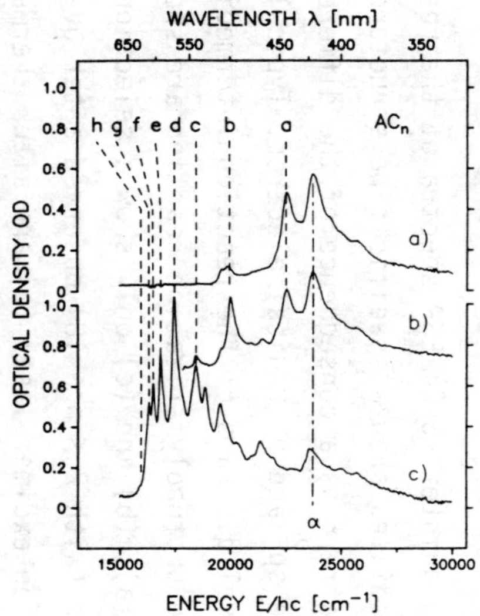

FIGURE 4

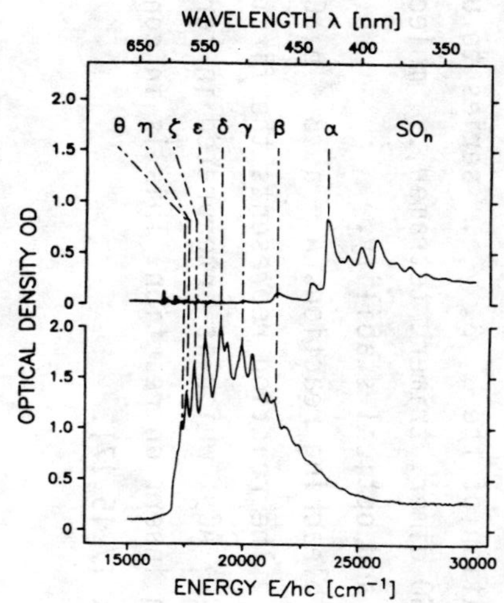

FIGURE 5

Polarized optical absorption spectra of the diradical (DR), the asymmetric carbene (AC) and the stable oligomer (SO) molecules. The dimer (n=2) spectra (A,a,α) are given by the upper part. The spectra with n > 2 are shown in the lower part.

and the correlation of the a, b, c,... photoproduct absorption series to asymmetric carbene AC intermediates is based on the ESR fine structure and hyperfine structure analysis and on their reaction kinetics. The correlation of the α, β, γ,... series to unreactive stable oligomer SO dimer, trimer, tetramer,... molecules is based on their thermal and optical stability.

The resonant optical bleaching reactions A → a, B → b, C → c etc. are shown in Figure 6. The reaction represents the first chain termination steps with $DR_n \rightarrow AC_n$, which is interpreted in terms of an intramolecular hydrogen insertion reaction. This is in contrast to earlier interpretations [7,15,17].

ESR spectra

The triplet (S=1) and quintet (S=2) ESR spectra of the reactive DR, AC and DC intermediates show a specific fine structure and hyperfine structure pattern as a consequence of the different number and position of the sp^2 and p_z radical electron. Typical ESR spectra and energy level diagrams of the reaction intermediates of the low-temperature photopolymerization reaction are shown in Figure 7. The spectra (a), (b) and (c) only show a selection of the numerous intermediates (with $2 \leq n \leq 18$) observed after UV-irradiation (308 nm) with an excimer laser (a) and after thermal annealing (b) and (c) of the diacetylene monomer crystal.

Three types of intermediates can be distinguished. The DR triplet states (a), which show a small fine structure splitting between the two Δm=1 ESR lines, the AC triplet states (c), which show a large fine structure splitting between the two Δm=1 ESR lines and the DC quintet states (b), which also show a large fine structure splitting between the four Δm=1 ESR lines. The triplet state arise from two, the quintet states from four S=1/2 unpaired electrons. As shown theoretically by Benk and Sixl [8], the DC quintet states are not pure quintet S=2 states but are quantitatively described by mixed singlet-quintets especially for the limit of large n.

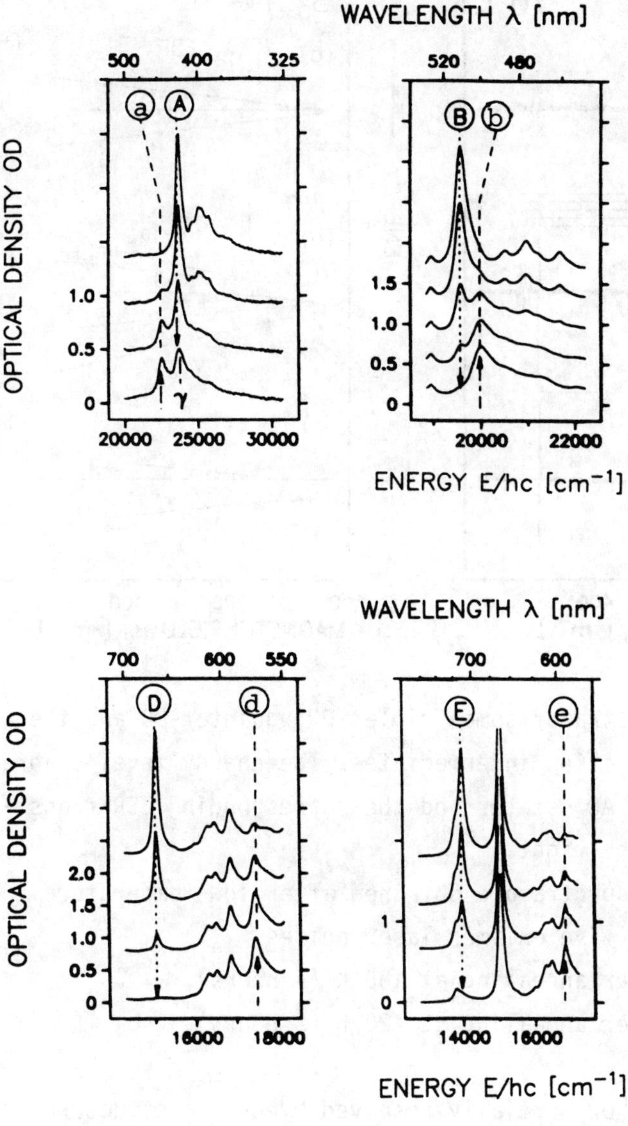

FIGURE 6 Resonant optical bleaching of the diradical molecules at T = 10 K. A → a (n=2), B → b (n=3), C → c (n=4), D → d (n=5), E → e (n=6), F → f (n=7).

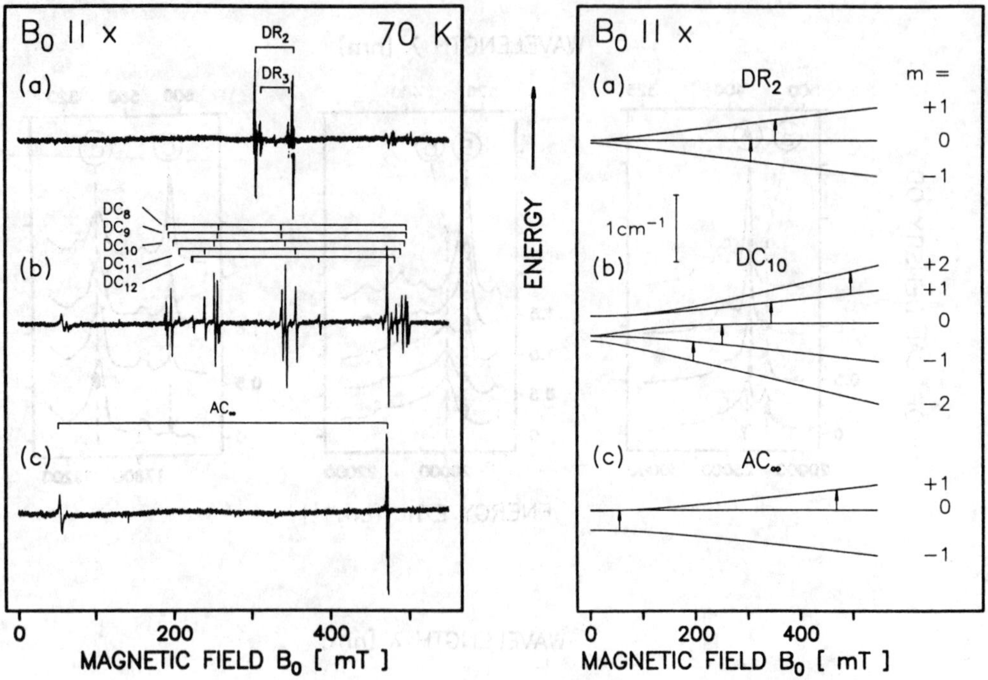

FIGURE 7 ESR spectra of some triplet-DR, quintet-DC and the final triplet-AC_∞ reaction intermediates. The energy level diagrams of the DR_2, DC_{10} and AC_∞ states and the corresponding ESR transitions are shown on the right.

(a) The DR_2 and DR_3 spectra are obtained after low-temperature irradiation with five excimer laser pulses.

(b) Same crystal after annealing at 100 K (4 hours).

(c) Same crystal after annealing at 120 K (2 hours).

All ESR lines show partially resolved hyperfine structure as shown in Figure 8, due to hyperfine interaction of the electron spin with the nuclear spins $I_1 = I_2 = 1/2$ of the two protons of the CH_2-rest group. In the case of the diradicals and dicarbenes two CH_2 groups of the two chain ends are involved in the hyperfine interaction. In the case of the asymmetric carbenes only one chain

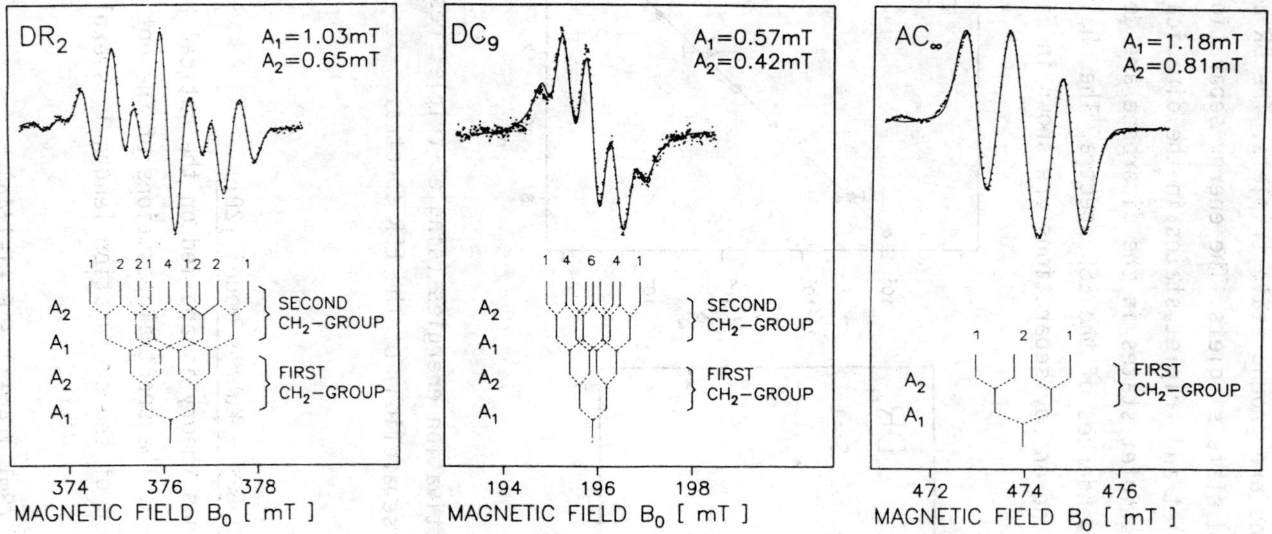

FIGURE 8 Hyperfine pattern of the DR, DC and AC reaction intermediates. The simulated hyperfine structure has been fitted to the experimental points. The hyperfine structure is due to the interaction of the radical electrons with two non-equivalent protons of the first and second CH_2-groups at the chain ends of the intermediates.

end is involved. The experimental hyperfine pattern is reproduced by a computer calculation.

The AC intermediates are ground state triplets. The DR and DC intermediates are ground state singlets. The energy separations ε_{ST} and ε_{SQ} of the singlet and triplet states in the diradicals and of the singlet and quintet states in the dicarbene are deduced from the temperature dependences of the ESR spectra. The chain length dependencies of the energy separations are shown in Figure 9.

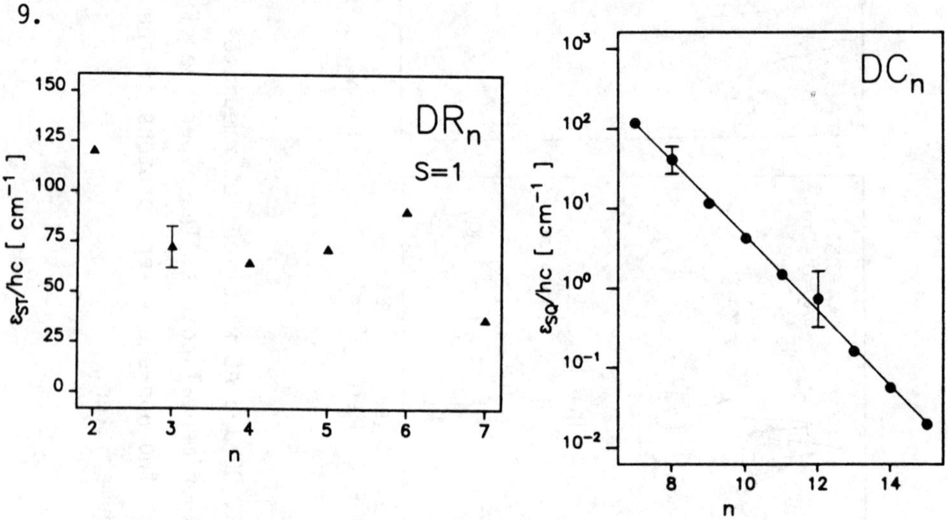

FIGURE 9   Thermal activation energies (singlet-triplet resp. singlet-quintet energy separation) of the ESR spectra.

THEORETICAL MODELS

One-Dimensional Electron Gas - Kuhn's model [20]

A simple electron gas theory is applied on the optical absorption series of Figure 3-5. The spectral positions of the optical absorptions as a function of the conjugation length l is explicitely given by

$$E_n = (h^2/8ml^2)(4n+1) + E_\infty(1-1/4n) \tag{1}$$

with $l = a_1 n + a_2$; h and m are Plancks constant and the electron mass, respectively. a_1 is the length of the repetition unit with $a_1(SO) = a_1(AC) = d(C \equiv C) + 2d(C-C) + d(C=C)$ and $a_1(DR) = d(C-C) + 3d(C=)$. a_2 describes the boundary conditions at the ends of the oligomers.

The zero-phonon line positions of the DR_n, AC_n and SO_n photoproducts of Figures 3-5 are summarized in Figure 10. The fit of the theoretical curves to the experimental points by variations of only a_2 and E_∞ is almost perfect and reflects the essential features of the photoproducts.

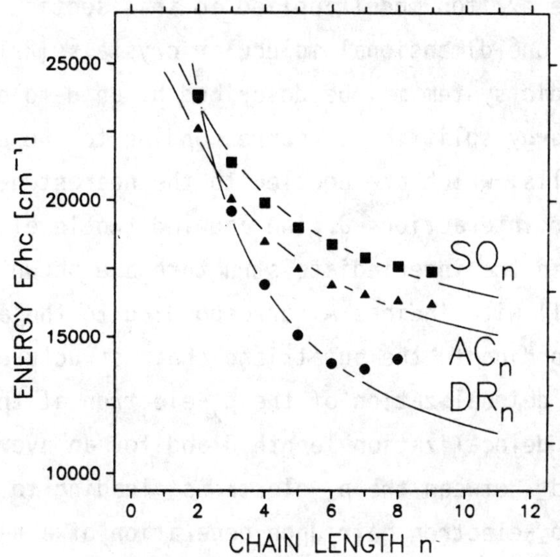

FIGURE 10 Zero-phonon absorption energies of the different SO, AC and DR photoreaction products. The points are obtained from the spectra of Fig. 3-5. The curves are obtained with Equ. (1).

The convergence energies E_∞ of the AC and SO series for $n \rightarrow \infty$ are closest to the absorption energy of the polymer chain (E_p=17000 cm^{-1}) and consequently are suspected of having related structures. However, the convergence of the DR series A, B, C,... is strongly red shifted in comparison with the absorption of the polymer. The red shift is due to the reduced bond alternation in the butatriene chain structure of the diradical molecules, consistent with the proposed chain structures of Figure 2.

Owing to the close relationship between the polymer absorption and the E_∞ value of the AC-series an acetylene structure of the final polymer chain is deduced. This is in accordance with the proposed acetylene structure of the long polymer chains, as deduced originally from the X-ray structure data [21].

One dimensional exciton Model [22]

The exciton model applied in this section leads back to earlier work in one-dimensional molecular crystals [22]. The excitation of a periodic system may be described by an n-fold two level system with energy splitting ε, corresponding to the excitation of the unit cells, which are coupled to the nearest neighbours by energy transfer interactions J. The coupled two-level systems of the finite chains in the intermediate structure are shown schematically in Figure 11 with indices A corresponding to the acetylene and B corresponding to the butatriene chain structure. The model accounts for the delocalization of the p_z-electron at the chain ends with an average delocalization length d and for an average interaction length d_π between the p_z-electrons, leading to recombination of the p_z-electron pair upon generation of a π-bond. With $d_\pi > \ell - 2d$ an "up" and "down" radical electron pair is formed as shown in the DC structure. The simple configuration of pure butatriene or pure acetylene structures is characterized by a degenerate energy level system described by the Hamiltonian

$$\hat{H} = \varepsilon \sum_{i=1}^{n} |\phi_i\rangle\langle\phi_i| + J \sum_{i=1}^{n-1} \left(|\phi_i\rangle\langle\phi_{i+1}| + |\phi_{i+1}\rangle\langle\phi_i|\right) \qquad (2)$$

$\varepsilon = \varepsilon_A$ or ε_B is the excitation energy of the acetylene or butatriene unit, $J = J_A$ or J_B is the energy transfer matrix element. Only nearest neighbour interaction is taken into account. The energy eigenvalues then are given by

$$E_{nk} = \varepsilon + 2J \cos \frac{k\pi}{n+1} \text{ with } k = 1...n, \qquad (3)$$

The excitation energy of the n-mer molecule therefore is split into n states. The optically allowed transition energy is given with k = n. The following energies and energy transfer matrix elements are obtained by a fit of the theoretical dependence (Equ. 3) to the experimental points of Fig. 11.

$$\varepsilon_B = 36\ 100\ \text{cm}^{-1}, \qquad \varepsilon_A = 31\ 080\ \text{cm}^{-1},$$
$$J_B = 12\ 050\ \text{cm}^{-1}, \qquad J_A = 7\ 080\ \text{cm}^{-1}.$$

The DR structures change to DC structures at a critical chain length given by $\bar{l} > 2d + d_\pi$, forming a p_z-radical electron pair.

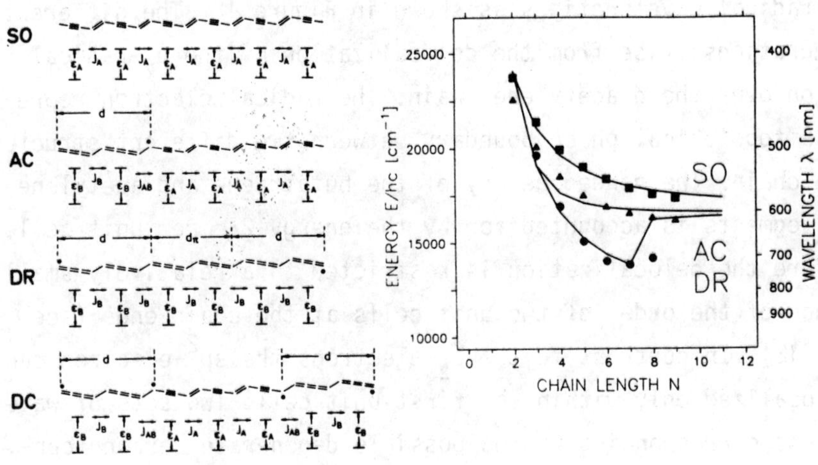

FIGURE 11 One-dimensional exciton model. The energy level scheme corresponding to the SO, AC, DR and DC structures is shown on the left. The resulting chain-length dependence of the absorption energies is shown on the right. The kink in the DR-curve is due to a DR → DC transition at about n=7.

Due to the presence of butatriene and acetylene chain structures diagonal and off-diagonal disorder is introduced into the Hamiltonian of the resulting DC as well as in the AC-states. Using the values ε_A, ε_B, J_A and J_B of the DR and SO dependencies and d = 2a, d_π = 3a (with unit cell length a) the best fits of the AC and DC-dependencies are obtained with $J_{AB} = 9\,770\ cm^{-1}$.

Radial Electron Model - Valence Bond Theory

The identification of the AC and DC intermediates in terms of pure carbene states $-\ddot{C}-$ as shown in Figure 2 is not correct. It is obvious from the ESR-fine structure splitting, that the fine structure constants D of the AC and DC states discussed below are only about half the value $D(-\ddot{C}-) = 0{,}7\ cm^{-1}$ of the pure carbene, which has been deduced from ESR measurements [23]. Therefore we have to conclude that about 50 percent of the wavefunctions are not carbene but diradical wavefunctions as shown in Figure 12. The different configurations arise from the delocalization of the p_z-radical electron over the diacetylene chain. The radical electron represents a topological phase boundary between two different structures of the chain. The nondegeneracy of the butatriene and acetylene chain segments is accounted for by the energy $2\cdot\varepsilon$ per unit cell. Therefore the delocalization is restricted to a relatively small distance of the order of two unit cells at the chain ends (see Figure 12). In contrast to the p_z-electrons the sp^2-electron can be delocalized only within the first unit cell. Two sets of wavefunctions corresponding to two possible degenerate carbene configurations $\psi_{1,1}$ and $\psi_{3,3}$ are used in the calculation.

The wavefunctions shown below represent a system of basis functions characterized by the positions of the two radical electrons i = 1,3 and j = 1,3,5,... (4n-5). The ground state electronic configuration ψ of the AC chain end is given by the linear combination

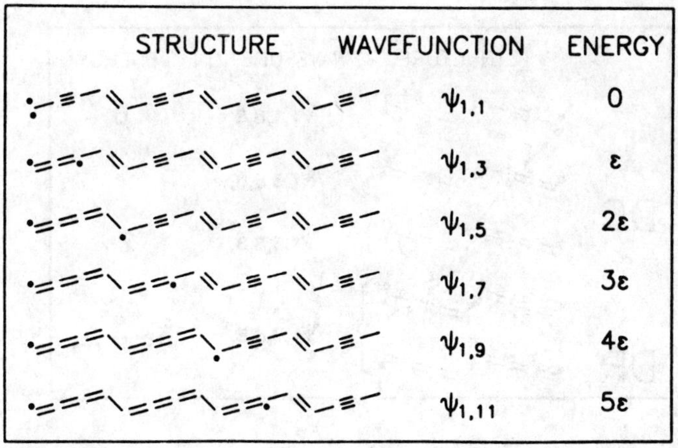

FIGURE 12 Basic configurations of the radical electron distribution in the AC intermediates. The delocalization of the p_z-electron is correlated with a linear increase in the energy.

$$\psi(AC) = \sum_{i,j} c_{i,j}\,\psi_{i,j} + \sum_{i} c_{i,t}\,\psi_{i,t} \tag{4}$$

with $$\langle\psi_{i,j}|\psi_{k,l}\rangle = \delta_{ik}\,\delta_{jl}.$$

At the unreactive chain ends we have to account for weak delocalization of the p_z radical electron into the rest group. The corresponding wavefunction is denoted by $\psi_{i,t}$.

The wavefunction of Figure 13 schows some dimer DC configurations characterized by the positions of four radical electrons $i = 1,3$; $j = 1,3,5,7$; $k = 8,6$; $l = 8,6,4,2$. The dicarbene configuration $\psi_{1,7,8,8}$ is equivalent with the diradical configuration $\psi_{1,8}$, due to recombination of the p_z-electrons forming a π-bond with energy ε_π. The ground state configuration ψ of the DC or DR chains is given by:

$$\psi(DR/DC) = \sum_{i,j,k,l} c_{i,j,k,l}\,\psi_{i,j,k,l} \tag{5}$$

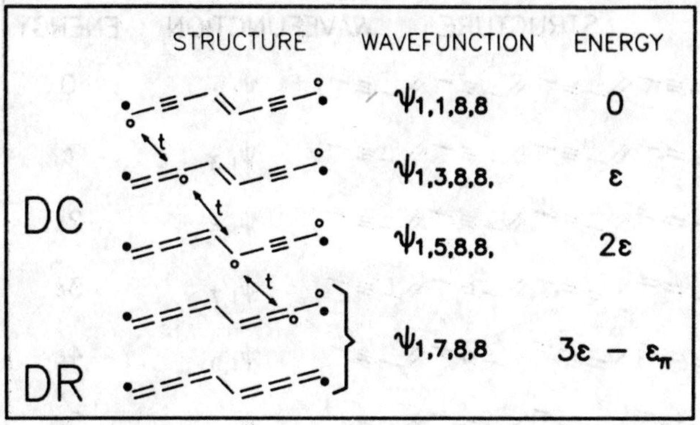

FIGURE 13 Basic configurations of the radical electron distribution in the DC and DR intermediates. The dicarbenes change to diradicals upon recombination of the p_z-electrons.

with $\langle\psi_{i,j,k,l}|\psi_{m,n,o,p}\rangle = \delta_{im}\,\delta_{jn}\,\delta_{ko}\,\delta_{lp}$.

The Hamiltonian of the AC and DR/DC states is given by

$$\hat{H} = \hat{H}_o + \hat{T} \qquad (6)$$

$\hat{H}_o$ accounts for the different energies of the individual configuration and of the special energy of the terminal group. An energy amount of ε is necessary for each transfer of the p_z-radical electron by two carben atoms, due to the increase of the high energetic butatriene chain length. For the transfer of the radical electron into the terminal group an energy of ε_t has to be taken into account.

$$\hat{H}_o(AC) = \sum_{i,j} \left(\frac{|i-j|}{2}\,\varepsilon|\psi_{i,j}\rangle\langle\psi_{i,j}| + \sum_i \{[(4n-5-i)/2]\varepsilon+\varepsilon_t\}|\psi_{i,t}\rangle\langle\psi_{i,t}| \right. \qquad (7)$$

$$\hat{H}_o(DR/DC) = \varepsilon\left(\frac{|j-i|}{2} + \frac{|l-k|}{2}\right)|\psi_{i,j,k,l}\rangle\langle\psi_{i,j,k,l}| \qquad (8)$$

The interaction between the different configurations is given by

$$\hat{T}(AC) = \sum_{i,j} t\{|\psi_{i,j}\rangle\langle\psi_{i,j\pm2}| + |\psi_{i,j}\rangle\langle\psi_{i\pm2,j}|\}$$

$$+ \sum_{i}\{t_t|\psi_{i,4n-5}\rangle\langle\psi_{i,t}| + t|\psi_{i\pm2,t}\rangle\langle\psi_{i,t}|\}. \tag{9}$$

The interaction t_t with the terminal group differs from that of the p_x and p_z electrons with $t = t_x = t_z$.

$$\hat{T}(DR/DC) = \sum_{i,j,k,l} t\{|\psi_{i\pm2,i,k,l}\rangle\langle\psi_{i,j,k,l}|$$

$$+ |\psi_{i,j\pm2,k,l}\rangle\langle\psi_{i,j,k,l}| + |\psi_{i,j,k\pm2,l}\rangle\langle\psi_{i,j,k,l}|$$

$$+ |\psi_{i,j,k,l\pm2}\rangle\langle\psi_{i,j,k,l}|)\}. \tag{10}$$

The energy matrix was diagonalized numerically. The spin distribution of the ground state is given by the projection coefficients $c_{i,j,k,l}$ of the eigenvector belonging to the lowest eigenvalue.

The probability for the appearance of the electronic configuration $|\psi_{i,j,k,l}\rangle$ is given by

$$|c_{i,j,k,l}|^2 = |\langle\psi_{i,j,k,l}|\psi\rangle|^2. \tag{11}$$

Figure 14 shows the probability density of the radical electrons for short chain intermediates, which is equal to the probability density of the unpaired spins. The only free parameters in the theory are given by the ratios t/ε and $\varepsilon_\pi/\varepsilon$. Figure 14 shows the best fit to the fine structure and hyperfine structure data of Figure 8. The essential result of the calculation is a very small probability density of the p_z-radical electrons in short chains with $2 \leqq n \leqq 5$. This is due to an almost perfect recombination of the p_z-radical electrons. In chains which are longer than $n = 5$ the p_z-radical electrons are well separated from each other. Therefore they are not able to recombine. The two p_z-radical electrons

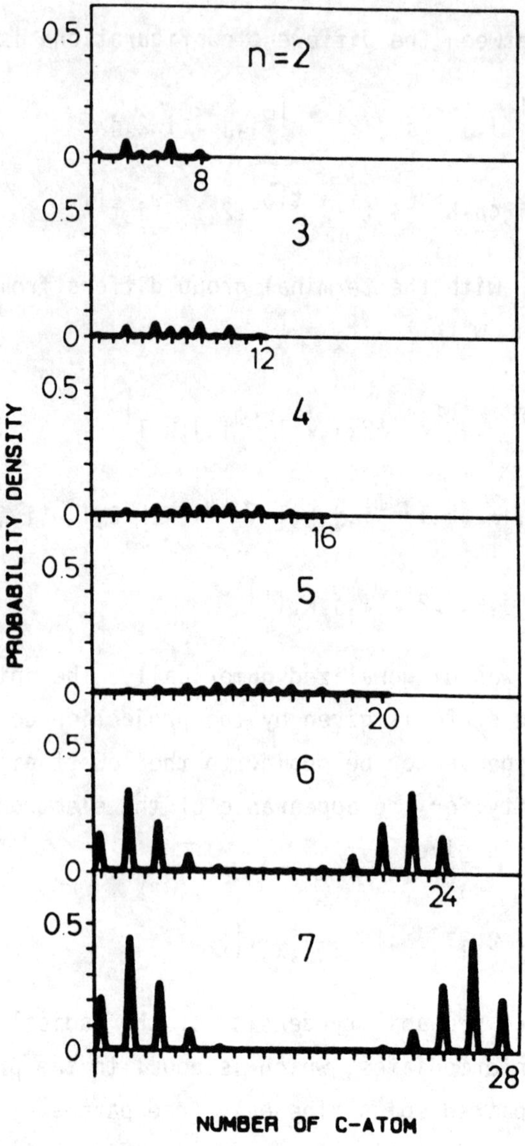

FIGURE 14 Probability density of the two p_z-radical electrons. The first four molecules are almost perfect diradicals, the hexamer and heptamer molecules have changed to dicarbenes. Note that the "up" p_z-electron on the left is located on odd carbon atom positions, whereas the "down" p_z-electron on the right is located on even carbon atom positions.

then are localized symmetrically on both chain ends. With $n \geqq 6$ the energy gain of the central acetylene structure is larger than the π-bond energy, which is gained by recombination of the p_z-electrons. Consequently the intermediates with $2 \leqq n \leqq 5$ are almost pure diradicals DR_n and with $n \geqq 6$ are almost pure dicarbenes DC_n.

The transition from dicarbenes DC_n to diradicals DR_n at short chain length corresponds to a structural phase transition of the polymer backbone from an acetylene to a butatriene chain structure (see Figure 2). Reducing the chain length the transition may be interpreted as recombination of the two p_z-radical electrons. Increasing the chain length the transition may be interpreted as generation of a p_z-radical electron pair. All effects are essentially described by only two parameters $\varepsilon_\pi/\varepsilon = 14.37$ and $t/\varepsilon = 2.5$, yielding $2\varepsilon \cong 0.4$ eV an $t \cong 0.5$ eV with a π-bond energy of $\varepsilon_\pi = 2.8$ eV. The resulting ε-value is in excellent agreement with theoretical calculations of Whangbo et. al. [24] and Karpfen [25].

A comparison of the polydiacetylene p_z-radical electron distribution function with that of the trans-$(CH)_x$ model system is shown in Figure 15. In the upper part the result of the soliton calculation of Su, Schrieffer and Heeger [26] is reproduced showing symmetrically shaped "up" and "down" electron densities on an infinitely long trans-$(CH)_x$ chain. The shape and finite extension of the soliton wavefunction in $(CH)_x$ is due to electron-phonon coupling. The probability density of the corresponding p_z-radical electrons in long polydiacetylene dicarbene structures is shown below. In contrast to trans-$(CH)_x$ the radical electrons now are localized at the ends of the polydiacetylene chains. As compared to trans-$(CH)_x$ the corresponding non-symmetrically shaped p_z-electron distribution is strongly compressed. These effects are due to the non-degeneracy of the butatriene and acetylene structures on either side of the p_z-radical electrons only.

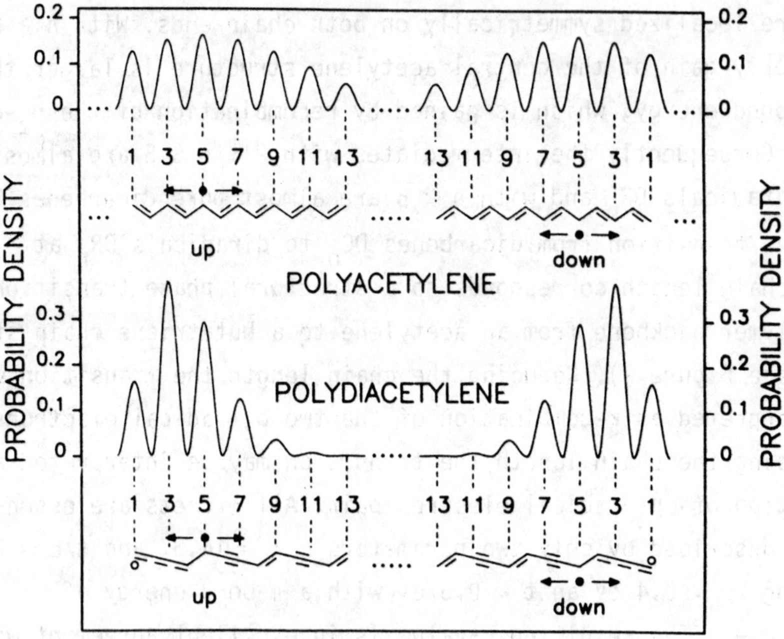

FIGURE 15 Comparison of the probability densities of the p_z radical electron distribution in trans-polyacetylene and in polydiacetylenes with finite chain length. Due to the non-degeneracy of the butatriene and acetylene structure the p_z-radical electrons are localized at the ends of the polydiacetylene chains.

Therefore the radical electrons are forced to the ends of the polydiacetylene chains. Consequently the shape of the radical electron wavefunction is predominantely determined by ε and the boundary conditions.

The ESR spectra of pure triplet and quintet states are described by the spin Hamiltonian

$$\hat{H} = g\mu_B \vec{B}_o \hat{\vec{S}} + D(\hat{S}_z^2 - \frac{1}{3}S(S + 1) + E(\hat{S}_x^2 - \hat{S}_y^2) \qquad (12)$$

with fine structure parameters D and E. $\hat{\vec{S}}$ is the spin operator with spin quantum number S = 1 for the triplet and S = 2 for the quintet states. A detailled discussion for mixed singlet-quintet states is given in refs [8-10].

The shape of the electronic wavefunction is essentially determined by the D-value with

$$D_{i,j} = \frac{3}{4} g^2 \mu_B^2 < \frac{r_{i,j}^2 - 3z_{i,j}^2}{r_{i,j}^5} > \qquad (13)$$

$r_{i,j} = (x_{i,j}, y_{i,j}, z_{i,j})$ is the distance vector of the two electrons i and j. The total D-value is given by

$$D(AC) = \sum_{i,j} |c_{i,j}|^2 D_{i,j} \qquad (14)$$

and by

$$D(DR/DC) = \frac{1}{2} \sum_{i,j,k,l} |c_{i,j,k,l}|^2 (D_{i,j} + D_{k,l}). \qquad (15)$$

The D(AC) and D(DR/DC) values of the different AC and DR/DC reaction intermediates as well as the hyperfine interaction constants A_1 and A_2 have been calculated for different chain length n [14]. The chain length dependences of the experimental and theoretical D(AC) and A_1(AC), A_2(AC) values are shown in Figure 16. The largest D value is obtained in the shortest unit, which is the monomer. The decrease of the D values with increasing chain length is due to the delocalization of the p_z radical electron into the dimer, trimer, tetramer ... units. The asymptotic dependence converges very rapidely at n = 4 to the final D(∞) value, which is identical to that of the DC molecules. The rapid convergence of the D-values is a clear indication of the short delocalization length of the p_z-electron, which is of the order of two monomer units. The parameters of the computer fits are $t/\varepsilon = 3.7, \varepsilon/\varepsilon_t = 1.6$ and $t_z/t = 0.7$.

The chain length dependences of the experimental and theoretical D(DR/DC) values and of the hyperfine constants A_1 and A_2(DR/DC) are shown in Figure 17. The best fit of the calculated D(n) dependence to the experimental data is obtained with the parameters of the probability density of the p_z-electron shown in Figure 14.

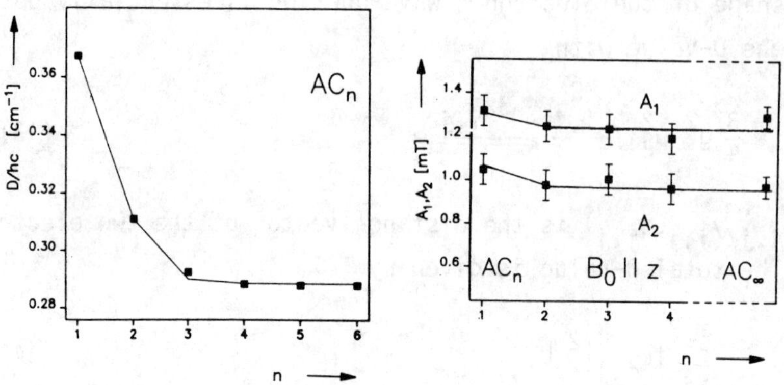

FIGURE 16 Fine structure constants D and hyperfine structure constants A_1, A_2 of the AC_n states as a function of the chain length n. The points are deduced from the ESR-spectra [5], the curve is calculated using the configurations of Figure 12.

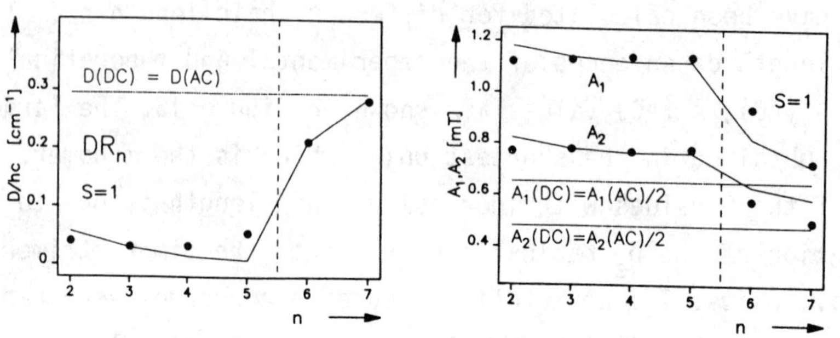

FIGURE 17 Fine structure constants and hyperfine structure constants of the DR_n/DC_n states as a function of the chain-length n. The dashed line separates the almost pure diradical states with $n \leqq 5$ from the mixed states witn n = 6 and from the dicarbene states with $n \geqq 7$. The points are experimental [13] the curves are calculated using the probability densities of Figure 14.

SUMMARY

The electronic structures of the DR, DC, AC and SO reaction intermediates in the polymerization reaction of diacetylene crystals are consistently interpreted in terms of their butatriene or acetylene structure and by their radical electron structure. The restricted delocalization of the p_z electron is essentially due to the energy difference of $2\varepsilon = 0.4$ eV per unit cell between the butatriene and acetylene chain structure. The chain length dependencies of the fine structure and hyperfine structure constants of the ESR spectra as well as the chain length dependencies of the absorption series have been described by theoretical model calculations. The complexity of the polymerzation reaction in diacetylene crystals has been demonstrated by the manifold of different reaction products observed in the experiments. The large amount of detailed information concerning the electronic structures and concerning the reaction mechanisms is essentially due to the crystalline structure of this model system and to the thermal stability of the reaction intermediates at low temperatures.

ACKNOWLEDGEMENTS

This work has been supported by the Deutsche Forschungsgemeinschaft and by the Stiftung Volkswagenwerk. The author thanks H. Gross, W. Neumann and R. Huber for efficient cooperation and W. Tuffentsammer for crystal preparations. Stimulating discussions with all co-workers as well as with E. Sigmund, G. Wegner, M. Schwoerer and H.C. Wolf are gratefully acknowledged.

REFERENCES

1. Wegner, G., Molecular Metals (W.E. Hatfield, Plenum Press, N.Y. 1979) p. 209 and Chemistry and Physics of One-Dimensional Metals (H.J. Keller, Plenum Press N.Y. 1977) p. 297.
2. Baughman, R.H., and R.R. Chance, Synthesis and Properties of Low-Dimensionals Materials (J.S. Miller and A. Epstein, Ac. of Science, N.Y. 1978) p. 705.
3. Enkelmann, V., Adv. in Polym. Sci. 63, (1984) 91.
4. Sixl, H., Adv. in Polym. Sci. 63, (1984) 49.
5. Bubeck, C., H. Sixl and H.C. Wolf, Chem. Phys. 32, (1978) 231.
6. Bubeck, C., W. Neumann and H. Sixl, Chem. Phys. 48, (1980) 269.
7. Bubeck, C., W. Hersel, H. Sixl and J. Waldmann, Chem. Phys. 51, (1980) 1.
8. Benk, H. and H. Sixl, Mol. Phys. 42, (1981) 779.
9. Huber, R.A., M. Schwoerer, H. Benk and H. Sixl, Chem. Phys. Lett. 78, (1981) 416.
10. Kollmar, C., H. Sixl, H. Benk, V. Denner and G. Mahler, Chem. Phys. Lett. 87, (1982) 266.
11. Neumann, W. and H. Sixl, Chem. Phys. 50, (1980) 273 and 58, (1981) 303.
12. Gross, H., W. Neumann and H. Sixl, Chem. Phys. Lett. 95, (1983) 584.
13. Sixl, H. and W. Neumann, Mol. Cryst. Liq. Cryst. 105, (1984) 41.
14. Sixl, H., W. Neumann, R. Huber, V. Denner and E. Sigmund, Phys. Rev. B. submitted (1984).
15. Hersel, W., H. Sixl and G. Wegner, Chem. Phys. Lett. 73, (1980) 288.
16. Gross, H. and H. Sixl, Chem. Phys. Lett. 91, (1982) 262.
17. Niederwald, H., H. Eichele and M. Schwoerer, Chem. Phys. Lett. 72, (1980) 242.
18. Gross, H., W. Neumann and H. Sixl, Chem. Phys. Lett. 95, (1983) 584.

19. Gross, H., H. Sixl, S.F. Fischer and E.W. Knapp, Chem. Phys. 84, (1984) 321.

20. Kuhn, H., J. Chem. Phys. 17 (1949) 1198. G.J. Exarhos, W.M. Risen jr. and R.H. Baughman, J. Am. Chem. Sox. 98, (1976) 481.

21. Kobelt, D. and E.F. Paulus, Acta crystallogr. B. 30, (1974) 232. E. Hädicke, E.C. Mez, C.H. Krauch, G. Wegner and J. Kaiser, Angew. Chemie 83, (1971) 253.

22. Hochstrasser, R.M. and J.D. Whiteman, J. Chem. Phys. 56, (1977) 4995. H. Benk, H. Haken and H. Sixl, J. Chem. Phys. 77, (1982) 5730. C. Kollmar and R. Huber, unpublished results.

23. Bernheim, R.A. and S.H. Chien, J. Chem. Phys. 65, (1976) 2023.

24. Whangbo, M.-H., R. Hoffmann and R.B. Woodward, Proc. R. Soc. Lond. A 366, (1979) 23.

25. Karpfen, A., J. Chem. Phys. C 13, (1980) 5673.

26. Su, W.P., J.R. Schrieffer and A.J. Heeger, Phys. Rev. B 22, (1980) 2099.

STRUCTURAL STUDY OF THE TOPOCHEMICAL POLYMERIZATION OF TWO DIACETYLENES

P.A. ALBOUY, J.N. PATILLON and J.P. POUGET

Laboratoire de Physique des Solides and LURE, Université de Paris-Sud, 91405 ORSAY , France.

Abstract

Some structural features of the mechanism of thermal polymerization of two diacetylenes, PTS and DNP, are reviewed. It is shown that the reaction proceeds homogeneously with polymer chains growing randomly in the crystal. An evaluation of the average polymer chain length has been obtained in both cases.

I - INTRODUCTION

The nearly defect free nature of diacetylene single crystals offers a unique opportunity for the study of the topochemical polymerization reaction. The goal of this paper is to summarize three years of study on some structural aspects of this polymerization process by X ray scattering techniques. We shall try to show what kind of information can be collected depending upon the price we are ready to pay for it. The results will concern two representative compounds, PTS ($R = -CH_2-O-SO_2-\Phi-CH_3$) and DNP ($R = -CH_2-\overset{NO_2}{\Phi}-NO_2$). Which display the same well known kinetics upon thermal annealing (1,2) : an induction period followed after a time to (conversion : about 15%) by an autocatalytic period (see Fig. 1).

In both compounds, the ratio k/k_o is of the order of 150. Furthermore, the activation energy is the same in both periods for both compounds. At the difference of PTS, DNP does not preserve its single-crystal arrangment up to the end of the reaction. Above 90 per cent conversion, there is an evolution toward a new phase made of micro-crystals of different lateral parameters.

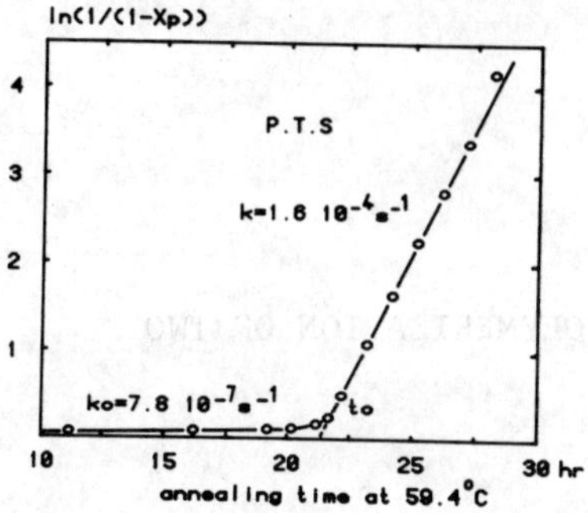

a

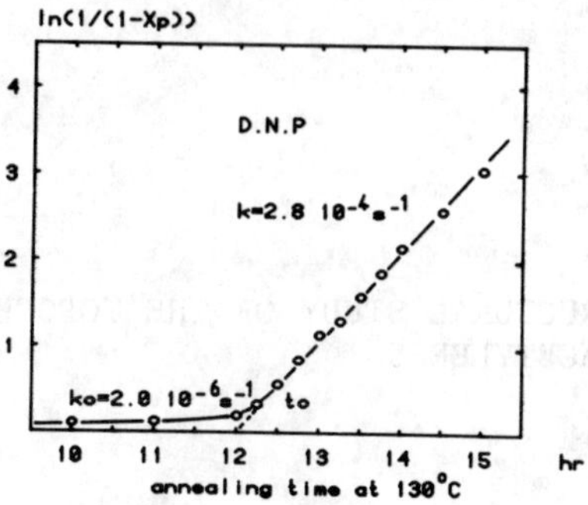

b

Fig. 1. a and b . $\ell n\ (1/(1-x_p))$ versus time for PTS and DNP. x_p is the polymer content. The kinetics is first order in both periods .(to = 21^{hr} 25 (a) and 12^{hr} (b)).

II - EXPERIMENTAL

a) Sample preparation

PTS monomer single crystals were grown from acetone solutions by slow evaporation at 0°C (3). DNP single crystals were grown by slow cooling of a saturated nitromethane solution from 90°C down to room temperature (2). DNP monomer crystals are practically insensitive toward light and do not polymerize markedly below 100°C. Both compounds were polymerized by thermal annealing at 59.4°C for PTS and 130°C for DNP (3).

b) X ray apparatus and techniques

We have used either conventional techniques like precession, Weissenberg or oscillating crystal methods to get the lattice parameters of the crystals, or the more unusual "monochromatic Laue method" for the study of diffuse scattering. In this latter technique, both the crystal and the film are kept fixed in a monochromatized X ray beam. Thus the same portion of the reciprocal lattice is continuously in reflection position, which permits the observation of very weak scattered intensities.

In the case of PTS which polymerizes rapidly under the X ray beam at room temperature, the crystal was cooled at 20 K in order to quench the polymerization. This was not necessary with DNP which is much less photoreactive.

X ray sources were either conventional generators, or the synchrotron radiation of LURE (Orsay) which has the advantage of combining an higher intensity with a greater resolution than conventional sources.

III - X RAYS SCATTERED BY POLYMER CHAINS

In fact, it is difficult to establish a simple and easy to handle formula in the general case of X rays scattered by a mixed (monomer polymer) crystal . We shall thus follow a step by step approach and discuss the underlying hypothesis and simplifications.

a) Case of an isolated polymer chain

Let us consider a polymer chain of N units as shown on Fig. 2 below. The scattered X ray amplitude for a scattering vector $\vec{q}$ is given by :

$$A(\vec{q}) = \sum_{k=1}^{N} F_p(\vec{q})\, e^{i\vec{q}.(k.d_z.\vec{z})}$$

, where $F_p(\vec{q})$ is the molecular structure factor of a polymer unit. It can be written :

$$A(\vec{q}) = F_p(\vec{q}) \sum_{k=1}^{N} e^{iq_z k d_z} = F_p(\vec{q})\, e^{+iq_z d_z \frac{N+1}{2}} \frac{\sin \frac{q_z N d_z}{2}}{\sin \frac{q_z d_z}{2}}$$

The observed intensity is : $I(\vec{q}) = |A(\vec{q})|^2 = |F_p(\vec{q})|^2 \dfrac{\sin^2 \frac{q_z N d_z}{2}}{\sin^2 \frac{q_z d_z}{2}}$

This quantity depends on the two lateral components q_x and q_y only through $F_p(q)$, which is a smoothly varying function. The right part of the formula is the well-known Laue function. It displays a maximum value at $q_z = \pm 2\pi\ell/d_z$ where ℓ is an integer, and half the width at half maximum (HWHM) of the main peaks is related to the chain length Nd_z by the Scherrer formula $\delta q_z^{1/2} \simeq \dfrac{0.888\pi}{N.d_z}$ (4).

To summarize, for a given orientation of the chain with respect to the incident X ray beam, the scattering is composed of a serie of cones having the chain direction as axis (see Fig.3). Each cone corresponds to scattering $\vec{q}$ vectors with the in-chain component $q_z = \pm 2\pi\ell/d_z$. Their intensity is smoothly modulated by the $\vec{q}$ dependence of the structure factor $|F_p(\vec{q})|$.

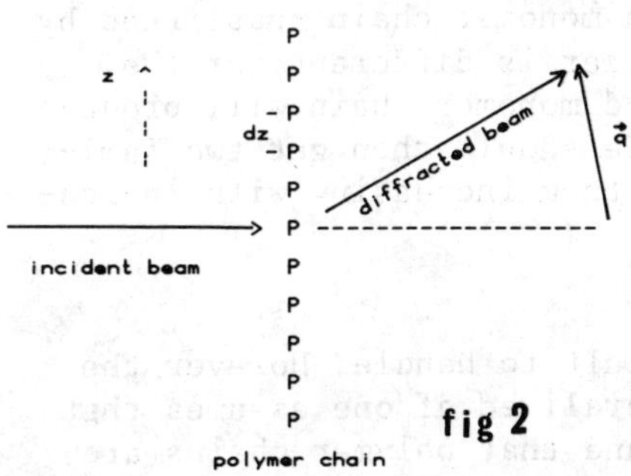

fig 2

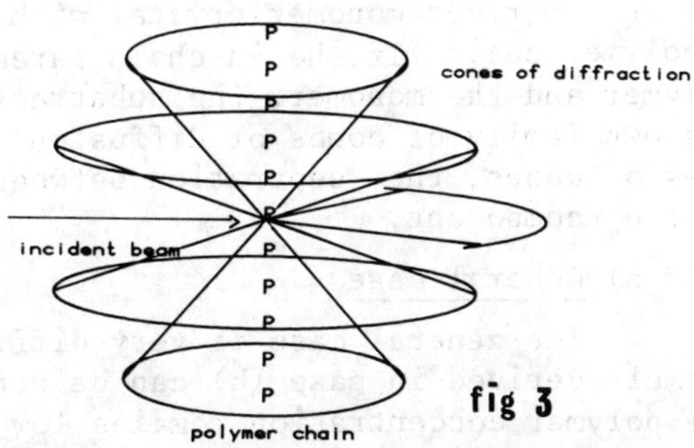

fig 3

b) Case of a polymer chain embedded in a monomer matrix :

For the sake of simplicity, we shall first assume that polymer and monomer have the same parameter d_z along the chain direction. The lateral parameters are d_x and d_y.

In this case :

$$A(q) = \sum_{\ell=-N_x/2}^{N_x/2} \sum_{m=-N_y/2}^{N_y/2} \sum_{n=-N_z/2}^{N_z/2} F_M(\vec{q})\, e^{i\vec{q}.[\ell d_x\vec{x} + md_y\vec{y} + nd_z\vec{z}]}$$

$$- \sum_{k=1}^{N} F_M(\vec{q})\, e^{i\vec{q}.kd_z\vec{z}} + \sum_{k=1}^{N} F_P(\vec{q})\, e^{i\vec{q}.kd_z\vec{z}}$$

The first two terms correspond to diffraction by the monomer (crystal size : $N_x d_x \times N_y d_y \times N_z d_z$) and the last one, to the diffraction by the polymer chain.

For large N_x, N_y, N_z : $I(\vec{q}) = |F_M(\vec{q})|^2 \dfrac{\sin^2 \frac{q_x N_x d_x}{2}}{\sin^2 \frac{q_x d_x}{2}} \dfrac{\sin^2 \frac{q_y N_y d_y}{2}}{\sin^2 \frac{q_y d_y}{2}} \times$

$$\frac{\sin^2 \frac{q_z N_z d_z}{2}}{\sin^2 \frac{q_z d_z}{2}} + |F_M(\vec{q}) - F_p(\vec{q})|^2 \frac{\sin^2 \frac{q_z N d_z}{2}}{\sin^2 \frac{q_z d_z}{2}} .$$

The first term represents the Bragg intensity. It displays intense maxima (Bragg reflections) for particular values of the $\vec{q}$ vector.

The second term represents a diffuse intensity which comes from the difference between the perfect monomer crystal and the real crystal. This is the same formula as case (a) with $|F_p(\vec{q})|^2$ replaced by $|F_M(\vec{q}) - F_p(\vec{q})|^2$.

The scattered intensity was calculated by decomposing the crystal in a perfect monomer crystal with a monomer chain substitued by a polymer chain. If the in-chain parameter is different for the polymer and the monomer, the substracted monomer chain will produce its own family of cones of diffusion. We should then get two families of cones, the separation between them increasing with increasing q_z component.

c) General case

The general case is very difficult to handle. However the formula derived in case (b) can be generalized if one assumes that the polymer concentration remains low and that polymer chains are randomly distributed in the matrix.

In that case : $I(\vec{q}) = |C_M F_M(\vec{q}) + C_p F_p(\vec{q})|^2 \times$

$$\frac{\sin^2 \frac{q_x N_x d_x}{2}}{\sin^2 \frac{q_x d_x}{2}} \; \frac{\sin^2 \frac{q_y N_y d_y}{2}}{\sin^2 \frac{q_y d_y}{2}} \; \frac{\sin^2 \frac{q_z N_z d_z}{2}}{\sin^2 \frac{q_z d_z}{2}}$$

$$+ C_M C_P \, |F_M(\vec{q}) - F_P(\vec{q})|^2 \sum_{N_z=2}^{\infty} P(N_z) \frac{\sin^2 \frac{q_z N_z d_z}{2}}{\sin^2 \frac{q_z d_z}{2}}$$

where $C_M (C_p)$ is the monomer (polymer) concentration and $P(N_z)$ the number of chains having N_z units.

The first term is the Bragg intensity coming from an average periodic lattice of molecular structure factor : $C_M F_M(\vec{q}) + C_P F_P(\vec{q})$, and the second term is the diffuse intensity coming from the difference between the average lattice and the real crystal.

IV - EXPERIMENTAL RESULTS

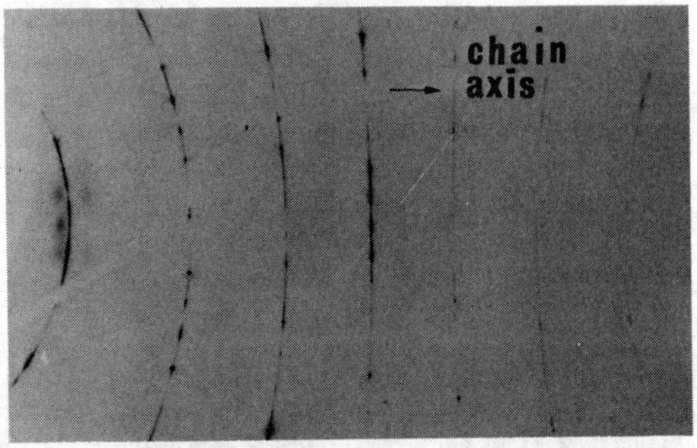

-A- Fig.4 : Monochromatic Laue pattern of a partially polymerized DNP crystal ($X_p \simeq 20\%$) taken at LURE. The polymerization axis is horizontal.

On Fig.4 the Bragg spots arrays of the average lattice clearly appear, and passing through them, one can see diffuse lines perpendicular to the polymer chain direction. The same kind of diffraction pattern is obtained if the crystal is rotated around the polymerization axis(5). These diffuse lines are thus produced by the intersection of cones of diffusion with the photographic plate. From the results of section III, one can see that this form of diffusion is directly related to the existence of parallel polymer chains. Different informations can be obtained from this diffraction pattern.

a) First we see that diffuse lines pass through the Bragg spots. This is verified throughout the reaction for both compounds. Thus monomer and polymer adopt the same lattice constant along the polymerization axis during the reaction. If it was not the case, one should observe double sheets, at least for large line index (see section III). In this respect our results differ from that of Ref.5.

The commonce lattice constant is shown as a function of conversion in Fig.5 for PTS and DNP.

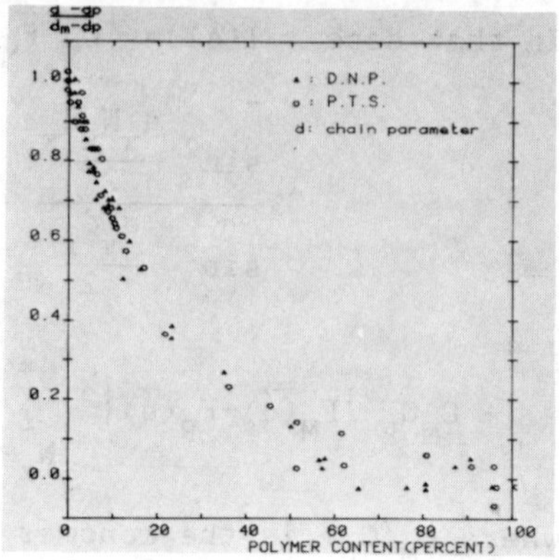

Fig.5. : Chain parameter as a function of polymer conversion for PTS and DNP (obtained from precession photographs).

The behaviour is the same for both compounds. In particular one starts from an elongation of 5% of polymer chains at the beginning of the reaction (6, 7).

b) Secondly, the modulation of the diffuse lines appears to be too strong to be accounted for by the only $|F_M(\vec{q}) - F_P(\vec{q})|^2$ factor. These intensity reinforcements indicate correlations between polymer chains along some directions. That is, a polymer chains favors the growth of other chains in its neighborhood. The concept of secondary chain initiation has already been considered in the litterature (8). In fact, this phenomenum is rather weak and the size of the aggregate should not exceed two or three chains (9). Thus, the polymer forms a quasi perfect solid solution in the monomer matrix.

-B-

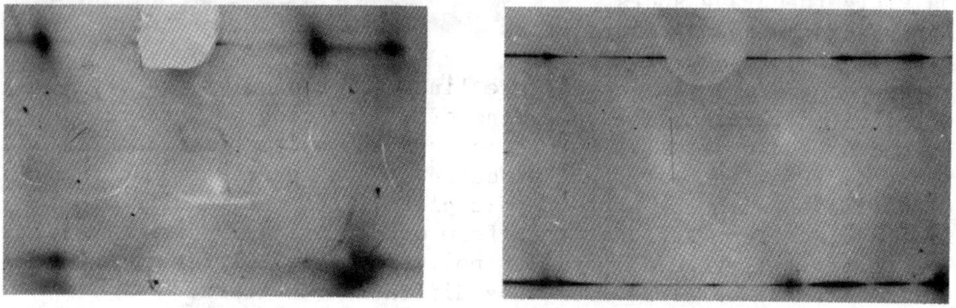

a b

Fig. 6 : Monochromatic Laue patterns of partially polymerized DNP crystals (LURE). (a) in the induction period and (b) in the autocatalytic period.

One can observe important differences between these two X ray patterns (Fig. 6 - a, b). In (b), Bragg spots are well-shaped,while in (a) they are strongly deformed. Furthermore, diffuse lines are broad in (a) and thin in (b).

The shape of the Bragg reflections reveals that intense and anisotropic stresses are acting in the crystal, probably in the vicinity of the polymer chains in strong extension, during the induction period. During the autocatalytic period, the reduction of the mismatch (see Fig.5) relaxes these stresses.

As for the width of the diffuse lines, we have seen in section III that this quantity is related to the length of the polymer chains. In the induction period, the observation of broad lines means that short chains are formed, while that of thin lines means that long chains have grown in the autocatalytic period (in Fig.6.b, the narrowness of the lines is limited by the experimental resolution). More quantitatively, we have analyzed the data in two steps.

a) In a first step, we have assumed that all the chains had the same length. In that case, the average polymer chain length $\bar{\ell}$ is related to the H.W.H.M. of the lines, $\delta q_z^{1/2}$, after correction for resolution, by the Scherrer formula : $\bar{\ell} = \dfrac{0.888\pi}{\delta q_z^{1/2}}$.

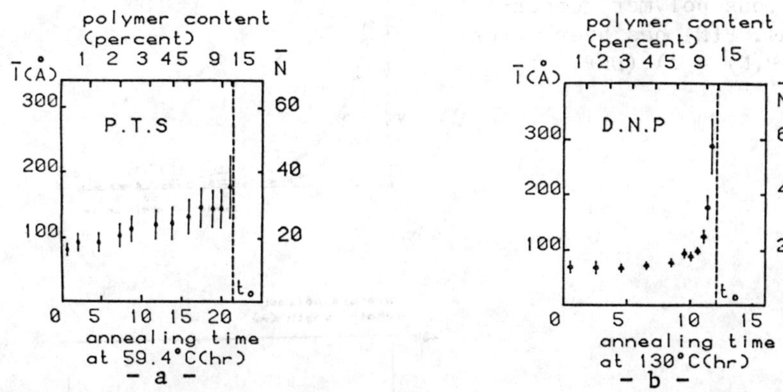

Fig. 7.a - b. Average polymer chain length as a function of annealing time for PTS (a) and DNP (b). The Scherrer formula has been used. The precision is improved in case (b) by the use of the synchrotron radiation.

In the induction period , short polymer chains developp , slightly longer in PTS (about 120 Å) than in DNP (about 70 Å). This difference could be related to the fact that PTS polymerizes more easily than DNP. On the other hand, an inferior limit of 400 Å for PTS and 700 Å for DNP can be given for the long polymer chains formed in the autocatalytic period. This compared well with other determinations (10).

b) The second step was an estimation of the chain length distribution function P(N) (see section III). For that, one makes the following remark : as the width of the lines remains always small before the distance between two lines, $\sin^2 q_z d_z/2$ can be approximated by $(q_z d_z/2)^2$. Thus :

$$I_{diffuse}(\vec{q}) \simeq C_M C_P |F_M(\vec{q}) - F_P(\vec{q})| \sum_{N_z} P(N) \frac{\sin^2 \frac{q_z N_z d_z}{2}}{(\frac{q_z d_z}{2})^2}$$

It has been shown by E. Bertaut (11) that in such a case P(N) is given by the second derivation of the Fourier transform of $I_{diff}(\vec{q})$, that is the line profile corrected for resolution. The mathematical treatment of the data is rather heavy and has been applied to DNP only (9).

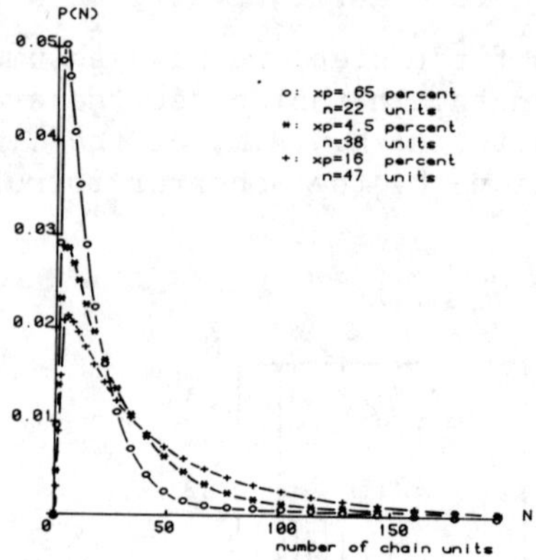

Fig.8 : Polymer chain length distribution function P(N) for various polymer concentrations. P(N) has been normalized($\Sigma P(N) = 1$) (DNP)

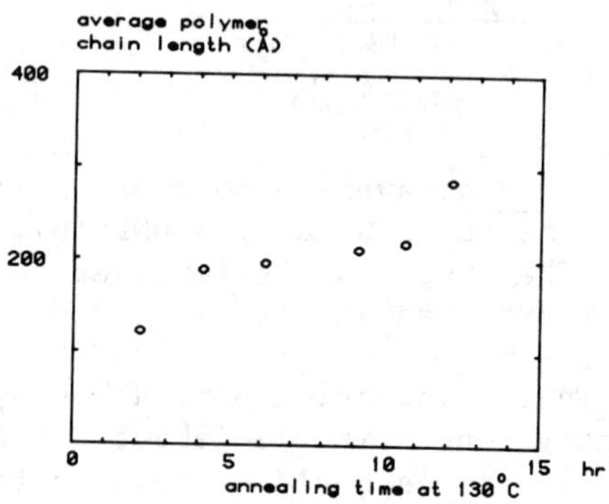

Fig. 9 : Average polymer chain length given by the function P(N) for different annealing times (DNP).

We see that the average length obtained from P(N) (Fig.9) agrees qualitatively well with by the results of the first step.On Fig.8 P(N) displays a maximum value at about 20 Å (4 units) which remains practically constant in the induction period. In regard to what is known about PTS (12), such chains seem to be very short. The fact that higher temperatures are required to polymerize DNP could account for this difference. On the other hand, short chains give rise to a foot of the line profile which is difficult to measure precisely.

V - CONCLUSION

A carefull analysis of the X ray diffuse scattering from two diacetylenes has directly shown that :

- The topochemical reaction is practically homogeneous
- In a mixed crystal, polymer and monomer share the same in-chain parameter. The first polymer chains are strongly elongated (5%) and produce a local stress field which deforms the crystal.
- First polymer chains are short (about 20 units) and grow much longer in the autocatalytic period.

Additional experiments could be proposed for the future. It has been observed in PTS that photoinduced polymer chains are short below 90 K and develop above (12). We have also observed this by X ray scattering : a monomer crystal was cooled down at 20 K, irradiated, warmed up and cooled again. A photograph was then taken at 20 K. Below a warm up temperature of 90 K, no noticeable diffuse lines appeared. Above 90 K, thin lines appeared (long chains). This kind of study could be made more quantitatively and extensively.

Acknowledgment

We thank P. Robin and R. Comès who participated to the first stages of this study. We also acknowledge useful discussions with J.P. Aimé, M. Bertault and M. Schott and the help of S. Megtert at the station D 16 of LURE.

References

1. Wegner G., in Chemistry and Physics of one dimensional Metal, 297 (Plenum Press N.Y., H.J. Keller Ed. 1977).
2. Kalyanaraman, P.S. et al. Makromol. Chem. 180 (1979) 1393.
3. Wegner G. Makromol. Chem. 85 (1971) 145.
4. Guinier A. in Théorie et Technique de la Radiocristallographie (Dunod, Paris, 1956).
5. Robin P. et al. J. Physique 41 (1980) 415.
6. Baughmann, R.H., J. Chem. Phys. 68 (1978) 3110.
7. Albouy, P.A. et al. J. Am. Chem. Soc. 104 (1982) 6556.
8. Bässler, M. in Advance in Polymer Science, 1 (Springer Verlag, 63, 1984).
9. Patillon J.N. Thesis (Orsay 1983).
10. Mondong R. and Bässler H., Chem.Phys.Lett.78 (1981) 371.
11. Bertaut, E.F. Acta Cryst. 3 (1950) 14.
12. Sixl, H. in Advance in Polymer Science, 49(Springer Verlag, 63, 1984).

MICROSTRUCTURES AND POLYMER CHAIN LENGTH IN DIACETYLENE SINGLE CRYSTALS

W. GÜTTLER AND M. SCHWOERER
Bayreuther Institut für Makromolekülforschung (BIMF) and Physikalisches Institut, Universität Bayreuth, POB 3008, D-8580 Bayreuth, Fed. Rep. of Germany

Abstract Microstructures of polymer in a TS-6 diacetylene monomer matrix were produced using electron beam lithography and UV-holography with a resolution of 300 nm in the chain direction and (for e-beam lithography) less than 40 nm perpendicular to it. From the experimental results an average polymer chain length of 200 nm (400 repeat units) is estimated for e-beam polymerization and of the same order of magnitude for photopolymerization at about 8 % conversion. It is supposed that the chain length is determined by the termination reaction during chain propagation and not by the initiation reaction.

1. INTRODUCTION

Since the discovery of the solid state polymerization of macroscopic diacetylene crystals by Wegner (1) information on the chain length was lacking for a long time due to the insolubility of most of the poly(diacetylenes) (2). It was only recently that data on chain lengths in these materials became available (3-7). The solid-state polymerization of diacetylenes can be initiated by heat, by UV and by ionizing irradiation. From the work on photopolymerization it is known that short chain oligomers are stable at temperatures below 100 K whereas long chains develop by irradiation at room temperature (8). In this paper we intend to quantify how long the chains grow at room temperature under UV and electron beam irradiation. Furtheron we want to prove whether the chains are short enough to allow the generation of submicron structures in diacetylene crystals, too. Submicron structures have already been produced in Langmuir-Blodgett films of diacetylenes by Garito and Singer (9).

The first attempt to determine the chain length in diacetylenes was made by Mondong and Bässler (3) using UV-lithography i. e. generation of microstructures by UV exposure and subsequent etching. The distortions of the microstructures were viewed under the electron microscope and yielded information about the chain length l. A value of l = 0.4 µm in TS-6 single crystals was given by the authors. A further attempt was made by Richter et al. using UV-holography (4) and by Niederwald et al. using electron beam lithography (7) in TS-6 single crystals. These results will be discussed in detail in this paper. Albouy et al. used diffuse X-ray scattering for the determination of chain lengths in TS-6 (5, and this volume). All the above mentioned methods are non standard methods for the determination of chain lengths. The only measurements with standard methods of molecular weight determination were reported by Wenz and Wegner (6) on soluble poly(diacetylenes), which became available very recently.

2. SAMPLE PREPARATION

Our experiments were performed on TS-6 single crystals (2,4-hexadiynylene-bis-(p-toluene-sulfonate also named PTS or TS). Monomer crystals were grown from solution as described by Wegner (1). They were cleaved along their (100) plane to give samples of some mm^2 area and a thickness of about 100 µm. The polymer content of the fresh crystals was about 0.01 %.

3. ELECTRON BEAM LITHOGRAPHY

Fig. 1 shows the schematics of the experimental procedure. A quadratic point lattice of polymer in TS-6 monomer was generated by irradiation with 30 keV electrons in a scanning electron microscope (SEM). The electron beam diameter at the sample surface was about 5 nm; however it increased inside the crystal by about one order of magnitude due to scattering. The penetration depth of 30 keV electrons is about 10 µm in this material.

After irradiation the crystals were etched with methanol in order to remove the monomer within a layer between 1 and 5 µm thickness. This yields a surface structure of the irradiation pattern, which can be looked at under the SEM. To avoid polymerization during observation in the SEM, which can lead to a peeling-off of the surface, the remaining monomer was thermally polymerized. Before observation in the SEM the sample was coated with gold (less than 10 nm).

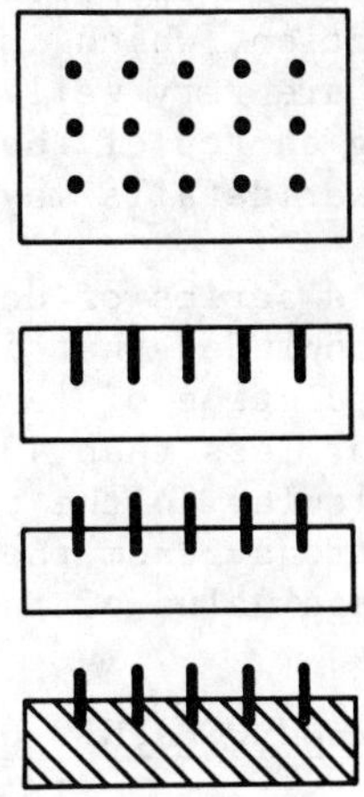

(1) Generation of a regular dot pattern of polymer by electron beam polymerization (top view)

(cross section)

(2) Etching

(3) Thermal polymerization of the remaining monomer

(4) Coating with gold

(5) Observation under the scanning electron microscope

Fig. 1: Generation of a dot pattern in a diacetylene crystal by electron beam lithography

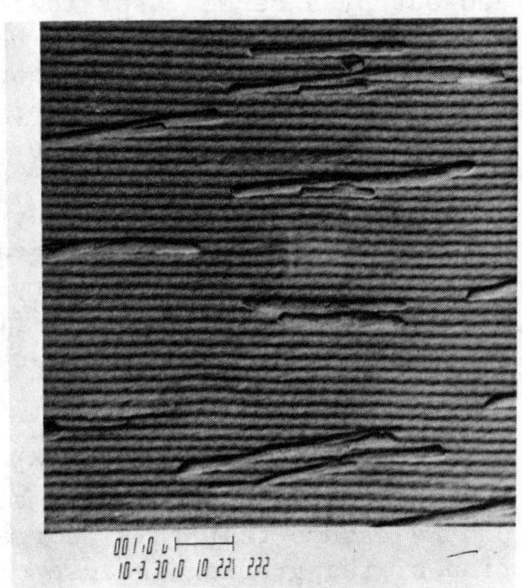

Fig. 2: Scanning electron micrograph of a dot pattern. Dot-to-dot distance is 200 nm in horizontal and vertical direction. The dots are nearly grown together along the chain direction.

Fig. 2 shows a dot pattern as it is viewed under the SEM. The dot-to-dot distance is 200 nm in horizontal as well as in vertical direction. The dots are almost grown together in the horizontal direction, which is the one-dimensional chain direction, whereas they are very well resolved vertically. We guess that the worms lying on top of the surface are artefacts of the preparation. Further details may be found in ref. (7).

From a series of dot patterns with varying dot-to-dot distances we concluded that it is possible to generate microstructures with a resolution of less than 300 nm parallel to the chain direction and of less than 40 nm perpendicular to it. The resolution perpendicular to the chains seems to be limited by scattering of the electrons. From the anisotropy of the resolution we estimate an average value of the chain length of 200 nm (400 repeat units).

4. UV-HOLOGRAPHIC GRATINGS

Microstructures in diacetylenes can also be produced by UV light. Fig. 3 shows the experimental set-up (4b). The frequency doubled light (257 nm) of an Ar^+ laser (514 nm) is split into two beams and directed onto the sample surface. By interference of the two beams a line pattern of polymer in monomer is created at the maxima of the interference fringes. We shall call this a holographic grating or shortly a hologram.

The hologram was read out by a He-Ne laser (633 nm). It should be pointed out that we get thin phase holograms, because the penetration depth of the exposing light is only about 1 micron and because diffraction is mainly due to the higher refractive index of the polymer at 633 nm and not to absorption.

Fig. 4 shows the maximum holographic efficiency η_{max} for different spatial frequencies $1/\Lambda$ (lines/mm). The holographic efficiency is defined as the first order diffracted intensity divided by the incident intensity of the reading beam. The diffraction efficiency is dependent on the exposure at 257 nm. It reaches a maximum at an exposure of 100 - 150 mJ/cm^2 and decreases to very low values above 250 mJ/cm^2 (4b,c). From fig. 4 it can be seen that the maximum diffraction efficiency decreases sharply above a spatial frequency of 1500 lines/mm, which corresponds to a grating distance Λ of 0.6 µm. We assume that this decrease is due to the limited resolution of the grating because of the polymer chains which lay perpendicular to the grating lines. So we estimate an average chain length of less than 0.6 µm. Nevertheless it is possible to generate holograms with a spatial frequency of up to almost 3000 lines/mm (Λ = 0.3 µm) and a holographic efficiency of about 1 %.

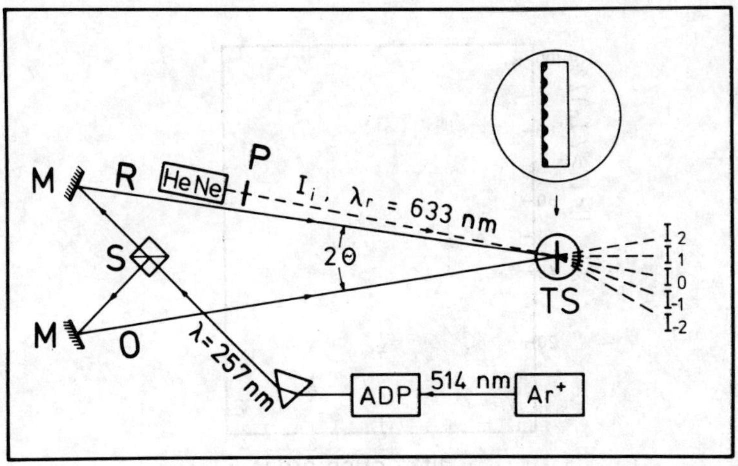

Fig. 3: Generation of holographic gratings (Ar^+: 18 W argon ion laser, ADP: frequency doubler, S: beam splitter, M: mirror, He-Ne: 2mW He-Ne laser, P: polarizer)

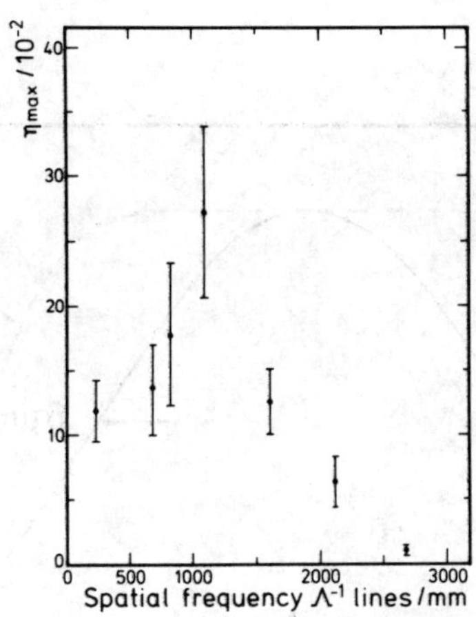

Fig. 4: Maximum holographic efficiency η_{max} for different spatial fequencies $1/\Lambda$

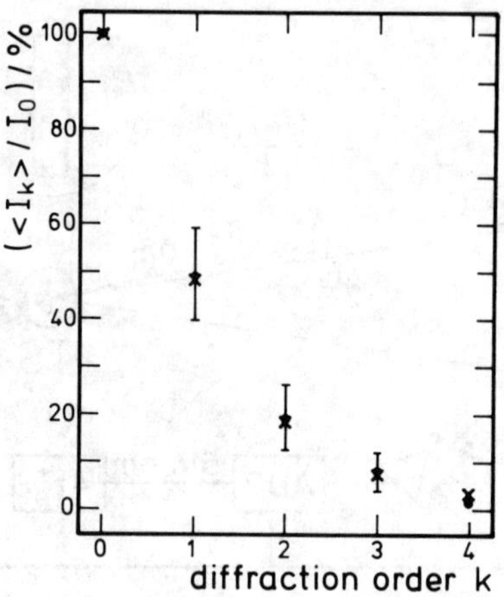

Fig. 5: Normalized higher-order diffraction intensities I_k / I_o at normal incidence of the He-Ne laser beam (k = 0...4). Dots are experimental values, crosses are calculated with the grating shape of fig. 6

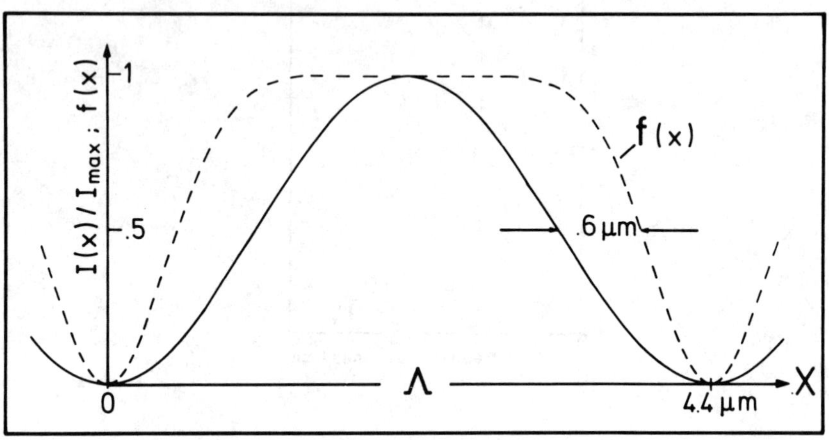

Fig. 6: Grating shape f(x) (broken line) and normalized UV intensity for a grating distance of 4.4 µm. f(x) is chosen to fit the higher-order diffraction intensities of fig. 5.

We also recorded the higher-order diffractions (fig. 5). These yield the approximate grating shape by evaluation of the Fraunhofer diffraction formula (4b,c). If we choose the grating shape to fit the data of fig. 5 we get the grating shape which is diplayed in fig. 6. Assuming that the displacement between the normalized light intensity and the grating shape is due to the polymer chains, which again lay perpendicular to the grating **lines, we estimate an average chain length of 0.6 μm (n = 1200) in agreement with the independent result which was deduced from fig. 4.**

It should be noted that these results are still somewhat preliminary because it was so far not possible to do the same experiments with a different orientation, namely, with the chains parallel to the grating lines. Measurements with the latter orientation should yield remarkably different data concerning figs. 4 and 5 if our estimate of the chain length is correct.

5. DISCUSSION

The most detailed investigations on chain lengths of poly(diacetylenes) were reported by Wenz and Wegner (6) on TS-12 (Dodeca-5,7-diyne-1,12-ylene- bis (p-toluene sulfonate)), which yields a solvable polymer. It was shown that the chain length increases with conversion from about 50 repeat units in the beginning to 2000 at 10 % conversion, and then stays constant.

To compare our chain lengths with values given in the literature it is therefore crucial to know the conversion, as was pointed out during this meeting by R.R. Chance. Given the polymer yield of P = 0,068 $\pm$ 0,010 at 257 nm (4b,c) we can calculate that the average conversion was 8 % within the penetration depth of the UV light of 0.5 μm at an exposure of E = 100 mJ/cm^2. This exposure was used for the data in figs. 4 and 5. So the conversion for our chain length data is near the autocatalytic regime and our estimate for the average chain length of 1200 repeat units is in reasonable agreement with the above cited values in TS-12 at this conversion. Measurements of the chain lengths at lower conversion are in progress (10).

Unfortunately we cannot give a value of the conversion for the electron beam polymerization. But from the similar value of the chain length (n = 400) it seems that also in the latter case we deal with conversions near the autocatalytic regime.

There seems to be a discrepancy between our chain length data during electron beam polymerization and photopolymerization and the ones during thermal polymerization which were measured by Albouy et al. (5) and which were also presented at this meeting. They report values for the average chain length of less than 50 repeat units in TS-6 and DNP at the approach to the autocatalytic regime and even smaller values for lower conversions. This might be due to experimental limitations of the diffuse X-ray diffraction method. The chain length was determined from the half width of the diffuse diffraction lines by the Scherrer formula. The longer the chains are, the smaller becomes the half width. From the experimental results it seems probable that longer chains were not resolved experimentally. So we guess that the discrepancy to our results could be ruled out by an improvement of the experimental resolution of the diffuse X-ray scattering.

Our values of the chain lengths are almost the same for photopolymerization and e-beam polymerization. This is perhaps surprising because the way of initiation is very different in both cases. So we conclude that the chain length is not influenced by the initiation but mainly by the termination reaction during chain propagation.

REFERENCES

1. G. Wegner, Polymerisation von Derivaten des 2.4-Hexadiin-1.6-diols im kristallinen Zustand. Z.Naturforsch. 24b (1969) 824

2. G. Wegner, IV. Polymerization of Bis-(p-Toluene Sulfonate) of 2.4-Hexadiin-1.6-diol. Makromol. Chem. 145 (1971) 85

3. R. Mondong, H. Bässler, Determination of the Chain Length in Polydiacetylenes by Scanning Electron Microscopy. Chem. Phys. Lett. 78 (1981) 371

4. a) K.H. Richter, W. Güttler, M. Schwoerer. Holographic Gratings on Diacetylene Single-Crystal Surfaces. Chem. Phys. Lett. 92 (1982) 4
   b) K.H. Richter, W. Güttler, M. Schwoerer. UV-Holographic Gratings in TS-Diacetylene Single-Crystals. Spatial Resolution and Sensitivity. Appl. Phys. A 32 (1983) 1
   c) K.H. Richter. Holographische Gitter auf Diacetylen-Einkristall-Oberflächen. Diplomarbeit, Universität Bayreuth 1983

5. P.A. Albouy, J.N. Patillon, J.P. Pouget. X-Ray Study of the Topochemical Polymerization of Two Diacetylenes. Mol. Cryst. Liq. Cryst. 93 (1983) 239

6. G. Wenz, G. Wegner. Kinetics of Radiation Induced Polymerization of a Diacetylene Studied by Analysis of the Molecular Weight Distribution. Mol. Cryst. Liq. Cryst. 96 (1983) 99

7. H. Niederwald, H. Seidel, W. Güttler, M. Schwoerer. Submicron Structures on Diacetylene Single-Crystal Surfaces by Electron Beam Irradiation. J. Phys. Chem. 88 (1984) 1933

8. M. Schwoerer, H. Niederwald, Photopolymerization of Diacetylene Single-Crystals. Makromol. Chem. (1985) (in press)

9. A.F. Garito, K.D. Singer
Laser focus 18 (1982) 59

10. H.-D. Bauer. Diplomarbeit, Universität Bayreuth 1985
(in preparation)

X-RAY TOPOGRAPHIC STUDIES OF THE SOLID-STATE POLYMERIZATION OF PTS [2,4-HEXADIYNE DIOL BIS(p-TOLUENE SULPHONATE)]

M. DUDLEY AND J.N. SHERWOOD

Department of Pure and Applied Chemistry, University of Strathclyde, Glasgow G1 1XL, Scotland, U.K.

D.J. ANDO and D. BLOOR
Department of Physics, Queen Mary College, University of London, London.

1. INTRODUCTION

Currently, there is widespread interest in the process of solid-state polymerization in single crystals of di-acetylene compounds. Much of this is centred round the question of whether or not lattice defects play any role in this process. Consequently, some attempts have been made to define the nature and properties of lattice defects in both monomer and polymer crystals.

Using etching techniques, Schermann, Wegner, Williams and Thomas (1) located the emergent ends of dislocations on (100) and (102) surfaces of PTS monomer crystals and demonstrated their potential role as nucleation centres for thermally induced polymerization. Partial confirmation of their assignment of Burgers vector and line direction to the dislocations has come from the transmission electron microscope studies of Young and Petermann(2) who identified edge dislocations of Burgers vector [010] emerging perpendicular to the (100) faces of thin specimens.

An alternative technique for this purpose is X-ray topography(3). This has the additional advantage that it can be used to differentiate between bulk and surface defects in large (cm^3) specimens of high quality. Hence it allows a judgement to be made of the role of both types of defect structure in subsequent processes.

To date, studies of solid-state polymerization using conventional X-ray topographic techniques have revealed gross defect structures with little or no resolution of individual defects(4,5). Progress in this field has been aided by the advent of the highly versatile White Beam Synchrotron Radiation Topography technique(6), which is tolerant of a greater degree of lattice distortion, and by the successful growth of large monomer single crystals of a quality which surpasses that of previously available samples(7).

Using the synchrotron beam as both the investigative probe and the driving force for radiation induced polymerization, we have examined the monomer to polymer conversion process and the defect structure in each species. Additionally, preliminary studies have been made of thermally induced polymerization.

2. EXPERIMENTAL

Crystals of highly purified PTS monomers were grown by self-nucleation in slowly evaporating saturated acetone solutions at 273K. Temperatures of this order were used to arrest polymerization during growth. X-ray topographs were taken on the White Radiation Camera at the SERC Synchrotron Radiation Source at Daresbury, UK.

3. RESULTS

3.1 Kinetic Experiments

Initial experiments were directed towards the definition of the polymerization process in bulk ($1x1x0.2cm^3$) crystals. A crystal was exposed with a (100) face normal to the beam and the polymerization process followed dynamically at ambient temperatures (293K) by successive (30 sec) exposures of Polaroid film in normal transmission. The resulting series of Laue photographs showed the gradual transition of the monomer single crystal to polymer single crystal over a period of 20 mins at a beam current of 120 mA. Simultaneously, the crystal changed from a pale red to a gold colour. Analysis of the films showed that the transformation followed the usual sigmoid variation of conversion versus time/accumulated dose. Curves of this type and others determined by Raman spectroscopy as a function of temperature were used to define the degree of polymerization of the crystals. A detailed account of these experiments has been published previously (8,9).

3.2 Structural Experiments

Structural experiments required longer (3 min) exposure time (at 260 mA beam current). At room temperature the first exposure allowed the imaging of the monomer crystal. Subsequent irradiation (12 min) completed the process. Simultaneously, the crystal ($1x1x0.1cm^3$) deformed to become concave to the beam reaching a maximum curvature of 1cm radius after 3 min. On continued irradiation the crystal relaxed back to planarity. This massive deformation arises because of the initial preferential polymerization in the surface regions of the monomer crystal due to the high absorption of X-rays. The relaxation occurs as the polymerization gradually propagates through the crystal on continued irradiation. The direction of curvature, around the 'c' direction, is consistent with the relative changes in lattice parameter on polymerization. Reduction of the temperature or attenuation of the soft components from the white radiation beam successfully resulted in prolonging the induction period of the radio-polymerization process by 10-100x. This allowed the detailed examination of the crystal at initial, intermediate and final stages of the polymerization process. It also allowed the examination of thermal polymerization without interference from radio-polymerization. The thermal processes could be achieved without crystal distortion.

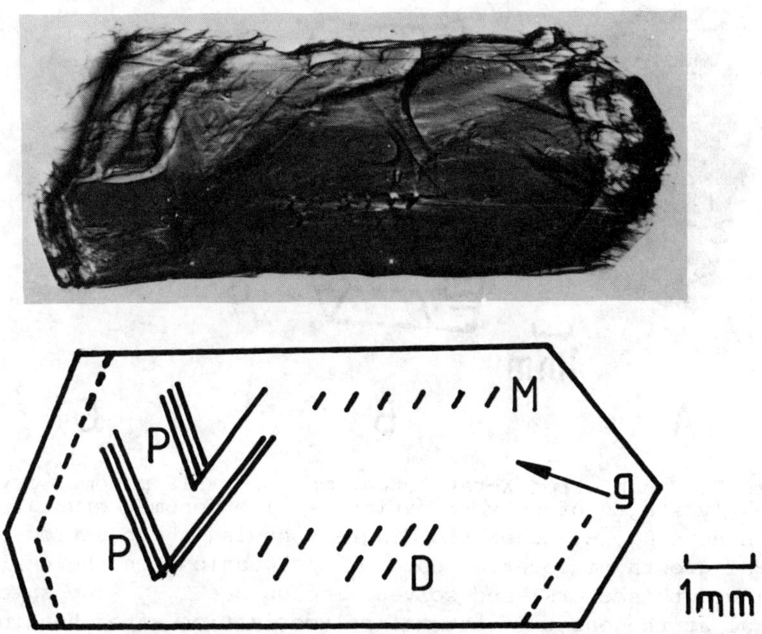

Figure 1. Synchrotron X-ray topograph of a PTS polymer crystal showing growth dislocations D, mechanically induced dislocations M and a planar twin boundary P (g $1\overline{12}$, λ = 1A (crystal plane (100))

Topographs of the monomer and polymer revealed that not only was the transformation a single crystal to single crystal process but that despite the considerable distortion observed, the defect structure of the polymer was equivalent to that of the monomer.

A typical topograph of the polymer crystal is shown in Figure 1. Various defects can be identified and characterised:

- growth dislocations in the {111} and (100) growth sectors
- slip dislocations; b[010]*
- growth sector boundaries
- twins; twin plane (122)*

(* in agreement with references 2 and 10)

These defects existed in the same geometry in the monomer and polymer crystal and hence are 'frozen in' during the radio-polymerization process. There was no visual evidence of enhanced polymerization in the vicinity of these defects or of dislocation multiplication. This apparent lack of enhanced reactivity at dislocations to high energy radiation was confirmed in kinetic studies of the γ-ray induced process.

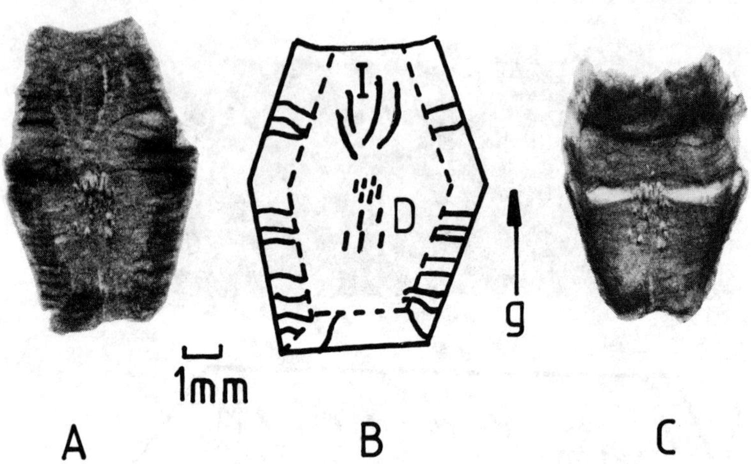

Figure 2. Synchrotron X-ray topographs of a PTS monomer crystal in the early stages of polymerization, A. The monomer crystal before heating, B. Schematic diagram showing the position of the central growth dislocation core D, dislocations in the peripheral {111} growth sectors and solvent inclusions I, C. The same crystal at the onset of thermal polymerization after heating to 70°C showing the dynamical strain lobe centred on the growth dislocation core and the loss of contrast in the outer growth sectors (crystal plane (100))

In contrast, preliminary experiments on thermal polymerization do show some evidence of the initiation of this reaction at strained regions in the crystal. Figure 2a shows a topograph of a PTS crystal prior to thermal polymerization. The resolution is low because the topograph was recorded on fast film (exposure 12 sec) to avoid lengthy exposure to the synchrotron beam. The external growth sectors are, however, clearly recognisable as is the central dislocation core. As the reaction initiates, considerable strain builds up in the central region and is evidenced by the large strain lobe which develops centred on this region (Figure 2b). We note that this is a dynamic effect due to the interaction of the strain with the incident X-rays and is not a structural phenomenon. Additionally, there is a marked deterioration of crystallinity in the more highly dislocated outer growth sectors and around solvent inclusions. On continued thermal polymerization, the crystal polymerizes fully to yield a single crystal polymer and the strain evidenced by the dynamical lobe becomes relaxed by cleavage of the crystal.

Thus we conclude that radio-polymerization of diacetylene monomers initiates at random sites in the lattice and initiation does not localise around dislocation cores. In contrast, and in confirmation of previous observations(1), the thermal polymerization does appear to initiate at the site of dislocation bundles and possibly other strained features in the lattice. Further experiments are in progress to define the present speculations and to examine the structural nature of the photolytic process.

4. REFERENCES

1. W. Schermann, G. Wegner, J.O. Williams and J.M. Thomas, J. Polym. Sci., Polymer Phys. Ed., 13, (1975) p 753.
2. R.J. Young and J. Petermann, Makromol. Chem., 182 (1981) p 621.
3. Characterization of Crystal Growth Defects by X-ray Methods, Eds. B.K. Tanner and D.K. Bowen, NATO ASI, series B, Physics Vol. 63, Plenum NY (1980),
4. J.M. Schultz, J. Mat. Sci., 12 (1977) p 2258.
5. R.G. Rosenmeier, R.E. Green Jr., and R.H. Baughman, J. Appl. Phys., 52 (1981) p 7129.
6. J. Miltat in Reference (3) p 401.
7. M. Dudley and J.N. Sherwood, to be published.
8. M. Dudley, J.N. Sherwood, D. Bloor and D.J. Ando, J. Mat. Sci. Letters, 1 (1982) 479.
9. M. Dudley, J.N. Sherwood, D. Bloor and D.J. Ando, Mol. Cryst. Liq. Cryst. 93 (1983) 223.
10. R.J. Young, D. Bloor, D.N. Batchelder, C.L. Hubble, J. Mat. Sci., 13 (1978) p 62.

ACKNOWLEDGEMENT

We gratefully acknowledge the financial support of SERC. We also wish to express our appreciation of the support of the staff of the SERC's Synchrotron Radiation Source in the performance of the Synchrotron X-ray experiments.

ELECTRONIC STRUCTURE OF CONJUGATED POLYMERS

R. Silbey

Department of Chemistry, M.I.T., Cambridge, MA 02139 USA

In this paper, I will review recent work on certain aspects of the electronic structure of conjugated polymers, and try to make contact with the work on the electronic structure of short oligomers. It is impossible to cover this vast subject except by ignoring a great deal. I therefore do not pretend to be exhaustive; instead I will cover a few topics of current concern and hope that these overlap sufficiently with the interests of people studying the properties of polydiacetylenes.

1 THE QUANTUM CHEMISTRY OF CONJUGATED SYSTEMS

First, I will briefly remind you of the standard methods for finding approximate wave functions for the states of conjugated molecules (1). We work in the Born-Oppenheimer approximation always, so that we fix the nuclear positions and write the electronic Hamiltonian as

$$H = \sum_i \{-\frac{1}{2}\nabla_i^2 - \sum_n \frac{Z_n}{r_{in}} + \frac{1}{2}\sum_{j\neq i}\frac{1}{r_{ij}}\} + \sum_{n>m}\frac{Z_n Z_m}{r_{nm}}$$

where Z_n is the charge on the nth nucleus, r_n is the position of that nucleus and r_i is the position of the ith electron. Of course, the Schrodinger equation cannot be solved for such a Hamiltonian if there is more than one electron, so we must resort to approximate methods. The usual first approximation is to assume that the electrons are each in one-electron orbitals which are solutions to the Hartree Fock (HF) equations (2). These treat each electron in the average field of all the other electrons, and yield one electron orbitals and energies. The HF approximation is often very good, especially for ground state closed shell systems. To go beyond HF, one must use multiconfigurational methods (3) or generalized valence bond methods[4] or pseudopotentials plus density functional methods (5). All these methods are time consuming and expensive and have not yet been applied to many polymeric systems.

In recent years, a pseudopotential-like method, the valence effective Hamiltonian method (VEH) (6) has also been used to obtain HF-caliber one electron orbitals for polymers with little computer time and expense.

Because all the ab-initio methods mentioned above are quite time consuming, semi-empirical methods have been developed which replace certain integrals appearing in the Hartree Fock equations by parameters (or in some cases, functions) chosen so that the results fit the spectral properties of certain model compounds. Some of these methods are the Huckel model, Hubbard model, Pariser-Parr-Pople (PPP), CNDO, MNDO, ··· (7). Since the first three of these methods are related and have played a significant role inthe work of the last few years, I will discuss them in more detail.

The PPP model is the most complete model in this hierarchy (of Huckel, Hubbard and PPP), and it begins with the HF equations for the one electron π-molecular orbitals of a conjugated molecule. Only π orbitals are considered so all the σ electrons are lumped together with the nuclear potential to form a "core" Hamiltonian, h_{core}. Each carbon nucleus has one p_z atomic orbital (for simplicity I consider planar systems), called u_n, n labelling the nucleus. Note that the molecular orbitals are linear coefficients in this linear combination we are trying to find. In the HF equations for the molecular orbitals there will be one electron integrals, Coulomb integrals and exchange integrals. All of these can be written as sums of integrals over atomic orbitals. The semi-empirical nature of the PPP method enters here: these integrals over atomic orbitals are parameterized by comparing the final results with the experimental results for the spectrum of benzene, for example. Thus, one electron integrals become

$$\langle u_n(1)|h_{core}(1)|u_{n'}(1)\rangle \equiv \beta(r_{nn'})$$

while two electron integrals are

$$\langle u_n(1)u_m(2)|\frac{1}{r_{12}}|u_p(1)u_q(2)\rangle = \delta_{np}\delta_{mq}\gamma(r_{nm})$$

The zero differential overlap (ZDO) approximation has been made here so that only two center integrals are considered. Note that

$$\gamma(0) \equiv U \neq 0$$

$$\gamma(r) \sim \frac{1}{r} \quad \text{as } r \to \infty$$

The PPP model keeps all the one and two center two electron integrals, solves for the HF molecular orbitals (within these approximations) and compares the result to experiment. The Hubbard model neglects all the γ except $\gamma(0)$ ($\equiv U$). The Huckel model neglects all the γ including U.

The PPP model has been found to be an excellent approximation for many small planar conjugated molecules such as benzene, naphthalene, etc. (7) A standard parameterization gives (in benzene)

$$U \sim 11 \text{ eV}$$
$$\beta(\text{nearest neighbor}) \sim 2.4 \text{ eV}$$
$$\gamma(r) = U(1+0.6r^2)^{-1/2} \qquad (r \text{ in Å})$$

Even though the PPP model has been successful, it was recognized quite early that using single determinational wave functions was overly restrictive, and to obtain better results, configuration interaction with other determinants should be included. Karplus and co-workers (8) have used this procedure to take the PPP model further than before.

Since the Hubbard and Huckel models are approximations to the PPP model, they are bound to be less reliable. However, when used cleverly, they can be qualitative and even semi-quantitative guides to the electronic structure.

The MNDO method pioneered by Dewar (9) has proved itself to be reliable and semi-quantitative for the geometries of small organic molecules and has been used recently (10) to study the defect in polyenes, as has the PPP method (11). The results are in reasonably good agreement.

Another way of dealing with these semi-empirical methods is to write down the Hamiltonian, in operator form, after all the approximations have been made. The PPP Hamiltonian is then

$$\hat{H} = \sum_{n,n'} \{\sum_{\sigma} \beta_{nn'}(a^{+}_{n\sigma}a_{n'\sigma}+a^{+}_{n'\sigma}a_{n\sigma})$$

$$+ \hat{n}_n \varepsilon_n \delta_{nn'} + \frac{1}{2} U_n \delta_{nn'} \hat{n}_n(\hat{n}_n - 1) + \frac{1}{2} \gamma_{nn'}(1-\delta_{nn'})(\hat{n}_n - 1)(\hat{n}_{n'} - 1)\}$$

$$\hat{n}_n = \sum_\sigma a^+_{n\sigma} a_{n\sigma}$$

and σ labels the spin of the electron. Once this many electron Hamiltonian has been written down, we do not have to solve the Hartree-Fock problem, but can attempt to solve the Schrodinger equation exactly. Recently, Soos and co-workers have attempted this within a valence bond approach (12) with good results.

Note that if all the $\gamma_{nn'}$ are neglected, the standard Hubbard Hamiltonian results with its correlation parameter, U. (By the way, the U term, although a two electron term, is contained within the standard HF approximation, so a quantum chemist would not label it a correlation term, but a Coulomb term). Note further that if U is neglected in addition to $\gamma_{nn'}$, the general Huckel Hamiltonian results.

This Hamiltonian is also a useful starting point for discussing electron phonon interactions. If we expand $\beta_{nn'}$ and $\gamma_{nn'}$ around the equilibrium geometry of the structure so that for example

$$\beta_{nn'} = \beta^0_{nn'} + \kappa(u_n - u_{n'}) + \cdots$$

where κ is the electron phonon coupling constant and u_n is the derivation of the nth nucleus from its equilibrium position. If $\Gamma_{nn'}$ and u_n are neglected, the Hamiltonian becomes the Su, Schrieffer, Heeger (13) (SSH) Hamiltonian when the vibrational Hamiltonian for the cores is added. Thus, the SSH model is identical to Huckel theory with σ bond elastic energy (1).

2 APPLICATION TO POLYENES

2.1 Electronic Structure for Regular Geometries

The polyenes have been studied in the Huckel and PPP models by physical chemists over the last thirty years. The early work is summarized by Salem (1). An important question was the origin of the optical gap which persists as the polyene becomes longer. If all the bonds were of equal length so that all the nearest neighbor β's were equal, then Huckel theory predicts that the optical gap would disappear as N, the number of double bonds in the even polyene, goes to infinity. Salem and Longuet-Higgins (1) noticed that if bond alternation were allowed, then the gap persisted as $N\to\infty$, and, in addition, the bond alternated structure had a lower total energy than the equal bond length structure. (This is an example of a Peierls instability (13).)

If within the Huckel model there are just two β's, β_{long} and β_{short}, (corresponding to a fixed bond-alternated geometry) given by

$$\beta_{long} = \beta(1+\delta)$$

$$\beta_{short} = \beta(1-\delta)$$

then the gap is given by $\Delta E=4\beta\delta$ as $N\to\infty$.

The optical gap for short chain polyenes was computed in the PPP model by Johnson and Peacock (14) who found that even in the equal bond case ($\delta=0$), there seemed to be a gap as $N\to\infty$.

The gap has also been computed in the Hubbard model in the equal bond length geometry by Lieb and Wu (15) who found a nonzero gap as long as $U\neq0$. Thus, the optical gap can be caused solely by the electron-electron interactions. In the alternating bond structure in the Hubbard model, no exact solution is known, but a meanfield result (16) leads to a gap, Δ,

$$\Delta = [\{\Delta(\delta=0)\}^2 + \{4\beta\delta\}^2]^{1/2}.$$

The exact optical gap in the PPP model for an infinite bond alternated structure is not known; however, it is clear that electron interactions can play an important role in the size of the gap.

Another important question is whether single determinational functions (i.e. within the Huckel or Hartree Fock approximations) describe the lowest excited states of polyenes correctly. In the last ten years (17), it has become clear from both experiment and theory that the lowest excited state of even polyenes with more than three double bonds is not the B_{1u} state reached by the absorption of light, but a second A_{1g} state accessible by two photon absorption. Thus the Hartree Fock model, although a very good description of the B_{1u} excited state, is not good at all for the lowest excited state. At the present time, the behavior of this A_{1g} state (and other such states) as the polyene grows longer is completely unknown, although a valence bond description of this state resembles a soliton-antisoliton pair of the SSH model, but within a rigid geometry.

Within the last year, Soos (12) has found the exact energies of the polyenes up to six double bonds within the PPP model, assuming a bond alternated geometry with δ=0.07 for the neutral molecules, and equal bond geometry for certain ions and radicals. This beautiful work is so new that the results are still being interpreted. However, I believe that Soos has opened a new era in the study of polyenes with this breakthrough. He finds that the PPP prediction (within the geometry mentioned above) for the vertical optical gap for tetraene is 4.56 eV (expt=4.4 eV), for pentaene 4.23 eV (expt=4.0 eV), and for dodecaene 4.0 eV (expt=3.7 eV), while the prediction for the $2A_{1g}$ vertical gap is 3.8 eV (expt 3.6 eV) and for pentaene 3.4 eV (expt=3.1 eV). An extrapolation to $N=\infty$ gives an optical gap of 2.8 eV (expt=1.8 eV)! How do we account for this discrepancy: (a) lack of bond alternation throughout the structure? (b) non-Franck Condon effects, i.e. relaxation of the B_{1u} state geometry? (c) σ-π interaction effects? Another important question: to what energy does the $2A_{1g}$ state go at infinite N? Where are the other A_{1g} states? Can we describe them as a band of soliton-antisoliton bound states?

Although the Soos work has raised important questions, it has also answered others with finality. For example, it is now absolutely clear that to describe the states of polyenes quantitatively requires electron-electron interactions (U and γ above). If they are left out, the excited state energies, spin densities in radicals, etc. are given incorrectly. Soos (12) has a careful discussion of all this.

2.2 Electron-Phonon Interactions

As Salem (1) points out, the electronic structure of spin 1/2 defects in long polyene chains was first discussed by Pople and Walmsley (18) in 1962. They treated the Huckel Hamiltonian for a bond alternated (i.e. dimerized) chain in which spin 1/2 defects were inserted by having two long (single) bonds in a row. This defect changes the phase of the alternation. They did not allow the structure to relax around the defect so that the domain wall was infinitely sharp. Using values of β current at that time, they estimated the creation energy of such a defect and thus the number of free spins in long polyene chains at room temperature. (From this they speculated on the semiconducting properties of such systems.) Two important features of this structure were clearly recognized in the paper: the defects could be pulled as far apart as one liked (i.e. degenerate ground states) and the easy movement of the defect (domain wall). However, these authors did not notice that the energy gain by allowing the defect to spread was large. In 1979, Su, Schrieffer, and Heeger examined this model again (13), and repaired this oversight.

By allowing the bond alternation to change continuously through the defect and taking $\beta(r)$ from the quantum chemists standard formulary, SSH showed that the optimized geometry was a soliton-like structure. That is, the difference in bond lengths in a unit cell varied as tanh (n/ℓ) where n numbers the unit cell (n=0 at the center of the defect) and ℓ is a parameter giving the width of the defect. SSH, using the Huckel model, found $\ell=7$ for the neutral and charged solitons (S_0, S_+, S_-). The seminal work of SSH led to an enormous interest in the theoretical description of this system.

The interaction between defects on a polyene chain were considered within the SSH model by Maki et al. (19), Bredas et al. (20) and Bishop et al. (21). The latter two groups noted that although like-charged defects do not bind together, a neutral and a charged defect will bind to form a polaron (or radical-ion) on the chain. Thus the complete SSH model must include the possible formation of polarons, and other defects (21).

Other authors have attempted to look at the effect of electron interactions (correlation) on these structures in a variety of ways. Some have added a Hubbard U term to the basic SSH Hamiltonian (22,23,24) and showed that as long as U<4,7 eV, the basic results of SSH are largely unaffected. The "true" value of U is not known with accuracy; however, it is likely to be close enough to the critical U to be disturbing. If U>4.7 eV, there is an undimerized spin density wave state which is lower in energy than the normal solution. Aslangul (24) has an interesting discussion on this. J. Fukutome and Sasai (11,25) have tried to solve the PPP Hamiltonian within the unrestricted Hartree Fock approximation, and find both SDW and CDW solutions, although the normal bond-alternated solution does seem to be the best description of the electronic structure here. As we pointed out above, Soos' (12) methods have not yet been applied to long polyenes.

Another approximate scheme, MNDO, has been applied to the study of defects (10,26) in long polyenes. This self-consistent method allows the inclusion of electron interactions in an approximate way and leads to predictions similar to those of the SSH model. For example, the neutral defect is studied by exaining a 41 carbon atom polyene with a center of symmetry. This forces the defect to be centered in the middle of the chain. The MNDO results show the standard tanh structure but with $\ell=3$ (instead of $\ell=7$ from the Huckel model). The charged defects were treated in the same manner and it was found that $\ell=5$ for the positive and $\ell<3$ for the negative solitons. Note however that the tanh form was found in all cases. These authors also discussed the structure of the orbital which is singly occupied in the neutral defect (i.e. mid-gap orbital). The results of the MNDO calculation are in good semi-quantitative agreement with the UHF results in the PPP model (25).

In spite of their simplicity and inexact character, the application of semi-empirical Hamiltonians and methods to the electronic structure of polyenes has proven to be an extremely fertile endeavor, allowing a better understanding of experiments and giving rise to simple models of immense utility.

3 QUANTUM CHEMISTRY OF POLYMERS

In the past few years, the work of Ladik, Andre, Karpfen, Suhai, Kerstesz and others (27) has led the way to a greater under standing of the electronic structure of polymers. Using ab-initio methods mainly, these authors have shown that high quality electronic structure calculations can be done on polymers in the same way as on semiconductors and metals. At the same time, semi-empirial methods have been applied to this problem with success. Duke et al. (28) have used CNDO methods, while Bredas et al. (29) have used valence effective Hamiltonian (VEH) methods.

The current status of ab-initio calculations on the ground state properties of polymers is given in the review by Karpfen.[30] The band structures and densities of states of polyacetylene (and polyethylene) are given in the crystal orbital method (Hartree Fock with periodic boundary conditions). Correlation effects have been treated by Suhai (31) in order to study the optical band gap. Although progress is surely being made in this field, the effort required per calculation is large, and it is important that approximate and semi-empirical methods be developed. One such method which has proven to be very useful is the valence effective Hamiltonian (VEH) method mentioned above. This is a method for obtaining molecular one-electron Hamiltonians from first principles and has been worked out by Nicolas and Durand (6,32). The effective Fock operator of the molecule is assumed to be the sum of the kinetic energy and the various atomic potentials in the molecule:

$$F_{eff} = -\frac{\Delta}{2} + \sum_{A} V_A$$

where V_A is the effective potential of atom A. For computational ease, simple nonlocal atomic potentials are chosen of the form of

Gaussian projectors:

$$V_A = \sum_{\ell} \sum_{m} \sum_{ij} C_{ij,\ell m} |\chi^A_{i\ell m}\rangle\langle \chi^A_{j\ell m}|$$

where the summations over ℓ and m define the angular dependence of V_A. The numerical coefficients $C_{ij,\ell m}$ are independent of m in the case of spherical symmetry, which we usually consider. The functions are normalized Gaussians:

$$\chi_{i\ell m} = N_i \exp[-\alpha_i r^2] Y_{\ell m}(\theta,\phi)$$

N_i is the normalization factor and $Y_{\ell m}$ denotes the usual spherical harmonics. Note that only 1s and 2p Gaussian Cartesian functions are used.

The parameterizations of the linear coefficients, C, and the nonlinear exponents, α, first require valence SCF calculations on model molecules by a theoretical pseudopotential method with an STO-3G minimal basis set and a double zeta basis set. The model molecules chosen to parameterize carbon and hydrogen atomic poten-tials were ethane, transbutadiene, and acetylene (29); for sulfur and carbon linked to sulfur, dimethyl sulfide and thiophene (29); for ni-trogen and carbon linked to nitrogen, dimethylamine and pyrrole (33). For each molecule, the Fock operator is constructed as:

$$F = \sum_{\upsilon} \varepsilon_{\nu} |\phi_{\nu}\rangle\langle\phi_{\nu}|$$

where the summation is over all occupied states; the valence orbitals ϕ_{υ} are taken from the minimal basis set calculation and the corresponding monoelectronic energies ε_{υ} from the double zeta calculation. The choice of this theoretical Fock operator leads to valence effective Hamiltonians providing double zeta accuracy for monoelectronic energies when solved with a minimal set. The parameterization of the atomic potentials is then determined by minimizing the quantity

$$\sum_{B} (F-F_{eff}|F-F_{eff})_{\text{molecule } B}$$

where the summation runs over the model molecules used for a given set of atomic potentials. $(F-F_{eff}|F-F_{eff})$ denotes the scalar product of $F-F_{eff}$ with itself in the subspace of the occupied valence orbitals. On the model molecules, standard deviations between the ε_{υ} energies produced by using the valence effective Hamiltonians and double zeta energies are of the order of 0.015 a.u., and in no case larger than 0.2 a.u. The position of the highest occupied orbital is especially well-produced - a result which lends confidence in obtaining good ionization potential estimates.

No information pertaining to the excited states is included in the atomic potentials. As a result, no special attention should in principle be given to the unoccupied levels. However, for the planar systems considered previously, surprisingly good agreement between experiment and theory has been obtained for the lowest optical energy transition (29).

The extension of the VEH method to polymer calculations is straightforward. The effective operator takes the form:

$$F_{eff} = -\frac{\Delta}{2} + \sum_{g} \sum_{A} V_A$$

the summations over g and A running, respectively, over the polymer unit cells and the atoms present in one cell. The band structure E(k) of the polymer, where k is a point in the first Brillouin zone of the polymer, is obtained from eigenvalues of the set of secular equations:

$$\underline{F}(k)\ \underline{C}(k) = \underline{S}(k)\ \underline{C}(k)\ \underline{E}(k)$$

$\underline{F}(k)$ and $\underline{S}(k)$ are the Fock and overlap matrices between Bloch functions and $\underline{C}(k)$ collects the coefficients of the linear combinations of Block functions that provide the crystalline orbitals.

The main advantages of the VEH technique are that it is completely theoretical and gives ab-initio double zeta quality results with negligible computer time, since only one-electron integrals need to be evaluated and SCF iterative cycles are completely avoided.

It must be pointed out that the VEH atomic potentials have not been parameterized for geometry optimization purposes and should be used with geometric parameters close to equilibrium. For systems whose geometries are experimentally unknown (as is the case of the majority of the large oligomers and polymers studied), we must make use of other techniques in order to obtain reasonable input geometries. Ab-initio techniques, even with small basis sets, rapidly become too expensive when large compounds are considered. As a result we often use geometries optimized with a method such as MNDO.

The results for polyacetylene, polydiacetylenes, polyphenylene, polypyrrole and other polymers using the VEH method have been published. A comparison between the ab-initio Hartree-Fock crystal orbital results and the VEH results on polyacetylene shows perfect agreement for minor details. It seems therefore that the VEH method is an inexpensive way to derive excellent one electron energies and wave functions for the ground state of polymers.

4 CONCLUSIONS

At the present time, ab-initio Hartree-Fock and VEH methods can yield reliable ground state one electron properties. Correlation energies require more effort and are not yet available for many systems.

Semi-empirical methods (MNDO, CNDO, PPP, and its variants Hubbard and Huckel) yield semi-quantitative information on both ground and excited state properties.

With the advent of supercomputers we will soon see tremendous advances in both areas. I expect that effects of impurities and dopants, interchain effects and solid state effects will become amenable to attack using these methods in the near future.

REFERENCES

1 L. Salem, Molecular Orbital Theory of Conjugated Systems (W. A. Benjamin, 1966).
2 J. C. Slater, The Self-Consistent Field for Molecules and Solids (McGraw Hill, NY, 1974).
3 H. F. Schaeffer, Methods of Electronic Structure Theory (Plenum, NY, 1977).
4 W. A. Goddard, Ann. Rev. Phys. Chem. 29, 363 (1978).
5 M. L. Cohen, Physica Scripta T1, 5 (1982).
6 G. Nicholas and Ph. Durand, J. Chem. Phys. 70, 2020 (1979).
7 R. Parr, Quantum Theory of Molecular Electronic Structure (Benjamin, NY, 1963).
8 K. Schulten, I. Ohmine, M. Karplus, J. Chem. Phys. 64, 4422 (1976); A. Lasaga, R. Aerni, M. Karplus, J. Chem. Phys. 73, 5230 (1980).
9 M. S. Dewar and W. Thiel, J. Am. Chem. Soc. 99, 4907 (1977).
10 R. Chance, D. Boudreaux, J. Bredas, R. Silbey, Phys. Rev. B27, 1440 (1983).
11 H. Fukutome and M. Sasai, Prog. Theor. Phys. 67, 41 (1982).
12 S. Ramesesha and Z. Soos, Syn. Met. 9, 293 (1984); J. Chem. Phys. 80, 3278 (1984).

13 W. Su, R. Schrieffer, A. Heeger, Phys. Rev. Lett. 42, 1698 (1979); Phys. Rev. B22, 2099 (1980).
14 L. Johnson and T. Peacock, Mol. Phys. 20, 849 (1971).
15 E. Lieb and F. Wu, Phys. Rev. Lett. 20, 1445 (1968).
16 A. Ovchinnihov, I. Ukrainskii, G. Kventsel, Sov. Phys. Usp. 15, 575 (1973).
17 B. Hudson, B. Kohler, K. Schulten, Excited Statesm vol. 6, edited by E. C. Lim (Academic Press, 1982).
18 J. Pople and S. Walmsley, Mol. Phys. 5, 15 (1962).
19 Y. Lin-Liu and K. Maki, Phys. Rev. B22, 5754 (1980); H. Takayama, Y. Lin-Liu, K. Maki, Phys. Rev. B21, 2388 (1980).
20 J. L. Bredas, R. Chance, R. Silbey, Mol. Cryst. Liq. Cryst. 77, 319 (1981); Phys. Rev. B26, 5843 (1982).
21 A. Bishop, D. Campbell, K. Fesser, Mol. Cryst. Liq. Cryst. 77, 253 (1981); Nucl. Phys. B200, 297 (1982).
22 S. Kivelson and D. Heim, Phys. Rev. B26, 4278 (1982).
23 K. Subbaswamy and M. Grabowski, Phys. Rev. B24, 2168 (1981).
24 C. Aslangul and D. St. James, Phys. Lett. 96A, 247 (1983); J. Physique 44, 953 (1983).
25 H. Fukutome and M. Sasai, Prog. Theor. Phys. 69, 1 (1983); 69, 373 (1983).
26 R. Chance, D. Boudreaux, J. L. Bredas, R. Silbey, Phys. Rev. B28, 6927 (1983).
27 a) Quantum Theory of Polymers, NATO-ASI Ser. vol. C39, Edited by J. Andre, J. Bredas, J. Delhalle, and J. Ladik (Reidel, Dordrecht, 1978); b) Recent Advances in the Quantum Theory of Polymers, Edited by J. Andre, J. Bredas, J. Ladik, G. Leroy and C. Moser (Springer, Berlin, 1980).
28 K. Kip, N. Lipari, C. B. Duke, B. Hudson and J. Diamond, J. Chem. Phys. 64, 4020 (1976); W. Ford, C. B. Duke and W. R. Solaneck, J. Chem. Phys. 77, 5020 (1982).
29 J. L. Bredas, R. R. Chance, R. Silbey, G. Nicholas and Ph. Durand, J. Chem. Phys. 75, 255 (1981); 77, 371 (1982); J. L. Bredas, R. Chance, R. Baughman and R. Silbey, 76, 3673 (1982); J. L. Bredas, R. Elsenbaumer, R. Chance, R. Silbey, J. Chem. Phys. 78, 5656 (1983); J. Bredas, D. Boudreaux, R. Silbey, R. Chance, J. Am. Chem. Soc. 105, 6555 (1983).
30 A. Karpfen, Physica Scripta, vol. T1, 79 (1982).
31 S. Suhai, Phys. Rev. B27, 3506 (1983); Phys. Rev. B29, 4570 (1984).
32 G. Nicholas and Ph. Durand, J. Chem. Phys. 72, 453 (1980).
33 J. L. Bredas, B. Themans, J. M. Andre, J. Chem. Phys. 78, 6137 (1983).

THEORETICAL EXCITON AND POLARITON SPECTRA OF POLYDIACETYLENES

S.Suhai

German Cancer Research Center, INF 280,
D-6900 Heidelberg, FRG

A review is given of first principles methods for the calculation of exciton and polariton spectra in polymer crystals. Correlation effects are treated using the polaron model, the interaction of electron and hole polarons is calculated with the help of the resolvent operator formalism. The important role of charge-transfer components in the singlet and triplet exciton and polariton spectra of polydiacetylenes is discussed.

1 INTRODUCTION

The visible spectrum of polydiacetylenes (PDA's) consists of a complex series of transitions starting with a strong absorption edge at about 2 eV (1-6). Absence of photoconductivity at the absorption maximum (3), observation of excitonic behavior in the resonance Raman spectra (4) as well as of surface polaritons (7) supported the suggestion of Bloor et al. that the first absorption peak is due to excitons in this polymer (1). Calculations of the exciton spectrum of PDA and of other structurally similar organic polymers both at semiempirical (8-11) and first principles (12) levels have undoubtedly shown that the bound electron-hole (e-h) pair is delocalized over several unit cells. On the other hand, the shape and the size of the experimentally observed excitonic electroreflectance response in PDA's could plausibly be accounted for (13-15), if the field-induced energy shift due to large charge-transfer components in the excitonic wave function had been taken into account.

Experimentally observed polariton spectra in polymer crystals have mostly been analyzed in terms of phenomenological models, treating the solid as a continuum (16-18). Microscopic treatments for molecular crystals have been proposed within the framework of the Frenkel-exciton model (19-22) using semiempirical Hamiltonians to represent the matter field.

The methods presented here extend previous work in two respects: First, we solve the eigenvalue problem of the interacting exciton-photon system, applying Green's-function formalism of charge-transfer exciton theory (12) supplemented with terms representing the radiation field. Second, we calculate all excitonic effects from first principles using ab initio Bloch- and Wannier-functions for one-electron orbitals and including electronic correlation effects on the spectrum with the help of the electron polaron model.

2 EXCITON-PHOTON INTERACTION IN POLYMER CRYSTALS

In the coupling region, where the electromagnetic field cannot be treated as a weak perturbation, its interaction with the electronic excitations of the solid has to be included in zeroth order. To calculate the energy spectrum of the new particles evolving from this interaction, we have to diagonalize the Hamiltonian of the interacting electron-radiation field system (23,24):

$$\hat{H} = \hat{H}_{el} + \hat{H}_{ph} \tag{1}$$

The electronic part, $\hat{H}_{el}$, is expanded in terms of the field operators $\Psi(\vec{x}), \Psi^+(\vec{x})$, obeying the usual fermion commutation rules ($\vec{x}$ represents both spatial and spin variables, $\vec{x}=\{\vec{r},\sigma\}$):

$$\hat{H}_{el} = \int \Psi^+(\vec{x})\left[-\frac{1}{2m_e}\left(\frac{\hbar}{i}\nabla - \frac{e}{c}\vec{A}\right)^2 + V_p(\vec{x})\right]\Psi(\vec{x})\,d^3\vec{x} + $$
$$+ \frac{1}{2}\iint \Psi^+(\vec{x})\Psi^+(\vec{x}')\frac{e^2}{|\vec{x}-\vec{x}'|}\Psi(\vec{x}')\Psi(x)\,d^3\vec{x}\,d^3\vec{x}' \tag{2}$$

$V_p(\vec{x})$ stands here for the periodic potential due to ions and $\vec{A}$ is the vector potential of the electromagnetic field containing only transverse photons ($\nabla\vec{A}=0$). $\vec{A}$ can be expanded in terms of the photon operators b and b^+, obeying Bose statistics (23):

$$\vec{A}=\sum_{\vec{k},j}\left(\frac{2\pi c^2\hbar}{V\omega_{\vec{k},j}}\right)^{1/2}\left\{\vec{e}_{\vec{k},j}\,e^{i\vec{k}\vec{r}}\,b_{\vec{k},j}+\vec{e}_{\vec{k},j}\,e^{-i\vec{k}\vec{r}}\,b^+_{\vec{k},j}\right\}, \tag{3}$$

where $\omega_{\vec{k},j}$ stands for the frequency of the photon with momentum $\vec{k}$ and polarization index j, $\vec{e}_{\vec{k},j}$ is the corresponding polarization vector and V is the normalization volume. In terms of the b's, the Hamiltonian of the free photon field has the form

$$\vec{H}_{ph}=\sum_{\vec{k},j}\hbar\omega_{\vec{k},j}\,b^+_{\vec{k},j}\,b_{\vec{k},j}\,. \tag{4}$$

Excluding non-linear optical phenomena by the neglect of the A^2-term in eq. (2) and taking into account the use of Coulomb gauge, we may identify the term responsible for the electron-photon interaction:

$$\hat{H}_{el,ph}=\int\Psi^+(\vec{x})\left[-\frac{\hbar e}{imc}\vec{A}(\vec{x})\nabla\right]\Psi(\vec{x})\,d^3\vec{x}. \tag{5}$$

To be able to evaluate the matrix elements of this operator for a polymer, we expand the fermion operators using as a basis the complete orthonormal set of Wannier spin orbitals (24):

$$\begin{aligned}\Psi(\vec{x})&=\sum_{\ell}a_{\ell v}\,w_v(\vec{x}-\vec{\ell})+\sum_{\ell}a_{\ell c}\,w_c(\vec{x}-\vec{\ell})\\ \Psi^+(\vec{x})&=\sum_{\ell}a^+_{\ell v}\,w^*_v(\vec{x}-\vec{\ell})+\sum_{\ell}a^+_{\ell c}\,w^*_c(\vec{x}-\vec{\ell}),\end{aligned} \tag{6}$$

with the notation $w_j(\vec{x}-\vec{\ell})=w_j(\vec{r}-\vec{R}_\ell)\chi_j(\sigma)$. $w_j(\vec{r}-\vec{R}_\ell)$ is the Wannier function (WF) belonging to either the conduction or the valence band (j=c or v, respectively). The WF's are localized around the unit cell and multiplied with the spin function χ. With the help of these expansions, the electronic Hamiltonian can be partitioned in the form (12):

$$\hat{H}_{el}=E_{HF}+\hat{H}_e+\hat{H}_h+\hat{H}_{e,h}\,. \tag{7}$$

In terms of the particle-hole operators a and d (defined as $a^+_\ell=a^+_{\ell,c}$ and $a_\ell=a_{\ell,c}$ for the creation and destruction of a particle in the conduction band and similarly

$d_{\ell}^{+}=a_{\ell,v}$ and $d_{\ell}=a_{\ell,v}^{+}$ in the case of a hole in the valence band), we can construct the symmetry-adapted wave function of an e-h pair with quasimomentum K and with an e-h separation of s unit cells in the Bloch form

$$|\Psi_{s,\vec{K}}\rangle = N_c^{-1/2} \sum_{\ell} \exp(i\vec{K}\vec{R}_{\ell})\, a_{\ell+s}^{+}\, d_{\ell}^{+}\, |\Phi_{HF}\rangle \equiv B_{s,\vec{K}}^{+}\, |\Phi_{HF}\rangle \tag{8}$$

where N_c stands for the number of monomer cells and $|\Phi_{HF}\rangle$ is the Hartree-Fock (HF) ground state of the polymer. Eq. (8) defines at the same time the e-h creation operator $B^{+}_{s,\vec{K}}$. The wave function of an exciton with momentum $\vec{K}$ is formed as a linear combination of e-h functions with different separation:

$$|\Psi_{\vec{K}}\rangle = \sum_{s} \Omega_{s,\vec{K}}\, |\Psi_{s,\vec{K}}\rangle \equiv B_{\vec{K}}^{+}\, |\Phi_{HF}\rangle \tag{9}$$

Applying the resolvent operator formalism (25-27), the exciton band levels $E_{\vec{K}}$ can be obtained as solutions of the Schrödinger equation

$$|\Psi_{\vec{K}}\rangle = \left[E_{\vec{K}} - (\hat{H}_e + \hat{H}_h)\right]^{-1} \cdot \hat{H}_{e,h} \cdot |\Psi_{\vec{K}}\rangle \tag{10}$$

Further details of the calculation of the Green's function and of the matrix elements of the e-h interaction in polymers are given in Ref.12. To obtain the exciton-photon interaction term, we substitute expansions (3) and (6) for the electromagnetic and fermion fields, respectively, into eq.(5). After having neglected terms, which describe either Compton-scattering or energy conservation violating processes, we make two further physically reasonable approximations:

i) Since the momentum of the photon capable of efficiently interacting with the exciton is much smaller than the size of the BZ, terms of the type $\exp\{i\vec{k}\vec{r}\}$ in eq.(3) can be substituted by $\exp\{i\vec{k}\vec{R}_{\ell}\}$

ii) The electronic properties of most polymers are strongly anisotropic, absorption is usually most efficient along the chains containing delocalized electrons. It seems reasonable, therefore, to simplify our treatment by restricting the calculations to one photon branch polarized parallel to the polymer axes, $\vec{e}_{\vec{k},\gamma} \rightarrow \vec{e}_{\vec{k}}$.

Finally, making use of the periodic symmetry, integrals of the type $\int w(\ell) \cdot w(\ell')$ can be handled as depending only on $|\ell-\ell'|$, and the interaction term is obtained in the form

$$\hat{H}_{exc,ph} = \hbar \sum_{\vec{k}} \sum_{s} [\, g^{c,v}_{s,\vec{k}} B^{+}_{s,\vec{k}} b_{\vec{k}} + g^{v,c}_{s,\vec{k}} B_{s,\vec{k}} b^{+}_{\vec{k}} \,], \tag{11}$$

where the coupling constants $g^{j,j'}_{s,\vec{k}}$ are defined by

$$g^{j,j'}_{s,\vec{k}} = -\frac{e}{m_e}\left(\frac{2\pi}{v_c \hbar \omega_{\vec{k}}}\right)^{1/2} \vec{e}_{\vec{k}} \int w^{*}_{j}(\vec{r}) \frac{\hbar}{i} \nabla w_{j'}(\vec{r}-\vec{R}_s)\, d^3\vec{r}. \tag{12}$$

($v_c = V/N_c$ is the volume of the unit cell). Analogously to eq.(4), the total energy of the exciton field can be written as

$$\vec{H}_{exc} = \sum_{\vec{k}} E_{\vec{k}} B^{+}_{\vec{k}} B_{\vec{k}} = \sum_{\vec{k}} \sum_{s} \sum_{s'} \hbar \nu_{\vec{k}} \Omega^{*}_{s,\vec{k}} \Omega_{s',\vec{k}} B^{+}_{s,\vec{k}} B_{s',\vec{k}}. \tag{13}$$

Now collecting the Hamiltonians from eqs.(4), (12) and (13), the total energy of the interacting exciton-radiation system is obtained in the form

$$\hat{H} = \hat{H}_{exc} + \hat{H}_{ph} + \hat{H}_{exc,ph} = \sum_{\vec{k}} \hbar \hat{h}_{\vec{k}}, \tag{14}$$

with

$$\hat{h}_{\vec{k}} = \sum_{s} \sum_{s'} \nu_{\vec{k}} \Omega^{*}_{s,\vec{k}} \Omega_{s',\vec{k}} B^{+}_{s,\vec{k}} B_{s',\vec{k}} + \omega_{\vec{k}} b^{+}_{\vec{k}} b_{\vec{k}} + \sum_{s} g^{c,v}_{s,\vec{k}} B^{+}_{s,\vec{k}} b_{\vec{k}} + \sum_{s} g^{v,c}_{s,\vec{k}} B_{s,\vec{k}} b^{+}_{\vec{k}}. \tag{15}$$

To facilitate the diagonalization of $\hat{h}_{\vec{k}}$, we build from the operator set $\{B,b\}$ the vector $\underline{C}$ defined by

$$C_i(\vec{k}) = B_{i-1-M,\vec{k}} \quad \text{for} \quad 1 \le i \le 2M+1$$

$$C_i(\vec{k}) = b_{\vec{k}} \quad \text{for} \quad i = 2M+2, \tag{16}$$

where M denotes the radius of the exciton (i.e. charge-transfer excitation components will be included from the reference cell, M=0, to neighboring cells -M,-M+1,...,M-1,M). Similarly, from the coefficients of the corresponding operators in eq.(15), we construct the matrix $\underline{\underline{A}}$:

$$A_{i,j}(\vec{k}) = \nu_{\vec{k}} \Omega_{i-1-M,\vec{k}} \cdot \Omega_{j-1-M,\vec{k}} \quad \text{for} \quad 1 \le i,j \le 2M+1$$

$$A_{i,j}(\vec{k}) = g^{c,v}_{i-1-M,\vec{k}} \quad \text{for} \quad 1 \le i \le 2M+1, \; j = 2M+2$$

$$A_{i,j}(\vec{k}) = g^{v,c}_{j-1-M,\vec{k}} \quad \text{for} \quad 1 \le j \le 2M+1,\ i = 2M+2$$

$$A_{i,j}(\vec{k}) = \omega_{\vec{k}} \quad \text{for} \quad i = j = 2M+2 \,. \tag{17}$$

It is easy to verify, using a term-by-term comparison that, expressed with the help of these quantities, operator $\hat{h}_{\vec{k}}$ takes the equivalent form

$$\hat{h}(\vec{k}) = \underline{C}^{+}(\vec{k})\, \underline{\underline{A}}(\vec{k})\, \underline{C}(\vec{k}) \,. \tag{18}$$

In seeking the eigenvalues of $\hat{h}$, we diagonalize first $\underline{\underline{A}}$:

$$\underline{\underline{A}}(\vec{k}) \cdot \underline{\underline{U}}(\vec{k}) = \underline{\underline{U}}(\vec{k})\, \underline{\underline{\Lambda}}(\vec{k}) \,, \tag{19}$$

where $\underline{\underline{\Lambda}}(\vec{k})$ is a diagonal matrix and $\underline{\underline{U}}(\vec{k})^{+} = \underline{\underline{U}}(\vec{k})^{-1}$. The transformation of our coefficient matrix C with this unitary matrix,

$$\underline{D}(\vec{k}) = \underline{\underline{U}}(\vec{k})\, \underline{C}(\vec{k}) \,, \tag{20}$$

provides us the eigenvector matrix of $\hat{h}$:

$$\hat{h}(\vec{k}) = \underline{D}^{+}(\vec{k})\, \underline{\underline{U}}(\vec{k})\, \underline{\underline{A}}(\vec{k})\, \underline{\underline{U}}^{+}(\vec{k})\, \underline{D}(\vec{k}) =$$

$$= \underline{D}^{+}(\vec{k})\, \underline{\underline{\Lambda}}(\vec{k})\, \underline{D}(\vec{k}) = \sum_{i=1}^{2M+2} \Lambda_{i,i}(\vec{k}) D_i^{+}(\vec{k}) D_i(\vec{k}) \,. \tag{21}$$

In this way we found new quasiparticles, the polaritons, corresponding to the eigenstates of the interacting exciton-photon field. Their energy spectrum is given throughout the BZ by $\Lambda_{i,i}(\vec{k})$.

3 EXCITON AND POLARITON SPECTRA OF POLYDIACETYLENES

As we can see from Equ. (15), four quantities are needed for the construction of the Hamiltonian of the coupled exciton-photon field, $\hat{h}_{\vec{k}}$:

i) The frequency dispersion of the excitons, $\nu_{\vec{k}}$;
ii) the weight factors of e-h pairs with different separation in the total exciton wave function, $\Omega_{s,\vec{k}}$;
iii) the dielectric constant of the crystal, ε, entering the photon dispersion, $\omega_{\vec{k}}$;
iv) the coupling constants $g^{c,v}_{s,\vec{k}}$ and $g^{v,c}_{s,\vec{k}}$, respectively.

The exciton frequencies throughout the BZ can be obtained from Equ. (10). In our previous work (12), we calculated them for two PDA polymers, PTS and TCDU, respectively.

One of the motivations behind that study was the comparison of two significantly different diacetylene backbones with acetylene-like (PTS) and with butatriene-like (TCDU) bonding patterns, respectively, from the point of view of optical properties. After substitution of expansions (8) and (9), Equ. (10) leads to a system of homogeneous linear equations of the form

$$\Omega_{r,\vec{K}} = \sum_{s}\sum_{t} G(\vec{R}_r, \vec{R}_s, E_{\vec{K}}) \cdot V(\vec{R}_s, \vec{R}_t, \vec{K}) \cdot \Omega_{t,\vec{K}}, \quad (22)$$

where G and V stand for the Green's function and e-h interaction matrix elements, respectively. The single particle energies $\varepsilon_c(\vec{k})$ and $\varepsilon_v(\vec{k})$ entering G, have been

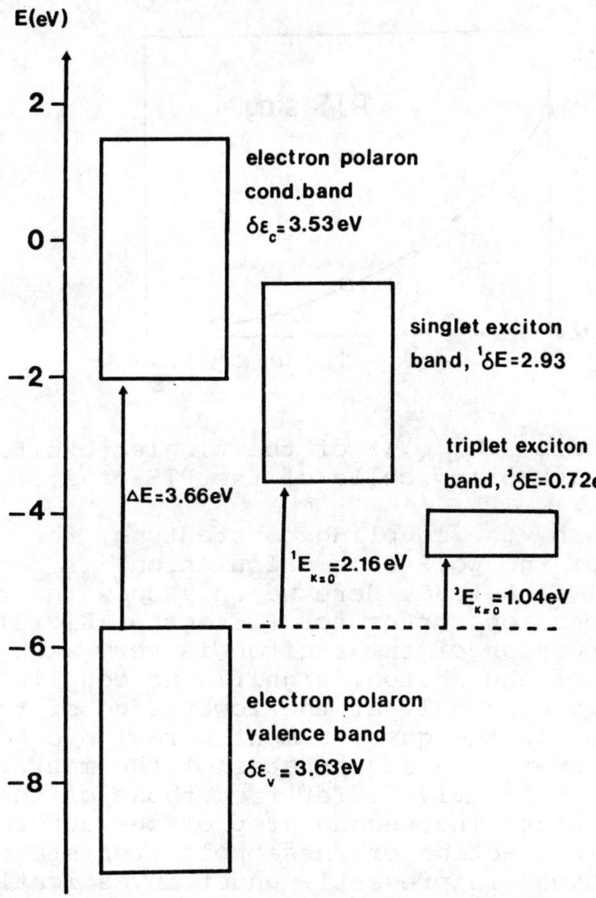

Fig.1. The first singlet and triplet exciton bands of the polydiacetylene PTS shown together with the original electron and hole polaron bands.

calculated both at Hartree-Fock (HF) and correlated levels using many-body perturbation theory and the electron-polaron model (28). The polaron corrections in G turn out to be substantial: they reduce the HF-exciton energies by about 2 eV. The dielectric screening in V, calculated using the Hermanson-Phillips scheme (29), causes a shift of about 0.4 eV in the opposite direction. As an example, Fig.1 shows the resulting exciton bands for PTS. The situation is very similar in TCDU, with the difference that the lower edge of the singlet exciton band is at 1.7 eV and its width is 3.5 eV. In both cases, the exciton wave function extends over 4-5 diacetylene units, as shown in Fig.2. The s=0 (Frenkel) component contributes only about 30% to the e-h binding energy.

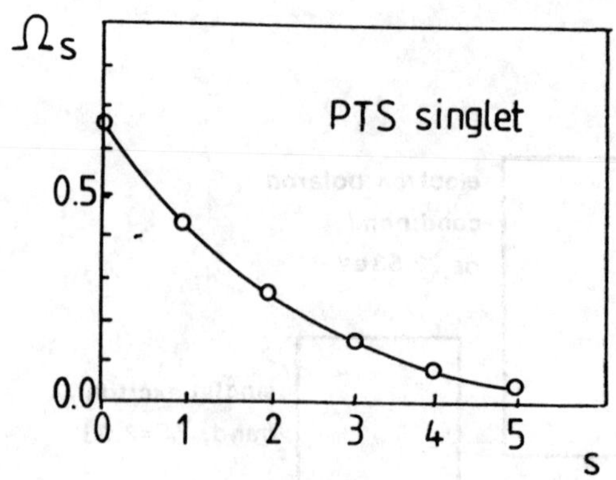

Fig.2. Components $\Omega_{s,K=0}$ of the singlet-exciton wave function in neighboring cells of the PTS crystal.

The evaluation of the coupling constants g, the most tedious part of the polariton calculation, is described in more detail in Ref.30. Here we only show the resulting polariton dispersion curves for PTS again (Fig.3). Since the dispersion of the exciton is very weak on the scale of that of the photon, significant coupling of the two particles occurs only at the lower edge of the Brillouin zone. In the quasi-momentum region of 0.005 to 0.015 π/a, however, the eigenstates of the matter+radiation field significantly differ from those of the non-interacting fields. The second step of the absorption process, the interaction of these polariton states with bulk phonon modes, is presently under investigation in our Laboratory. Future extension of this work will include also the corresponding surface states.

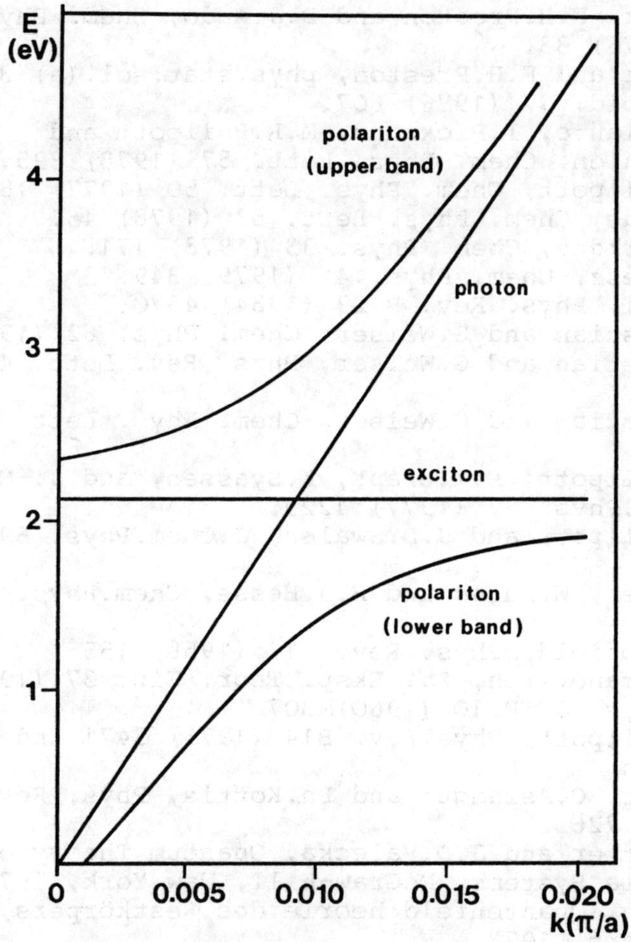

Fig.3. Polariton dispersion resulting from the interaction of an incoming photon with the lowest lying singlet exciton of PTS.

REFERENCES

1 D.Bloor. D.J.Ando, F.H.Preston and G.C.Stevens, Chem. Phys. 24 (1974) 407.
2 D.Bloor, G.C.Stevens, P.J.Page, and P.M.Williams, Chem. Phys. Lett. 33 (1975) 61.
3 B.Reimer and H.Bäßler, phys.stat.sol.(a) 32 (1975) 435.
4 D.N.Batchelder and D.Bloor, Chem.Phys.Lett. 38 (1976) 37.

5 D.Bloor, F.H.Preston and D.J.Ando, Chem. Phys. Lett. 38 (1976) 33.
6 D.Bloor and F.H.Preston, phys.stat.sol.(a) 37 (1976) 427; ibid. 37 (1976) 607.
7 A.Brillante, I.Pickrand, M.R.Philpott and J.D.Swalen, Chem. Phys. Lett. 57 (1978) 395.
8 M.R.Philpott, Chem. Phys. Lett. 50 (1977) 18.
9 N.A.Cade, Chem. Phys. Lett. 53 (1978) 45.
10 D.R.Yarkony, Chem. Phys. 33 (1978) 171.
11 M.Kertész, Chem. Phys. 44 (1979) 349.
12 S.Suhai, Phys. Rev. B 29 (1984) 4570.
13 L.Sebastian and G.Weiser, Chem. Phys. 62 (1981) 447.
14 L.Sebastian and G.Weiser, Phys. Rev. Lett. 46 (1981) 1156.
15 L.Sebastian and G.Weiser, Chem. Phys. Lett. 64 (1979) 396.
16 M.R.Philpott, P.M.Grant, K.Syassen, and J.-M.Turlet, J.Chem.Phys. 67 (1977) 4229.
17 M.R.Philpott and J.D.Swalen, J.Chem.Phys. 69 (1978) 2912.
18 G.Weiser, W.Fuchs and H.J.Hesse, Chem.Phys. 52 (1980) 183.
19 J.J.Hopfield, Phys. Rev. 112 (1958) 1555.
20 V.M.Agranovitch, Zh. Eksp. Teor. Fiz. 37 (1959) 340; Sov.Phys.-JETP 10 (1960) 307.
21 M.R.Philpott, Phys.Rev. B14 (1976) 3471 and refs. therein.
22 M.Orrit, C.Aslangul and Ph.Kottis, Phys. Rev. B25 (1982) 7263.
23 A.L.Fetter and J.D.Valecka, Quantum Theory of Many-Particle Systems, McGraw-Hill, New York, 1971.
24 H.Haken, Quantenfeldtheorie des Festkörpers, Teubner, Stuttgart, 1973.
25 M.Lax, Phys. Rev. 94 (1954) 1391.
26 G.F.Koster and J.C.Slater, Phys. Rev. 95 (1954) 1167.
27 Y.Takeuti, Prog. Theor. Phys. 18 (1957) 421.
28 S.Suhai, Phys. Rev. B27 (1983) 3506.
29 J.Hermanson and J.C.Phillips, Phys. Rev. 150 (1966) 652.
30 S.Suhai, J.Chem.Phys. (in press).

AB INITIO STUDIES ON POLYYNES AND POLYDIACETYLENES: STRUCTURE AND HARMONIC FORCE FIELD

Alfred Karpfen

Institut für Theoretische Chemie und
Strahlenchemie der Universität Wien
A-1090 Wien, Währingerstraße 17,
Austria

Abstract: Ab initio crystal orbital calculations have been perfomed on the equilibrium geometries and on parts of the harmonic force field of polyyne and polydiacetylene. Results obtained on the polymers are compared with conventional MO calculations on oligoynes and on divinylacetylene.

1 INTRODUCTION

Organic polymers with fully conjugated carbon backbones have attracted the interest of both experimentalists and theoreticians in the last years. Among them polyacetylene due to its remarkable changes in transport properties upon doping and polydiacetylenes because of their nearly perfect single crystal structures are the most prominent examples. The theoretical description of the electronic structure of these polymers has in the past mostly been restricted to band structure type calculations. The development of ab initio methods and recent technical improvements currently allow a reliable description of a number of ground state properties within the framework of the Hartree Fock approximation including the characterization of energy minima on the energy hypersurface and moreover offer also the possibility for a methodically consistent comparison with molecular and cluster-type calculations. For very recent reviews see (1-3). In this contribution the gradual change of bond lengths and harmonic force constants upon polymerization is demonstrated on the example of the polyynes. Oligoynes ranging from acetylene to tetraacetylene are compared with the infinite polyyne. A few methodical discussions on the convergence of crystal orbital calculations on polyyne are included as well. Finally the structure of polydiacetylene is compared with the geometry of divinylacetylene and the relative stability of valence-and rotational isomers of polydiacetylenes is investigated.

2 POLYYNES

The prototype of a conjugated polymer is the linear, infinite chain of carbon atoms. For a recent survey of experimental work on polyynes, often also termed carbynes, see (4). Only a few ab initio calculations have been performed on polyynes in the past (5-7). Here we present data considerably improved with respect to lattice summations than the previously published calculations.

Table 1: Ab initio crystal orbital calculations on polyyne 7s3p basis (8), $r_{C\equiv C}$=2.27 bohr, r_{C-C}=2.57 bohr.

N	thresh.	total energy	I.P.	B.W.	ref.
5	0.005	-75.5997	.3117	.2669	
6	0.005	-75.5997	.3117	.2669	
7	0.005	-75.5998	.3117	.2669	
5	0.003	-75.6002	.3114	.2668	this work
6	0.003	-75.6002	.3115	.2669	
7	0.003	-75.6002	.3115	.2668	
6	0.002	-75.6005	.3112	.2669	
6	0.001	unsuccessful			
7	0.001	-75.6013	.3107	.2665	
8	0.001	-75.6014	.3107	.2664	
4	0.01	-75.6007	.348	.273	(6)
3	0.02	-75.6389	.460	.211	(7)

N: number of neighbors, thresh.: Threshold for eigenvalues of $S^{-1/2}(k)$ below which the corresponding eigenvectors are discarded, I.P.: ionization potential, B.W.: π-band width.

In table 1 we show total energies, ionization potentials and π-band widths as a function of the number of neighbors and the overlap threshold applied in order to demonstrate the internal accuracy of our crystal orbital calculations and therefore also the direct comparability to standard molecular calculations. The latter is fur-

Table 2: Eigenvalues of oligoynes. All values in hartree.
7s3p/3s, $r_{C\equiv C}$=2.27, r_{C-C}=2.57, r_{C-H}=1.995 bohr.

C_2H_2	C_4H_2	C_6H_2	C_8H_2	polyyne	comment
			-.333	-.311	
		-.344			
	-.364		-.416		degenerate
-.405		-.447			
	-.492		-.491		π-levels
		-.525			
			-.540	-.577	
-.676	-.705	-.717	-.722	---	σ-
-.759	-.729	-.725	-.725	---	C-H
			-.792	-.784	
		-.810			
---	-.852		-.848		σ
		-.888			
			-.907	-.964	
			-1.036	-1.049	
		-1.035			
	-1.035		-1.071		
-1.040		-1.082			σ
	-1.101		-1.110		
		-1.125			
			-1.137	-1.168	
	-75.6005	-75.6024	-75.6027	-75.6014	ΔE[a)]

a) $E_{C_{n+2}H_2} - E_{C_nH_2}$

nished in table 2 on the example of cluster-type calculations on oligoynes. We compare eigenvalues and energy differences per C_2-moiety with band edges and with the total energy of polyyne. Computed equilibrium geometries of oligoynes and of polyyne are displayed in figure 1. Corresponding in-line harmonic force fields are compiled in table 3.

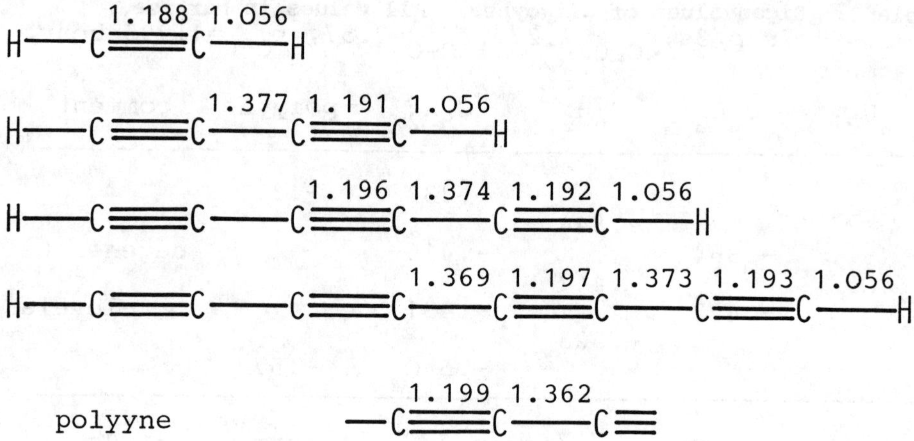

Figure 1: Equilibrium geometries of oligoynes and polyyne as obtained with a 7s3p/3s basis. All values in Å.

Turning first to the convergence behavior of the crystal orbital calculations (see tables 1 and 2) we observe that our best value for the total energy per C_2 is about 1 kcal/mol above the value extrapolated from cluster calculations. The main source of this discrepancy lies in the unavoidable use of a threshold value in the Löwdin orthogonalization in order to escape near linear dependence problems caused by the very small intercell distances in polyyne. The agreement between CO- calculations on polyyne and MO- calculations on oligoynes is, however, sufficiently close to reject a recent criticism by Teramae et al. (7) on earlier polyyne studies by the author (6).

Inspecting the changes in bond lengths upon polymerization we note:

i) C-H bond lengths remain practically unaffected,

ii) gradual increase of triple bond distances and simultaneous decrease of single bond distances, and seemingly smooth convergence to the polyyne structure,

iii) overall the geometrical changes are small if diacetylene is chosen as reference molecule.

A similarly consistent picture is found for the harmonic force constants. Triple bond force constants decrease and single bond force constants increase upon polymerization. Of particular significance are the large

Table 3: Harmonic C-C stretching force constants of oligoynes and polyyne. All values in mdyn/Å.

Acetylene	20.1						
Diacetylene	19.5						
	0.39	8.0					
	-0.18	0.39	19.5				
Triacetylene	19.36						
	0.44	8.05					
	-0.24	0.43	18.98				
	0.08	-.09	0.43	8.05			
	-0.02	0.08	-0.24	0.44	19.36		
Tetraacetylene	19.24						
	0.46	8.06					
	-0.26	0.45	18.84				
	0.10	-.12	0.51	8.13			
	-0.04	0.09	-0.31	0.51	18.84		
	0.02	-.02	0.09	-.12	0.45	8.06	
	0.00	0.02	-0.04	0.10	-0.26	0.46	19.24
Polyyne	17.66						
	0.55	7.47					
	-0.82	0.55	17.66				
	-	-.80	0.55	7.47			

values of the coupling force constants between nearest and next nearest neighbor bonds, a typical feature of conjugated systems. In case of force constants the convergence of cluster quantities to the polymer values is less clear-cut. This may partly be due to the use of analytical gradients for the molecular calculations versus pointwise scanning of the energy surface for the polymer and subsequent evaluation of force constants via polynomial fits. However, the main factor which complicates a direct comparison comes from the restriction in geometry-variations because of periodic boundary conditions. Force constants derived with the aid of crystal orbital calculations are necessarily Fourier transformed force constants and therefore contain always a number of small coupling constants. Vibrational frequencies of oligoynes and phonon dispersion curves of polyyne will be published elsewhere.

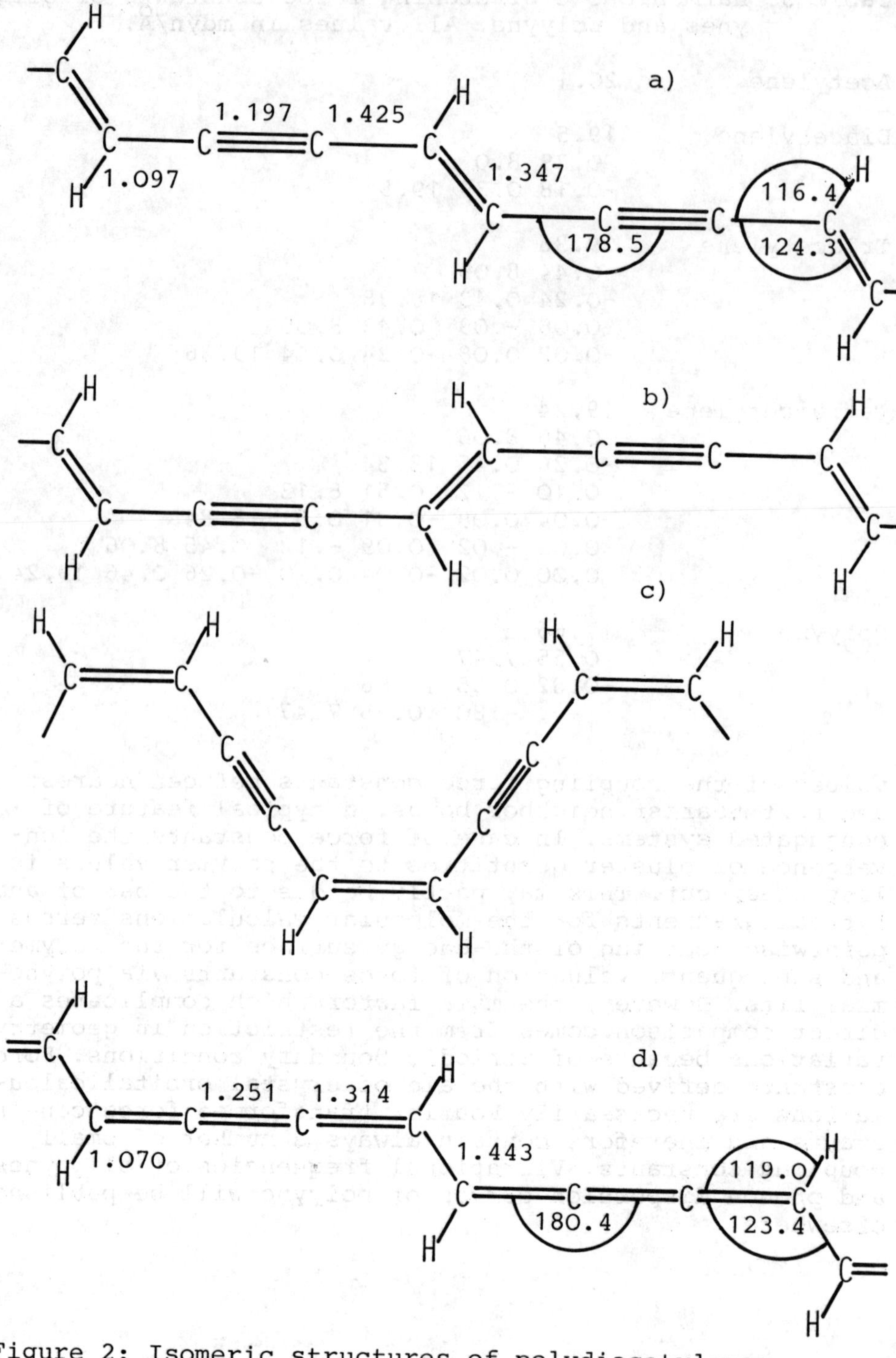

Figure 2: Isomeric structures of polydiacetylenes.

3 POLYDIACETYLENES

Although polydiacetylenes have already been the subject of several ab initio studies (9-13) we present here improved CO-calculations which as for the polyyne case discussed earlier are considerably better converged than the previous literature data and where a more extended geometry optimization has been carried out than in (11). The polydiacetylene isomers treated in this work are shown in fig. 2. Structure 2a corresponds to the syn-conformation with the acetylenic PDA-type valence structure as found in practically all PDA-crystals, whereas 2d has a butatrienic PBT backbone. Structures 2b and 2c are rotational isomers of 2a. Geometry optimization has been performed for 2a and 2d only. Equilibrium bond distances and bond angles are also shown in fig. 2. The relative stabilities of structures 2a - 2d are given in table 4. For the purpose of comparison the isolated molecule divinylacetylene has also been optimized at the same level of sophistication (see fig. 3). From these data the following conclusions may be drawn: i) Structures 2a - 2c are of nearly the same stability. This is in accord with recent electron diffraction data (14) where syn- and anti-conformer of divinylacetylene were found to be of similar stability and that the molecule behaves nearly as a free rotator. This high flexibility of divinylacetylene with respect to rotation around single bonds may be of importance for solution properties of polydiacetylenes (15-18). With the bulky substituents the planar syn-conformation is fixed in the crystal. For isolated chains rotation around single bonds is hindered only by steric interactions between substituents on the same chain and therefore the torsional potential in the vicinity of the syn-conformation might be quite flat resulting in considerable flexibility for a single PDA-chain as well.

Table 4: Relative stability of polydiacetylene isomers. 7s3p/3s basis, all values in kcal/mol.

structure	$\Delta E/C_4H_2$
2a,PDA	0.0
2b	0.05
2c	-0.1
2d,PBT	11.1

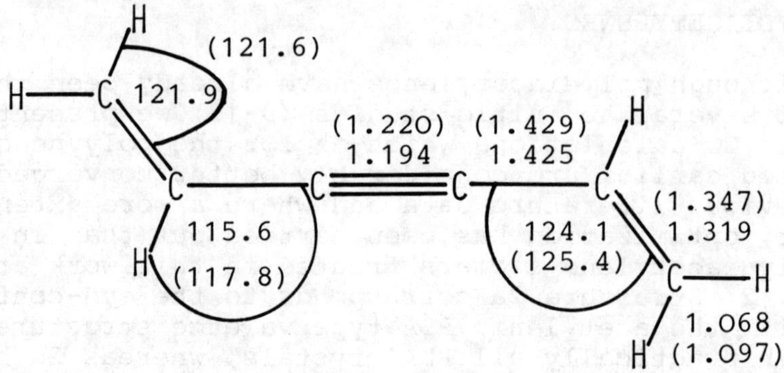

Figure 3: Equilibrium geometry of divinylacetylene as obtained with a 7s3p/3s basis. Bond distances in Å, bond angles in degrees. Experimental values in brackets taken from (14).

ii) In agreement with our earlier results (11) the PDA backbone is much more stable than the PBT structure. This agrees both with X-ray structure investigations on polydiacetylene crystals and with spectroscopic investigations on small oligomer radicals (19) which upon increasing the chain length convert from initially butatrienic structures to PDA-type structures. Transitions from PDA to PBT in longer chains are therefore very improbable.
iii) The computed polymer geometry is rather close to the geometry of divinylacetylene.

In table 5 part of the harmonic force field of a PDA polymer corresponding to carbon-carbon stretching coordinates only is shown. Qualitatively the picture

Table 5: Partial harmonic force field of a PDA polymer as obtained with a 7s3p/3s basis. All values in mdyn/Å.

	C≡C	C-C	C=C	C-C
C≡C	18.9			
C-C	0.3	6.4		
C=C	-0.3	0.6	10.9	
C-C	-	-0.1	0.6	6.4

emerging from these calculations is compatible with the polyyne case treated in the preceeding chapter. The triple bond force constant is somewhat larger and the single bond force constant is significantly smaller than in polyyne in line with the trends observed in computed equilibrium geometries. In both cases diagonal stretching force constants are considerably overestimated (by about 10 to 20%) as a consequence of the use of the Hartree Fock approximation. The finding that long range coupling force constants are non-negligible in these two and probably also in other conjugated polymers should, however, persist.

Acknowledgement: The author wishes to thank Prof. P. Schuster for his interest and continuous support. Ample supply with computer time by the Interfakultäres EDV-Zentrum Wien are gratefully acknowledged.

References:

1. Kertesz, M. Electronic Structure of Polymers, Adv. Quantum Chem. 15 (1982) 161-214.
2. Karpfen, A. Ab Initio Calculations on the Ground State Properties of Polymers, Phys. Scripta T1 (1982) 79-87.
3. Karpfen, A. Ab Initio Studies on the Structure and Phonon Spectra of Simple Polymers, Quantum Chemistry of Polymers-Solid State Aspects, eds. J.Ladik and J.M.André, NATO-ASI-Series C123 (Reidel,Dordrecht, 1984) p. 33-55
4. Heimann, R.B., J.Kleiman and N.M.Salansky. Structural Aspects and Conformation of Linear Carbon Polytypes (Carbynes), Carbon 22 (1984) 147-156.
5. Kertesz, M., J.Koller and A.Azman. Ab Initio Hartree Fock Crystal Orbital Studies II. Energy Bands of an Infinite Carbon Chain, J.Chem.Phys. 68 (1978) 2779-2782.
6. Karpfen A., Ab Initio Studies on Polymers I: The Linear Infinite Polyyne, J.Phys. C12 (1979) 3227-3237.
7. Teramae, H., T.Yamabe and A.Imamura. Ab Initio Effective Core Potential Studies on Polymers, Theoret. Chim. Acta 64 (1983) 1-12.
8. Huzinaga, S., Approximate Atomic Functions I (Canada, University of Alberta,1971)
9. Kertesz, M., J.Koller and A.Azman. On the Electronic Structure of Polydiacetylenes as Studied by the Ab Initio Crystal Orbital Method, Chem.Phys. 27 (1978)

273-280
10. Kertesz, M., J.Koller and A.Azman. Electronic Structure of Polydiacetylene Backbones, Chem.Phys.Lett. 56 (1978) 18-19.
11. Karpfen, A. Ab Initio Studies on Polymers IV: Polydiacetylenes, J.Phys.C13 (1980) 5673-5689.
12. Suhai, S. Ab Initio Hartree Fock Crystal Orbital Calculations on Polydiacetylene Backbones, Chem.Phys. 54 (1980) 91-101.
13. Brédas, J.L., R.R.Chance,R.Silbey,G.Nicolas and Ph. Durand. A Nonempirical Effective Hamiltonian Technique for Polymers: Application to Polyacetylene and Polydiacetylene, J.Chem.Phys. 75 (1981) 255
14. Almenningen, A.,E.Gogstad,K.Hagen,H.Schei,R.Stolevik, O.Thingstad and M.Traetteberg. Conformation and Molecular structure of 1,5-Hexadiene-3-yne (Divinylacetylene) and Perchloro-1,5-Hexadiene-3-yne as Determined by Gas Phase Electron Diffraction and Molecular Mechanics Calculations, J.Mol. Struct. 116 (1984) 131-140.
15. Patel, G.N., R.R.Chance and J.D.Witt. A Planar-Non-Planar Conformational Transition in Conjugated Polymer Solutions, J.Chem.Phys. 70 (1979) 4387-4392.
16. Lim, K.C., C.R.Fincher,jr. and A.J.Heeger. Rod to Coil Transition of a Conjugated Polymer in Solution, Phys. Rev. Lett. 50 (1983) 1934-1937.
17. Müller, M.A., M.Schmidt and G.Wegner. A Single Chain Rod to Coil Transition of a Conjugated Polymer in Solution? Caveat Emptor, Makromol. Chem. Rapid Comm. 5 (1984) 83-88.
18. Wenz, G., M.A.Müller, M.Schmidt and G.Wegner. The Structure of Polydiacetylenes in Solution, Macromolecules, (1984) in press.
19. Bubeck, C., H.Sixl and W.Neumann. ESR Analysis of the Initial Reaction Products in the Solid State Photopolymerization of Diacetylenes, Chem. Phys. 48 (1980) 269-275.

STRUCTURAL DEPENDENCE OF THE LONGITUDINAL ELECTRIC POLARIZABILITY OF FINITE POLYENE CHAINS : AN AB INITIO STUDY

V.P. Bodart, J. Delhalle, J.M. André

Laboratoire de Chimie Théorique Appliquée
Facultés Universitaires Notre-Dame de la Paix
Rue de Bruxelles, 61 B-5000 Namur (Belgium)

and

J. Zyss

Centre National d'Etudes des Télécommunications
196, rue de Paris 92220 Bagneux (France)

ABSTRACT :

ScaledSTO-3G calculations on the dependence of the longitudinal polarizability on bond- and chain-length modifications in compounds modelling polydiacetylene and polyene are reported. Values of the polarizability undergo noticeable variations with these parameters suggesting possible enhancement of polarizabilities and hyperpolarizabilities through controlled geometry adjustements.

1. INTRODUCTION

Some of the future applications in optical communication systems shall probably be based on the large second- and third-order nonlinear optical responses such as recently observed in several organic molecules and polymers (1). Nonlinear optical susceptibilities depend on the detailed electronic and geometric structure of the basic entities - atoms, molecules, macromolecules, etc. - of which the material is made. The electric polarization, $\vec{\mu}$, of these entities together with their mutual structural arrangement account for the medium response. In the framework of the Bloembergen's model (2), and considering homogeneous electric fields, $\vec{\mu}$ is represented as a power series expansion in terms of the electric field strengths E_I,

$$\mu_i = \mu_{io} + \alpha_{ij}E_j + \beta_{ijk}E_jE_k + \gamma_{ijkl}E_jE_kE_l + \ldots \qquad i,j,k,l = x,y,z \tag{1}$$

where μ_{io}, α_{ij}, β_{ijk} and γ_{ijkl} respectively are the components of the permanent dipole, $\vec{\mu}_o$, the linear $\underline{\underline{\alpha}}$ second-order $\underline{\underline{\beta}}$, and third-order $\underline{\underline{\gamma}}$, polarizabilities.

Disregarding possible unfavorable arrangements - e.g. symmetry, disorder, etc. - in which the basic entities organize to form the material, larger values of the above-mentionned tensor-components will correspond to higher linear and/or nonlinear responses of the medium. In organic compounds, large values of α_{ij}, β_{ijk} and γ_{ijkl} are found associated with the existence of conjugated π-systems. Finding the structural and constitutive attributes of the most responsive ones is in itself a research of fundamental and applied interest. In this context, quantum mechanical calculations, presently capable of predicting semiquantitatively α_{ij}, β_{ijk} and γ_{ijkl} of model systems that can acceptably simulate relevant aspects of actual situations, find a natural field of application.

In a previous work on the assessment of calculated static polarizabilities of conjugated chain molecules obtained after scaling of ab initio minimal basis set results (3), we observed a rather high sensitivity of α_{zz}, the longitudinal component, on geometry changes. In view of the fact that large structural variations can actually be produced in polydiacetylene crystals (4) - e.g. by application of hydrostatic pressure, mechanical deformation or stresses transmitted by side groups - we found worth studying in a more systematic way the importance of structural modifications on α_{zz}. This has at least relevance in the use of controlled geometry readjustments to enhance $\underline{\underline{\alpha}}$, $\underline{\underline{\beta}}$ and $\underline{\underline{\gamma}}$ components, and is also an indication of the important influence of electron-phonon coupling on the magnitude of polarizability related effects.

To our knowledge, changes in bond distances and their consequences on the values of $\underline{\underline{\alpha}}$-components have not been systematically investigated by self-consistent methods. In this paper, we first simulate the structural changes observed in polydiacetylenes (4) to evaluate their influence on α_{zz} and then proceed along this line with results of a prototypical study of the variation of α_{zz} as a function of the alternation degree in the C=C and C-C bond lengths for the series of $H-(CH=CH)_n-H$ polyenes where n=1 up to 10.

Under the name of PTS, TCDU etc., we shall refer to model structures corresponding to the actual polymers in which the side-groups have been replaced by hydrogen atoms.

2. METHODOLOGY

In spite of the appealing computational simplicity of semiempirical methods, basically designed for a narrow range of specific properties, we have prefered ab initio approaches for the sake of a better overall consistency in the computed results. Unfortunately, good quality ab initio calculations of $\underline{\underline{\alpha}}$ on relevant structures become rapidly prohibitive from a computational viewpoint. To overcome this limitation and, as first suggested by Chablo and Hinchliffe (5), we have chosen to scale minimal basis set results in order to produce values closer to the experimental ones.

All the calculations reported in this work were done on a DEC 20-60 using a modified version of the GAUSSIAN 80 series of program (6). Standard STO-3G minimal basis set (7) was considered. Polarizabilities were calculated by the finite-field SCF method of Cohen and Roothaan (8) which is virtually equivalent to the analytic Coupled Hartree-Fock scheme. A term $\vec{\mu}\vec{E}$, describing the interaction between the electric field, $\vec{E}$, and the molecule is added to the unperturbed molecular Hamiltonian, H_o; $\vec{\mu}$ is the total dipole moment of the molecule. At the Hartree-Fock level, the electric field appears explicitly in the one-electron part of the modified Fock operator, $F(\vec{r})$,

$$F(r) = F_o(\vec{r}) + e(\vec{r}.\vec{E}) \qquad (2)$$

where $F_o(\vec{r})$ is the molecular Fock operator in absence of electric field.

The numerical approach to the calculation of electric polarizabilities requires much higher accuracy of the total SCF energy than in standard applications. The convergence threshold on the density matrix elements was set up as 10^{-5}-10^{-6}. In this work we are only interested in the longitudinal component, α_{zz}, of the polarizability tensor $\underline{\underline{\alpha}}$, the z-axis being directed along the chain. The value of α_{zz} is obtained from the first derivative of the field-dependent dipole moment $\vec{\mu}(E)$, with respect to E_z, in the limit of zero-field. In practice this is done numerically using a

$$\alpha_{zz} = [\mu_z(E_z) - \mu_z(-E_z)] / 2E_z \qquad (3)$$

Polarizability calculations were carried out using Eq. (3) for four values of the external field, -0.002,-0.001,+0.001,+0.002 a.u. The Richardson extrapolation procedure (9) was applied to check the numerical stability of the results, a point which remains difficult in the finite field approach.

3. α_{zz} OF 1,5,9-DECATRIENE-3,7-DIYNE VERSUS BOND LENGTH VARIATION

PTS and TCDU polydiacetylenes exhibit different bond lengths in the polymer backbone, see Table 1; the first one adopts an acetylenic structure whereas the second one tends toward a butatrienic configuration. This difference in the geometries induces significant variations on the longitudinal polarizabilities, as can be seen in Table 1 for the oligomer in C10 (i.e. the 1,5,9-decatriene-3,7-diyne). We report also in this table the results for the full acetylenic and butatrienic structures as obtained by theoretical optimization on the infinite chain (10). In summary, we can say that α_{zz} substantially increases upon going from an acetylenic to a butatrienic structure. This conclusion suggests that geometry modifications represent a way to raise (hyper)polarizabilities. It is therefore interesting to estimate theoretically the response of α_{zz} to bond length changes in model molecules.

Table 1 : Geometries and STO-3G calculated longitudinal polarizabilities of polydiacetylene oligomers in acetylenic (PDA and PTS) and butatrienic (TCDU and PBT) structures. Distances in Angstroms, angles in degrees.

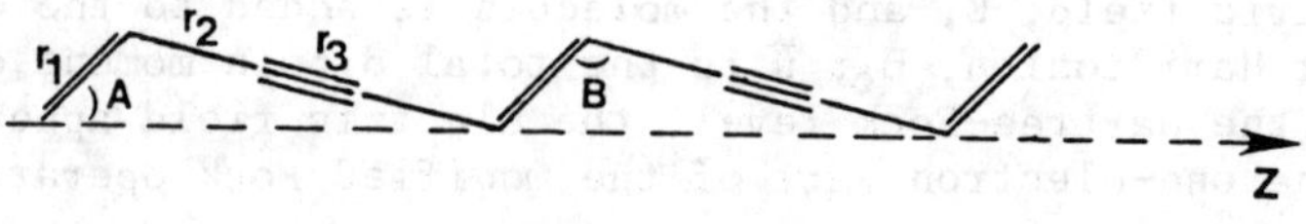

	r_1(Å)	r_2(Å)	r_3(Å)	A(°)	B(°)	α_{zz}(a.u.)
PDA(a)	1.321	1.425	1.194	43.2	123.9	183.7
PTS(b)	1.36	1.43	1.21	44.5	121.9	205.5
TCDU(c)	1.42	1.38	1.24	39.3	127.7	302.7
PBT(a)	1.444	1.319	1.248	41.9	123.7	412.4

(a) optimized geometry (10)
(b) experimental geometry (11)
(c) experimental geometry (12)

4. INFLUENCE OF THE ALTERNATION DEGREE ON α_{zz} OF POLYENE CHAINS OF INCREASING LENGTH

Another factor which is known to affect the values of the polarizabilities is the size of the chain. For regular polyene chains, both free-electron and the Hückel theories predict a behaviour of the longitudinal polarizabilities like L^3 (where L is the total chain-length) (13,14). Accordingly, the longitudinal polarizability per unit cell would grow as the second power of the chain-length, L^2, and diverge in the limit of an infinite chain. However, to our knowledge, nothing has been done beyond this level of the free-electron and Hückel theories to study the evolution of the longitudinal polarizability per unit cell of chains of increasing length and having different geometrical structures in their unit cell.

We have tested these two influences (geometry and chain-length) using the same procedure as above which, within its limits, has the merit of including interelectronic interactions in an average way and allowing the electron charge distribution to relax self-consistently as the size of the system increases. Calculations of α_{zz} have been made for four different alternation degrees while keeping the total chain-length identical in order to properly discriminate between length and alternation effects. The reference structure (hereafter denoted by A) is taken from the unit cell geometry of the infinite polyene theoretically optimized by Karpfen (15). Its alternation, Δr, is measured by the difference, 0.10 Å, between the C-C and C=C bond lengths, respectively 1.446 and 1.346 Å, see also Table 2. The other three situations, respectively denoted by B, C and D in Table 2, have been obtained by reducing the original alternation degree by a factor of 2, 4, and 8, respectively.

Table 2 : Geometries of the polyene chains used for the evaluation of the polarizability with respect to the chain-length and to the alternation degree $\Delta r = r_2 - r_1$.

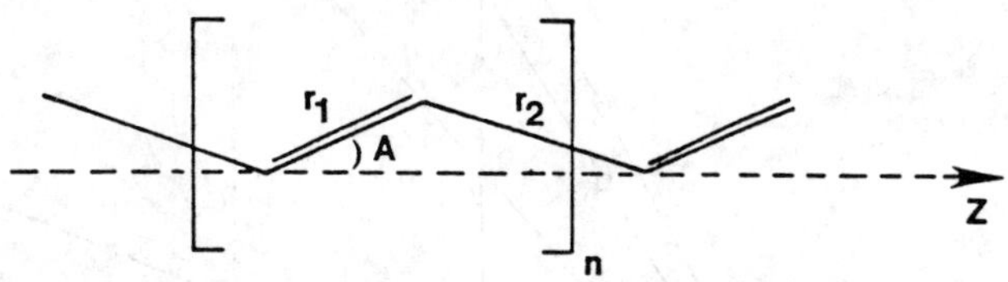

geometry	r_1(Å)	r_2(Å)	A(°)	Δr(Å)
A	1.346	1.446	28.4	0.1
B	1.371	1.421	27.9	0.05
C	1.3885	1.4085	27.6	0.025
D	1.38975	1.40225	27.5	0.0125

The evolution of the polarizability per unit cell(α_{zz}') with respect to the chain-length (up to 20 carbon atoms) is depicted in the figure. A change of slope is easily identified in each curve, it corresponds to the limit beyond which the parabolic behavior predicted by the free-electron theory($\alpha_{zz}' \# L^2$) starts to break down to reach a saturation regime in the limit of large n. Moreover, the shape of the curves noticeably depends upon the alternation degree.

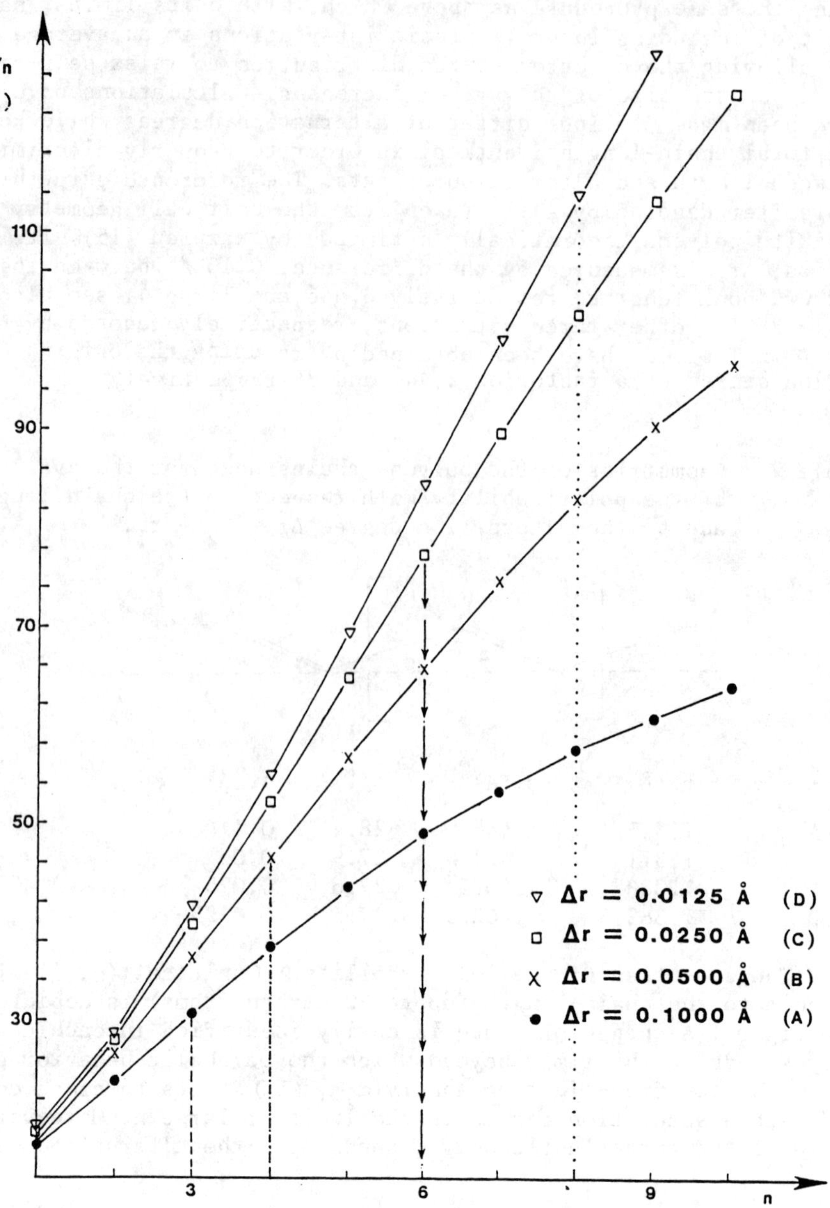

Figure : Evolution of the longitudinal unit-cell polarizability α_{zz}' of polyenes with respect to the chain-length (n) in four different geometries.

The values given in Table 3 allow a more quantitative appraisal of the trends. For each geometry are listed :

- α'_{zz} : the longitudinal polarizability divided by the number n of unit cells in the chain.
- $\Delta\alpha_n$: the increment of polarizability between the n- and the (n-1)-polyene chain.

Table 3 : Longitudinal unit-cell polarizabilities α_{zz}' of polyenes with respect to the chain-length and to the geometry; $\Delta\alpha_n = \alpha_{zz}'(n) - \alpha_{zz}'(n-1)$

n	A		B		C		D	
	α_{zz}'	$\Delta\alpha_n$	α_{zz}'	$\Delta\alpha_n$	α_{zz}'	$\Delta\alpha_n$	α_{zz}'	$\Delta\alpha_n$
1	17.34	0.0	17.91	0.0	18.20	0.0	18.34	0.0
2	23.52	6.2	26.34	8.4	27.91	9.7	28.73	10.4
3	30.70	7.2	36.33	10.0	39.57	11.7	41.34	12.6
4	37.34	6.6	46.50	10.2	52.03	12.5	55.09	13.8
5	43.45	6.1	56.46	10.0	64.75	12.7	69.40	14.3
6	49.08	5.6	65.95	9.5	77.50	12.8	84.29	14.9
7	53.31	4.2	74.82	8.9	89.72	12.2	98.99	14.7
8	57.34	4.0	83.02	8.2	101.86	12.1	113.90	14.9
9	60.73	3.4	90.39	7.4	113.57	11.7	128.36	14.5
10	64.04	3.3	96.87	6.5	124.25	10.7	142.39	14.0

As already intuited from the figure, $\Delta\alpha_n$ increases up to a given value of n, n_{max}, and then decreases. It is interesting to notice that n_{max} shifts toward higher values when the alternation degree is reduced, see also Table 4.

Table 4 : n_{max} with respect to the alternation degree Δr for a polyene chain

Δr(Å)	n_{max}
0.1	3
0.05	4
0.025	6
0.0125	8

Our results qualitatively agree with the free-electron and Hückel predictions in the sense that to a more regular structure, i.e. smaller alternation degree, corresponds a shape of α'_{zz} closer to a parabola. From these observations, n_{max} can be used as a rough measure of the electronic delocalization in the chain or equivalently of the mean free path of the electrons along the carbon backbone.

5. CONCLUSIONS

In this brief account of our recent theoretical results we have emphasized the role of the structure on the longitudinal static electric polarizabilities. On the one hand structural alterations, enforced chemically or mechanically and ranging from 0.05 to 0.12 Å, are predicted to yield important changes in α_{zz}. This suggests the possibility of enhancing the value not only of $\underline{\underline{\alpha}}$ but also $\underline{\underline{\beta}}$ and $\underline{\underline{\gamma}}$ by controlled tuning of the geometry. On the other hand it raises some warnings in the use of bond additivity schemes when applied to polarizabilities and hyperpolarizabilities considering their high sensitivity upon relatively small bond length differences. A more comprehensive and detailed study of these effects will form the subject matter of a forthcoming work.

6. ACKNOWLEDGMENTS

The authors are indebted to Dr. J.G. Fripiat for the implementation of the finite field method in the Gaussian 80 series of programs and to Prof. G. Berthier (ENSJF - Paris) and Mr. G. Baudoux for useful discussions. They acknowledge with appreciation the support of their collaboration under NATO grant # 130.83. One of us (VPB) is grateful to the Institut pour l'Encouragement de la Recherche Scientifique dans l'Industrie et l'Agriculture (IRSIA, Belgium) for financial support.

7. BIBLIOGRAPHY

1.a. B.L. Davydov, L.D. Derkacheva, V.V. Dunina, M.E. Zhabotinskii, V.F. Zolin, L.G. Korenava, M.A.S. Samokhina, JETP Lett. 12, 16 (1970).
b. J. Zyss, J. Chem. Phys. 71, 909 (1979).
c. Nonlinear Optical Properties of Organic and Polymeric Materials (Edited by D.J. Williams, ACS Symposium Series, Washington, 1983).
2. N. Bloembergen, Nonlinear Optics, Benjamin, N.Y., (1965).
3. V.P. Bodart, J. Delhalle, J.M. André, J. Zyss, Can. J. Chem. (in press).
4. Y. Tokura, Y. Oowaki, T. Koda, R.H. Baughman, Chem. Phys. 88, 437 (1984).
5. A. Chablo, A. Hinchliffe, Chem. Phys. Lett. 72, 909 (1980).
6. J.S. Binkley, R.A. Whiteside, R. Krishnan, R. Seeger, D.J. Defrees, H.B. Schlegel, S. Topiol, L.R. Kahn, J.A. Pople, QCPE 13, 406 (1981).
7. W.J. Hehre, R.F. Stewart, J.A. Pople, J. Chem. Phys. 51, 2657 (1969).
8.a. H.D. Cohen, C.C.J. Roothaan, J. Chem. Phys. 43, 534 (1965).
b. J. Zyss, G. Berthier, J. Chem. Phys. 77, 3635 (1982).
9. H. Rutishauser, Numer. Mathematik 5, 48 (1963).
10. A. Karpfen, J. Phys. C. 13, 5673 (1980).
11. D. Kobelt, E.F. Paulis, Acta Crystallogr. B30, 232 (1973).

12. V. Enkelmann, J.B. Lando, Acta Crystallogr. B34, 2352 (1978).
13. K.C. Rustagi, J. Ducuing, Opt. Commun. 10, 258 (1974).
14.a. P.L. Davies, Trans. Faraday Soc. 48, 789 (1952).
b. H.F. Hameka, J. Chem. Phys. 67, 2935 (1977).
15. A. Karpfen, J. Petkov, Theor. Chim. Acta 53, 65 (1979).

ELECTRICAL TRANSPORT AND DOPING OF POLYDIACETYLENES

H.Bässler

Fachbereich Physikalische Chemie, Philipps-Universität Marburg,
Hans-Meerwein-Straße, D-3550 Marburg, FRG

A review is given on generation and transport of charge carriers in crystalline polydiacetylenes. Emphasis is placed on separating experimental facts related to both intrinsic photogeneration and time dependent transport from model considerations. The problem of crystal doping is briefly addressed.

1 INTRODUCTION

Covalent bonding in conjunction with a large energy gap between highest filled and lowest empty molecular orbitals implies that the backbone of a polymer molecule should exhibit features characteristic of an 1D semiconductor or insulator. The weak van der Waals forces acting between neighboring chains in a real 3D polymer network should not alter the situation significantly. Nevertheless, the electrical performance of conventional polymers is that of a 3D hopping system. The prime reason is that the band gap associated with an aliphatic backbone, e.g. a polyethylene chain, is close to 9 eV (1). Therefore any pendant group or impurity will rapidly capture charge carriers, even if initially generated on the backbone chain, and subsequent motion occurs via hopping among weakly coupled substituents. If the backbone were conjugated and, consequently, gap energies were reduced, as, for instance, in polyacetylene, structural disorder would still preclude observation of 1D band motion on a macroscopic scale.

In crystalline polydiacetylenes (PDA) both limitations seem to be suspended at a time: (i) In all systems hitherto known the highest filled molecular orbitals and lowest unoccupied MO's are those of the chain and (ii) polymerization within a crystalline phase produces fully elongated polymer chains. Since in the crystal nearest neighbor chains are still 7.5 Å apart, PDAs were shortly after their discovery (2) regarded as prototypical model substances for an 1D system. A great deal of work has been devoted on elucidating the mechanisms by which those systems conduct an electrical current. The aim of this article is to summarize the present state of understanding of how carriers are generated and transported. An attempt will made to separate experimental facts with straightforward consequences and interpretations involving model assumptions that can be subject to conjecture.

2 ELECTRONIC TRANSPORT

2.1 Experimental Facts

Of central importance for the understanding of charge carrier generation in a semiconductor or insulator is the knowledge of its energy level structure. Photoemission work by Murashov et al.(3) allowed locating the top of the valence band of the chain in PTS at an energy $I_c = 5.5 \pm 0.1$ eV below vacuum, in good agreement with theoretical predictions (4,5). It also showed that I_c decreases in course of the polymerization process proving that the I_c value quoted above must be a property of the polymer chain. It should be noted that inferring I_c from an optical threshold is superior to analysing the kinetic energy distribution of electrons photoemitted in course of an ESCA experiment (6,7), since ambiguities due to spurious contact potentials and solid state line broading effects are avoided.

The analog to photoemission is electron injection from a well-defined contact to the polymer. From the observation that metals with low work function like Na, Ca, NaK, are able to inject electrons into DCH without energetic restriction, whereas injection from Mg ($\phi \simeq 3.7$ eV) is impeded by an energy barrier of 0.24 eV, it has been concluded, that the bottom of the conduction band in DCH must be at $A_c \lesssim 3.4$ eV below vacuum (8). Unfortunately, no electron injection into PTS was detectable, and neither could hole injection be identified unambiguously, contrary to earlier claim (9,10). This is probably due to severe trapping by oxygen or defects produced by surface degradation of PTS or chemical impurities. However, making the reasonable assumption, that I_c is similar for PTS and DCH chains, a band gap $E_g = I_c - A_c \gtrsim 2.1$ eV follows.

In view of the experimental uncertainties involved in determining I_c and A_c the above estimate for E_g is too close to the optical absorption threshold (≈ 2 eV) (11,12) to argue a priori that the dominant polymer absorption be not a transition between valence and conduction band of the conjugated polymer chain. Photoelectric action spectra were necessary to disregard this possibility. Results from several laboratories (13-17) consistently show that the threshold for intrinsic photocarrier generation is some 0.4 eV above the absorption edge whose excitonic nature is now no longer in question (4,18-20). That overall optical absorption and photocarrier production are not directly related is demonstrated by fig.1 which shows both types of spectra for PDA Langmuir-Blodgett films (16). Although not single-crystalline, the advantage of this system is that the total optical density of the sample is $\ll 1$. Therefore no absorption corrections are required for determining the true spectral response function. The most important experimental results are: (i) For a variety of PDA systems the photoelectric threshold varies between 2.2 and 2.6 eV, (ii) the quantum efficiency ϕ of free carrier production is much less than unity and increases with both excess photon energy and electric field (see fig.2), (iii) ϕ is temperature-activated with an activation energy $E_a \simeq 55$ meV. In evaluating ϕ the implicit assumption has been made that all photo-generated charge carriers reach the electrodes, i.e. no additional channels for carrier loss are operative, and once a carrier has been discharged it will not be replenished by subsequent electrode injection. This is the unity gain case in semiconductor terminology. Since contacts employed in these studies were found to act as perfect carrier sinks yet inhibit injection, the latter assumption appears uncritical. At low light intensity bimolecular carrier recombination can also be excluded, albeit not at arbitrarily low electric fields. Absence of monomolecular carrier loss by defect-catalysed recombination ("Shockley Reed" recombination) has been demonstrated by the ingenious cut experiment of Donovan and Wilson (21). They showed that deliberate introduction of recombination centers by laser-etching of the surface of a PTS crystal in a stripe-like fashion reduces the photoelectric gain in proportion

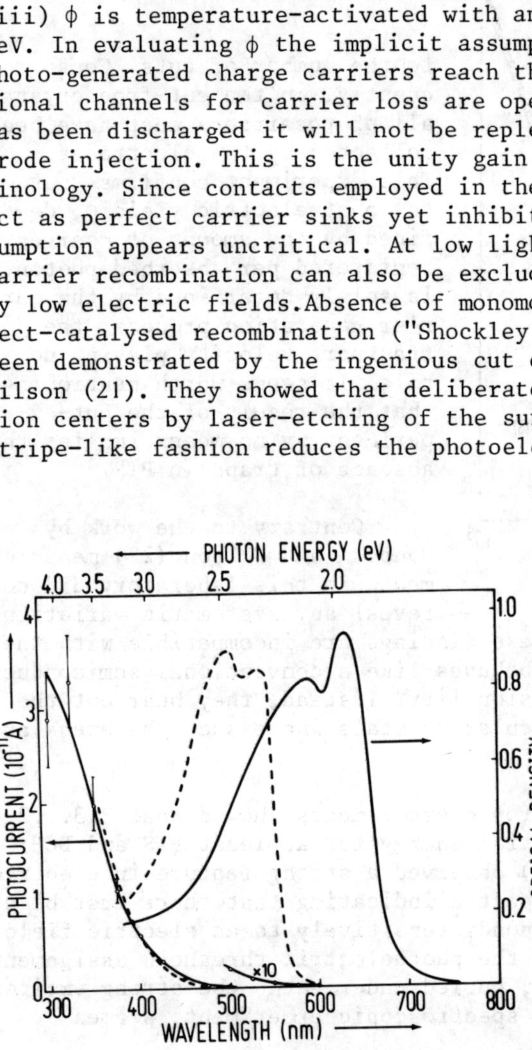

Fig.1

Photoconduction and absorption spectra of two different structural modifications of a PDA Langmuir-Blodgett film (polymerized tricosa-10,12-diin acid). From ref.(16).

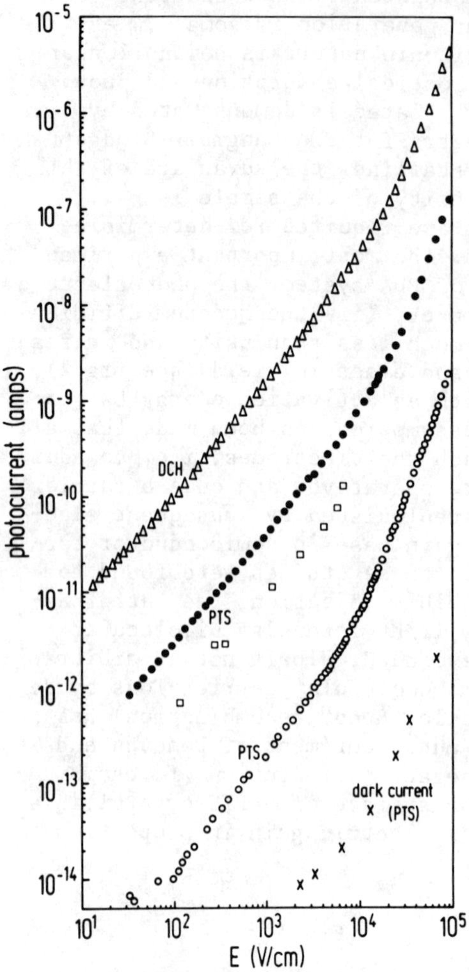

Fig.2

Photocurrent plotted against applied electric field for DCH (open triangles) and PTS (full and open circles were obtained under irradiation with a high-pressure Hg lamp and a tungsten lamp, respectively). Crosses indicate the dark current in PTS (from ref.(35)). Squares denote peak photocurrents measured by Donovan and Wilson (21) (arbitrary scale).

to the number of cuts. Consequently, in the cut-free crystal all photocarriers must have been collected by the electrodes. This experiment confirms that the photoelectric yield ϕ, defined by the amount of charges registered per absorbed photon, is solely determined by the carrier generation process, the transport velocity playing no role. It seems worth mentioning that the result of the cut-experiment by no means implies the absence of traps in PTS.

Contrary to the work by Donovan and Wilson (22) measurements in this laboratory did not reveal any systematic variation of E_a with photon energy. These findings are incompatible with the notion that crystalline PDA behaves like a conventional semiconductor, albeit of reduced dimensionality. Instead, they bear out the close relationship with molecular crystals where such phenomena are commonly observed (23).

A variety of opto-electronic experiments showed that 2.3...2.4 eV is a distinguished excitation energy for at least PTS and DCH: (i) Sebastian and Weiser (24) observed a strong feature in electric field modulated reflection spectra indicating that there must be an optical transition which responds sensitively to an electric field. Since the energy agrees with the photoelectric threshold assignment to a band to band transition, buried underneath the strong exciton transition in a conventional spectroscopic experiment, seemed straightforward. (ii) Resonant Raman work aimed at elucidating the role of adsorbed oxygen in PTS led Poole and Batchelder (25) to conclude that presence of an O_2 molecule next to a PDA chain induces an optical transition near 2.4 eV. (iii) Irradiating a biased PTS crystal with a 1.06 μm pulse of a Nd laser produces a strong photocurrent signal whose intensity dependence changes from I^2 to I^3 at a photon dose of 6×10^{25} cm^{-2} s^{-1} (26). Obviously, two photon absorption (TPA) generates a state at 2.34 eV, living for some 10 ps, which can either dissociate into a free carrier pair at low yield or can do so at larger yield if it absorbs another photon before decaying non-radiatively. Remarkably, a TPA coefficient is calculated which agrees with optical work (27,28) ($\beta = 5\times10^{-48}$ cm^4 s photon^{-1} (repeat unit)$^{-1}$ at 2.35 eV) provided the assumption is made that the carrier yield is the same irrespective of whether the carrier delivering state is populated via one- or two-quantum absorption. Interestingly, Tokura et al. (29) recently assigned an electroreflection signal observed in TCDU crystals near 2.4 eV to a transition to the dipole forbidden ^1A_g exciton state of the chain, in zero field accessible only via two quantum absorption. They concluded on field-induced transfer of oscillator strength by mixing with the ^1B_u exciton. If correct, this interpretation should also apply to the electroreflection studies of Sebastian and Weiser (24) although it is hard to reconcile the symmetric line shapes reported in their work with the concept of Tokura et al.

Since resonance integrals of covalently bonded carbon atoms are of order 1 eV, highest filled and lowest unoccupied orbitals of the chain must form bands several eV wide (30,31). On the other hand, interchain resonance integrals can at best be of order 10 meV or less. Therefore motion of charge carriers is expected to be anisotropic, band-like along the chain and incoherent in perpendicular direction. A simple $\mu = (e/m^*)\tau$-approach predicts $\mu_{\parallel} \approx 10^3$ $cm^2/(Vs)$ for $m^* \approx 0.1$ electron masses and a scattering time $\tau = 10^{-13}$ s. The expected anisotropy has meanwhile been verified. Both photo- (32) and darkcurrent (9) measurements give $\mu_{\parallel}/\mu_{\perp} \simeq 10^3$ for motion parallel and perpendicular to the chain. However, it is most likely that the true microscopic anisotropy is much higher since PDA crystals do not consist of infinitely long chains. From molecular weight studies Wenz and Wegner (33) concluded that in a soluble PTS homolog ("TS-12") one chain does on average not comprise more than 1000...2000 units. Therefore jumps between a chain end and either an adjacent intact chain or the head of a neighboring chain will more or less control the macroscopic transport properties and the true 1D character of the motion may be partially lost.

The question whether or not the charge carrier mobility along a PDA chain is in fact comparable to that in an inorganic semiconductor has attracted the attention of both experimentalists and theoreticians. The following experimental facts have been established: (i) From the time interval over which bimolecular charge carrier recombination was observed in a 120 μm thick PTS crystal following homogeneous two photon absorption (34), it follows that the carrier drift velocity must be $<10^5$ cm s^{-1} at a field of 2.5×10^4 V cm^{-1}, corresponding to a mobility of $\mu < 5$ cm^2 (Vs)$^{-1}$. (ii) Under similar experimental conditions yet reduced pump intensity Reimer and Bässler (35) observed a current transient carrying a shoulder whose position varied approximately inversely with field as expected for a time of flight (TOF) signal. Applying the conventional TOF analysis $\mu \approx 5$ cm^2 (Vs)$^{-1}$ at 295 K has been derived. (iii) Donovan and Wilson (DW) (21) compared the field dependence of the photogeneration efficiency and the peak photocurrent following a N_2-laser flash and found that both dependences are virtually identical (see fig.2). However, if the number of carriers produced by the light pulse varies as $\phi(E)$, the photocurrent measured before either bimolecular recombination or discharge at the contacts sets in, should be $i_p(E) \sim \phi(E) v_d(E)$ where $v_d(E)$ is the field dependent drift velocity. If $i_p(E)/\phi(E)$ is independent of E, v_d must be saturated, i.e. μ must be $\sim E^{-1}$. Data evaluation gave $v_d \simeq 2\times10^5$ cm s^{-1}. The maximum value of μ depends on the field range over which v_d is constant. Accepting the DW claim that $E^{crit} < 1$ V cm^{-1} would give $\mu > 2\times10^5$ cm^2 (Vs)$^{-1}$ for the zero field mobility, an astonishingly large value for an organic material. (iv) Analysis of electron dark-currents injected from a NaK contact into DCH in terms of trap-free space charge limited conduction led Spannring and Bässler (SB) (8) to conclude that the electron mobility must be 6000 cm^2 (Vs)$^{-1}$ up to a field of 100 V cm^{-1}. Although Wilson (37) has pointed out that these results are consistent with the DW concept of field-saturated motion provided that field-dependent trapping is invoked, a cautionary note

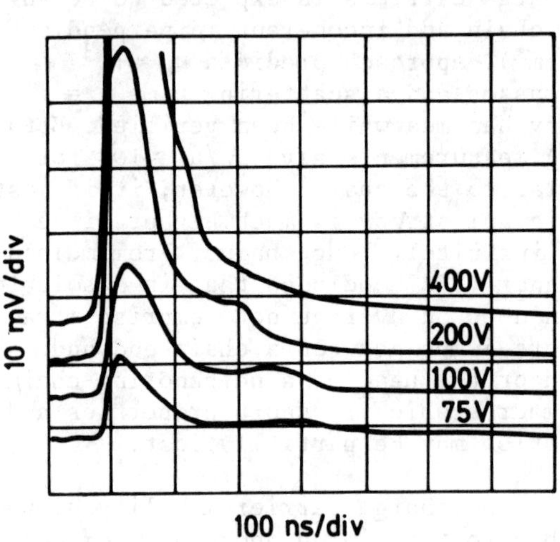

Fig.3

Oscilloscope traces of photocurrent transients recorded with a 120 μm thick PTS crystal in sandwich conformation. The crystal had been prepared by one side solvent etching of a native TS-6 monomer crystal and subsequent thermal polymerization. Photocurrents were excited through a semitransparent Ag-contact by 1.06 μm pulses of a Nd-laser at a dose of about 10^{18} photons cm^{-2}. Carrier generation has been assumed to originate from defect ionization at the polished sample side. Parameter is the applied voltage (from ref.(35)).

is necessary. Subsequent experiments with electrolytic contacts revealed a doping effect (38) (see below) and one may speculate that in the SB experiments metal atoms might have intercalated the DCH lattice along dislocation lines thereby having caused doping. In this case the SB data evaluation were fallacious.

Of considerable importance for interpreting charge transport in PDAs is Donovan and Wilson's (21) observation that photocurrent transients excited by a short laser pulse display long lasting tails. The decay follows an $i(t) \sim t^{-\alpha}$ law over five orders of magnitude in time with α being close to unity. Validity of this decay law has recently been confirmed by Movaghar et al. (39) (fig.4). By comparing carrier drift and yield data it has also been established that within experimental accuracy v_d must be virtually independent of field within the entire time domain of the experiment. From work on disordered materials it is known that decay laws of this form are generated if the rate determining transport parameter, for instance, the carrier hopping rate, is not a constant but is subject to a distribution that reflects the disorder of the material (40). In a single crystalline material the only conceivable type of disorder is

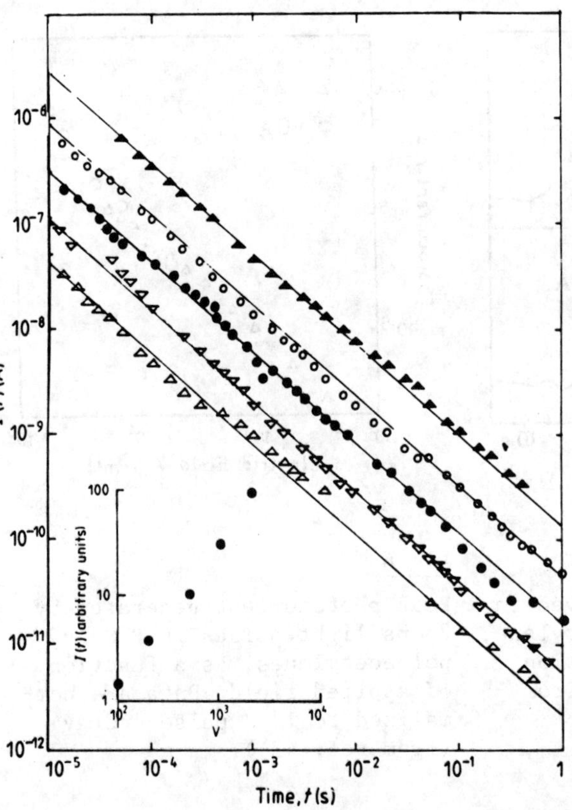

Fig.4

Decay of the photocurrent $I_{ph}(t,E)$ in a PTS single crystal, following a 2.7 eV laser pulse of duration 2 µs. ▲ 2 kV, ○ 1 kV, ● 500 V, ▽ 200 V, △ 100 V. The inset shows the voltage dependence of the photocurrent. From ref. (39).

related to defects. The conclusion that emerges from these decay time studies is that defects have a profound influence on charge transport in PDAs although they appear to be high quality single crystals. Keeping in mind that the density of chain ends is of order 10^{-3} this is not a surprising result. It is worth mentioning that introduction of additional defects, e.g. by ion bombardement or via poor crystallographic quality, gives rise to pronounced trapping effects characteristic of an 1D system (41,42).

Recently, Baumann et al.(43) have measured time-resolved photoconductivity in PDA-TS-6 crystals as well as polyacetylene excited by 25 ps pulses of a Nd:YAG laser ($\hbar\omega$ = 2.3 eV). The response time of the detector was 200 ps. The transient signal shown in fig.5 reveals a fast initial peak with instrument-limited pulse-shape followed by a slower decaying tail. The field dependence of the peak height (fig.6) parallels that of the carrier generation process and is in accord with what Donovan and Wilson have found on a 20 ns time resolution. The quantum efficiency associated with the fast photocurrent peak is 1.5×10^{-5} times the dc-quantum efficiency measured at $\hbar\omega$ = 2.7 eV.

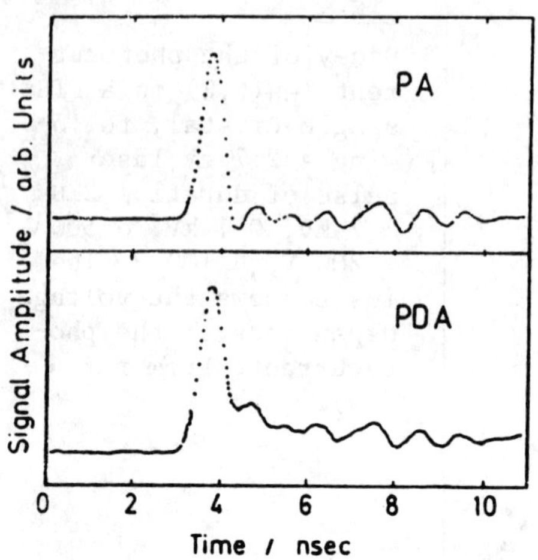

Fig.5

Photocurrent pulses observed in a PTS crystal and polyacetylene, respectively after excitation with 25 ps laser pulses (from ref.(43)).

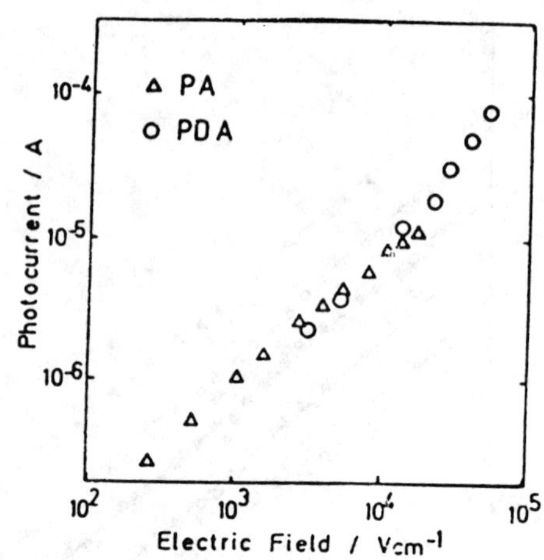

Fig.6

Peak photocurrent generated by 25 ps light pulses in PTS and polyacetylenes, as a function of applied field. Data are normalized to 1 μJ pulse energy (from ref.(43)).

2.2 Interpretations

In this section concepts developed for interpreting the experimental results on charge carrier generation and transport in crystalline PDA will be critically examined. The fact that the quantum yield ϕ for intrinsic photocarrier generation is much less than unity and depends on both temperature and field can in principle have two explanations: (i) Production of a fully dissociated electron-hole pair is a one step process like in an inorganic semiconductor. However, the driving optical transition is very weak, buried underneath the strong exciton transition, and only acquires oscillator strength via the perturbing action of an applied field, i.e. it vanishes in the $E \to 0$ limit. (ii) The concept worked out by Onsager in 1938 (44) to explain dissociation of a short-lived ion pair applies. It assumes that photon absorption initially generates a coulombically bound electron-hole pair (efficiency ϕ_o) which can either recombine or dissociate with the probability ϕ^{esc} into a free electron-hole pair in course of a thermally activated diffusive random walk, i.e. $\phi = \phi_o \phi^{esc}$. There are two arguments against the first alternative: Figure 2 indicates that at low electric fields $\phi \sim E$, whereas in a centrosymmetric system like a PDA-chain the perturbing action of an electric field on an optical transition must be symmetric in E, i.e. at low E ϕ should vary as E^2 if the field-independent term vanishes. Secondly, it is hard to explain why the optical transition should be associated with an activation energy and why this energy be independent of photon energy. (Even photoionization via defect excitation is associated with an activation energy of 55 meV.). On the other hand, Onsager's theory of geminate recombination in its 1D version (45,46) can explain the field dependence of ϕ up to fields of about 10^4 V cm^{-1} in a both consistent and quantitative fashion and provides a physically sensible explanation for E_a. It is the coulombic binding energy of the electron-hole pair before diffusive motion starts. An initial pair separation of $r_o = 65$ Å is calculated for a dielectric constant $\varepsilon = 4$ measured for PTS along the chain direction (47).

Albeit operationally straightforward, interpretation of photocarrier production in terms of geminate recombination theory creates implications for subsequent interpretatory work that are not straightforward. The most uncritical of these is that deviation from an ideal 1D behavior should manifest itself in a deviation of $\phi(E)$ from linearity at low fields. Recall that $\phi(E) \sim E$ behavior with no zero field intercept is a consequence of diffusion on a linear chain. If interchain jumps are prohibited a particle will with certainty return to the origin of its motion in the absence of a drain field (45,46,48). Observation of $\phi(E) \sim E$ down to fields where bimolecular recombination can reliably be excluded and where internal fields are still negligible, leads to the conclusion that the true transport anisotropy of a PTS or a DCH crystal must be $\gg 10^4$, much larger than the anisotropy following from macroscopy transport studies (9,32).

The second consequence derives from the fact that above 10^4 V cm^{-1} experimental yield data increase faster with field than predicted by theory and do not approach saturation (36). This indicates breakdown of the assumption, implicitly contained in the Onsager treatment, that the initial pair yield ϕ_o be independent of field. Such an independence would, in fact, be expected if the only primary process that produces e...h pairs were a weak band to band transition competing with the strong exciton transition above E_g. In this case that fraction of photons which choose the band to band absorption route should produce electron-hole pairs with unit efficiency and no further field-enhancement of ϕ_o should occur, contrary to experimental evidence. (At the fields employed field ionization can safely be ignored.). To explain the high field effect one has to invoke an excitonic state whose population is forbidden in first order but gains intensity via field-induced symmetry breaking and couples to the partially dissociated electron-hole pair state. Remembering that two photon absorption at a double quantum energy of 2.34 eV produces charge carriers at the same yield as does one photon absorption at the same energy, one is tempted to speculate that the intermediate is the parity forbidden ^1A_g exciton state. It therefore appears that two pathways exist for generation of geminate e...h pairs in PDAs. At fields below 10^4 V cm^{-1} a weak and field-independent one quantum band to band transition prevails. At higher fields the ^1A_g exciton state, which is almost degenerate with the states near the bottom of the conduction band of the PDA chain, becomes partially allowed and accounts for the experimentally observed E^2 term in the expression for the geminate pair yield (36) and, perhaps, for some high field electroreflection features (29). Since this state is likely to be responsible for the spectroscopic oxygen effect reported by Poole and Batchelder (25) one would also expect a strong oxygen effect on the photoconduction yield as well. Although qualitatively observed, a quantitative investigation of this effect is still lacking.

Irrespective of some still unresolved details of the primary process that leads to generation of electron-hole pairs one can determine the effective absorption coefficient, called α_b, that determines the photoelectric geminate pair yield in the low field limit, i.e. at fields where ϕ_o is constant. The result is shown in fig.7 for DCH. α_b has a singularity close to the energy where electroreflectance measurements of Sebastian and Weiser (24) locate the onset of the transition between delocalized valence and conduction band states of the chain. The feature near 3.5 eV is due to charge transfer from the carbazole substituent to the polymer chain (49).

The third non-trivial conclusion to be derived from the successful application of Onsager's theory is that the dissociative

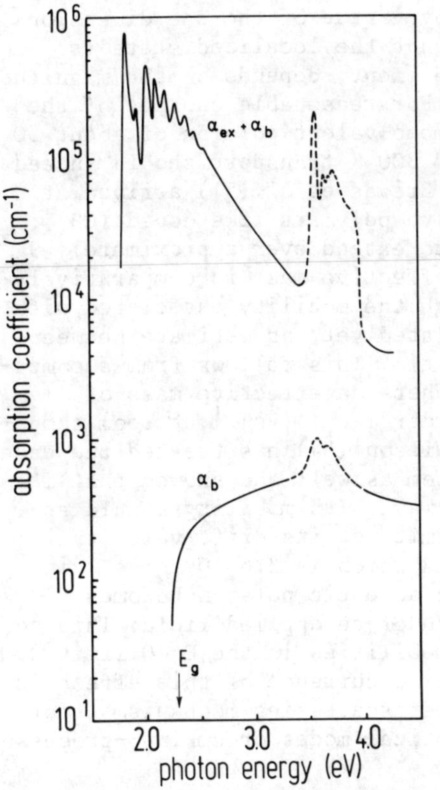

Fig.7

Total absorption coefficient $(\alpha_{ex} + \alpha_b)$ of DCH obtained by Kramers-Kronig analysis of 8 K reflection data. The broken portion of the spectrum reflects transitions within the carbazole chromophore. α_b is the profile of the ionizing transition; the broken portion reflects sensitised carrier generation via excitation of the sidegroup (49). E_g indicates the band gap measured by electroreflection at 2 K (from ref.(36)). The free carrier yield is $\phi = \phi^{esc}\alpha_b(\alpha_b + \alpha_{ex})^{-1}$ where ϕ^{esc} is the Onsager escape probability in 1D.

motion of the electron-hole pair is diffusive within the coulombic well. Considering that $r_o \simeq 65$ Å and the coulombic capture radius is $r_c \simeq 150$ Å at 300 K, there appears to be no way of avoiding the conclusion that the scattering length λ of a carrier is significantly less than 100 Å. This is incompatible with a band mobility of order 10^5 cm^2 (Vs)$^{-1}$. Recall, however, that this figure is a numerical construct, derived from the observation that the carrier drift velocity saturates at a value $\sim 2\times10^5$ cm s^{-1}, virtually independent of field. Within a coulombic potential well, the electric field is of considerable magnitude, $E(r=100$ Å$) = 3.6\times10^{-4}$ V cm^{-1} for $\varepsilon = 4$. Thus an average effective mobility only of order 10 cm^2 (Vs)$^{-1}$ would come out for motion of the electron-hole pair. This is compatible with the strong scattering requirement. Therefore the diffusive character of electron-hole pair motion only requires strong phonon-scattering to guarantee that the carriers be in thermal equilibrium with the surrounding lattice yet no direct conclusion regarding diffusion of an excess carrier in the low field limit can be drawn.

A mechanism that would always reduce the mobility of a carrier, be it free or coulombically bound to a countercharge, as compared to the value expected for a wide-band semiconductor is polaron formation. In fact, by calculating ground state and lowest-lying excited states of a PDA chain using a tight-binding Hamiltonian which includes electron-phonon interactions yet no electron-electron repulsion, Cade and Movaghar (50) showed that extra holes or electrons on a PDA chain are localized by virtue of the one-dimensional character of the system. Whether or not the localized state is stable at the temperature of an experiment, depends on the magnitude of the electron-phonon interaction. For reasonable choices of the constants binding energies for the mono-valent polaron of about 50 meV are calculated, implying that at 300 K transport should indeed be polaronic rather than band-like. Brédas et al.(51) arrived at very similar conclusion for conductive polymers like doped $(CH)_x$. Since the deformation is estimated to extend over approximately 9 repeat units of the PDA chain, the effective mass is comparatively small, ∿0.6 electron masses. Although the mobility associated with an optic polaron has not been calculated yet, an estimate between 10 and 100 cm^2 $(Vs)^{-1}$ appears realistic. This follows from a comparison with crystalline anthracene, where an effective mass of 11 m_e (52) corresponds to a mobility of order 1 cm^2 $(Vs)^{-1}$ at room temperature (53). In a series of papers Wilson (54) has treated the case of an acoustic polaron in an 1D system as well and showed that its motion should be that of a solitary wave. Its mobility should essentially be infinite in the $E \to 0$ limit yet its drift velocity should saturate at the speed of sound which is 3.6×10^5 cm s^{-1} in PTS. The physical reason is that the acoustic polaron becomes heavier as it travel under the action of an applied field. This concept is able to explain ultra-high mobilities in the $E \to 0$ limit that decrease as E^{-1} at finite E. However, occurrence of this effect is subject to the condition that no other scattering mechanisms, for instance, scattering by low lying optical modes or umklapp-processes, are operative.

It appears that the essential clue to solution of the problem of carrier motion along a conjugated chain is provided by the concept of defect-controlled motion in an 1D system. It has been developed by Movaghar et al.(39) on the basis of Alexander et al.'s (55) model of diffusion along an 1D chain with nearest-neighbor hopping rates W distributed according to the function

$$\rho(W) = (1 - \alpha)\, W_o^{-(1-\alpha)}\, W^{\alpha} \qquad (0 < W \le W_o,\ 0 < \alpha < 1) \qquad (1)$$

Movaghar et al.(56) have shown that this distribution function for the event times leads to a breakdown of the linear response theory in the long time domain, which means that drift current and electric field are no longer linearly related. Instead

$$j(t) \underset{t\to\infty}{\sim} 2ea\eta\, |2\eta t|^{-\alpha}\, W_o^{1-\alpha}\, C(\alpha) \qquad (2)$$

is obtained. Here a is the intersite spacing, $C(\alpha)$ a numerical factor, and η is a measure of the field (in the low field limit $\eta = eEa/2kT$ where a is the hopping distance). Thus for $\alpha \approx 1$, j is independent of field E and decays linearily with time, which is what has been observed with PTS (30,39) (see fig.4). The qualitative explanation for this peculiar behavior is that in an 1D system the electric field has two counteracting effects on transport: it makes carriers drift faster but also makes them reach strongly scattering defects (barriers) faster with no way to avoid scattering by making a detour around the defect. In terms of chemical kinetics it is the scattering effect exerted by the defect - assumed to be field-independent - that controls the transport process rather diffusion to reach the defect. One therefore deals with the case of a non-diffusion limited reaction and no information on the transport process before a carrier reaches the defect can be extracted from transport data collected within a time domain where the scattering effect is already operative.

The only modification required for applying this concept to PDA, known to be crystallic yet also known to contain chain ends, which by necessity generate barriers for a moving carrier, as well as defects (57), is to assume that only a small fraction x of all sites is subject to the above distribution law for jump rates, the rest, $1 - x$, being associated with a discrete jump rate W_o (fig.8). The analytic treatment shows that after a time $t_c = (2\eta x W_o)^{-1}$, which is the time a carrier needs to reach the first defect, the current $j(E,t)$ follows eq.(2), is multiplied by x^{-1}. Both time and field dependence of the decay curves of fig.4 are in full accord with this theory, giving $\alpha = 0.85$ and, since $v_d \sim E^{1-\alpha}$, predicting a drift velocity that is almost saturated with field. Observation that for short times, $20 < t < 100$ ns decay function of photocurrent transients is closer to an exponential (35) is not at variance with this concept. It indicates that the distribution function $\rho(W)$ for defect-controlled jump rates contains a discrete peak superimposed onto the distribution given by eq.(1).

This theory also predicts that the mean square distance moved by a carrier in a time t is $\langle R(t)\rangle \sim (2\eta W_o t)^{1-\alpha}\, a/x$. The time required by a carrier to traverse a sample of length L_o is therefore

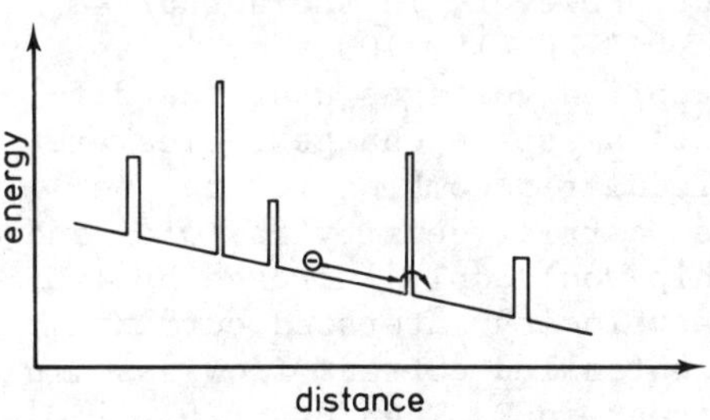

Fig.8

Schematic view of an electron travelling along the field direction in an 1D system where a fraction x of the hopping sites carries a repulsive potential which a carrier has overcome either thermally or via tunnelling. Barriers are subject to a statistical distribution leading to a distribution $\rho(W)$ of barrier crossing rates.

$$t_o \simeq (2\eta W_o)^{-1}(L_o x/a)^{1/(1-\alpha)} , \tag{3}$$

i.e. t_o scales inversely with electric field like in a conventional time of flight experiment. One therefore expects that in a sufficiently thin sample a TOF signal, albeit dispersive, might be observable. It is gratifying to note that the data from fig.3 can indeed be interpreted on these premises in a quantitative fashion using the parameter set $\alpha = 0.8$, $W_o = 10^{14}\ s^{-1}$, and $x = 3\times10^{-4}$. However, in this interpretation the arrival signal of a group of charge carriers does no longer carry information on the true micromotion of a carrier on an unperturbed chain. The $(L_o x)^{1/(1-\alpha)}$ factor in eq.(3) also implies that observation of a TOF signal in PDA be extremely sensitive to sample dimension and defect concentration. With increasing sample thickness the kink in the transient that signals arrival of the carrier packet will be obscured by the dispersive character of the transport process.

Since the peak values of photocurrent transients observed in the Donovan-Wilson experiment (21) on a 20 ns time scale display the same field dependence as do the slowly decaying tails of the transients it seems straightforward to conclude that one unique rate determining transport mechanism is operative over the entire time domain of the experiment. The conclusion is that drift currents measured in PDA solely reflect scattering and, perhaps, trapping with subsequent release by defects. Since the microscopic mobility μ_f enters $j(E,t)$ only via the factor $W_o^{1-\alpha}$ - note that $\mu_f = (ea^2/kT)W_o$ - no quantitative information on μ_f can be drawn from absolute current measurements. The only quantity which gives a handle on the mobility μ_f of a carrier travelling along an intact PDA chain is the time $t_c = (2\eta x W_o)^{-1} = a(xE\mu_f)^{-1}$ after which a carrier encounters a trap and which should therefore establish a criterion for the applicability of the model. Inserting $t_c < 20$ ns, $x = 10^{-3}$ which follows from the concentration of chain ends, $a = 5$ Å yields $\mu_f > 2.5\times10^3\ E^{-1}\ cm^2\ (Vs)^{-1}$. If the carrier velocity were saturated down to a field as low as $E^{crit} = 1\ V\ cm^{-1}$ as DW claim, $\mu_f > 2.5\times10^3\ cm^2\ (Vs)^{-1}$ would come out. However, in the author's opinion a more conservative estimate, $10 < E^{crit} < 100\ V\ cm^{-1}$, appears appropriate since (i) at low applied voltages spurious differences in contact potentials as well as space charge fields are no longer negligible and (ii) bimolecular recombination may become important. In this case the free carrier density is $\sim\phi^{1/2}$ and the exponents in a $j_p \sim E^n$ relationship would only increase by 1/2 when passing from recombination free velocity saturated current flow to conventional recombination controlled current flow assuming a carrier generation rate that is linear in E. Then $\mu_f > (25...250)\ cm^2\ (Vs)^{-1}$ would follow.

Major achievement towards understanding charge transport along a conjugated polymer chain is provided by the recent sub-ns photoconduction studies of Baumann et al.(43). Unless the fast current spike observed in these studies (see fig.5) is an experimental artifact, it indicates that free chain motion is terminated after less than 200 ps at $E = 5\times10^3$ V cm^{-1}. Knowing that the photocarrier yield is $\phi_o\phi^{esc}$ s, where s is the carrier schubweg and $\phi_o(2.7 \text{ eV})/\phi_o(2.3 \text{ eV}) \simeq 3...4$ (15), data quoted in ref.(43) yield s = 2000±400 Å or 500±80 repeat units for the distance travelled by a carrier while being in a high mobility state. It seems straightforward to associate s with the length of an individual PTS chain (58) and to conclude that the fast current spike is attributable to free chain motion and that barrier-controlled transport starts once a carrier has reach a chain end. From $t_c < 200$ ps a free chain mobility $\mu_f > 20$ cm^2 $(Vs)^{-1}$ follows.

In summary it appears fair to conclude that there is no need to invoke a mobility greater than about 100 cm^2 $(Vs)^{-1}$ for carrier transport along a defect-free PDA chain, compatible with optic polaron transport, yet experiments available to date cannot rule out existence of an ultra-high mobility either. The experiment that clarifies whether or not the drift velocity of a free carrier is saturated with field as predicted by the acoustic polaron model (54,59) needs still to be done. If performed in the time domain it requires ps-photoconduction work under conditions where the transit time of a carrier along an individual chain exceeds the response time of the circuit. Experiments done on a ns-time scale will always reveal barrier- or trap-controlled transport with pronounced 1D-features. High frequency ac-photoconduction studies would be extremely useful to answer the fundamental question about the nature of the transport process of an excess carrier on a conjugated perfect 1D chain.

3 DOPING EFFECTS

With the exception of $(SN)_x$, all polymeric systems with partially conjugated backbone show bond-alternation and have therefore finite energy gaps between highest filled and lowest empty chain states preventing metallic behavior. With porous materials like polyacetylene which provide easy diffusion paths for dopant molecules to penetrate the lattice, this handicap can be circumvented by exposure to vapors or solutions of highly reducing or oxidising agents acting as electron donors or acceptors (60). Apart from their chemical instability one inherent disadvantage of those systems is structural disorder which inevitably causes carrier transport to be determined by interchain jumps. Therefore the macroscopic carrier mobility cannot be and, in fact, is not (61) much larger than 1 cm^2 $(Vs)^{-1}$, which is a representative value for a molecular transport system where hopping sites are coupled by weak van der Waals interaction (53). Even assuming that the dopant concentration is comparable to the concentration of monomer units, the maximum conductivity cannot be much in excess of 10^3 $(\Omega cm)^{-1}$. With the idea in mind that the mobility in crystalline PDA is orders of magnitude larger, it seemed straightforward to search for dopable materials. Unfortunately, nature seldom grants two advantages at a time: The densely packed structure of PDA crystals prevents dopant penetration into the bulk and doping of the monomer before polymerization destroys the ability to undergo the topochemical solid state reaction. These facts suggest that doping, if possible at all, may only be successful for fibrous samples of polymerized PDAs with small sidegroups. The attempts by Bloor et al.(62) in this direction were greeted with only moderate success. Despite uptake of J_2 by the 1 OH PDA derivative upon exposure to iodine vapor the conductivity did not exceed 10^{-6} $(\Omega cm)^{-1}$.

A considerably larger doping effect was discovered accidentally in this laboratory (38). Since the vacuum-referenced work function of an aqueous Ce^{3+}/Ce^{4+} contact is 5.9 eV (23) it should oxidize a PDA crystal with $|I_c|$ = 5.5 eV (3), i.e. it should form an ohmic hole injector. By the same token, the J_3^-/J^- redox couple should behave similarly. It was indeed observed, that the dark conductivity of a DCH crystal increased up to a level of 10^{-3} $(\Omega cm)^{-1}$ if contacted with an electrolyt on one side and a pure H_2O contact on the other side (see fig.9). However, with a J_3^-/J^- contact the effect occurred at the wrong polarity and was subject to a slow

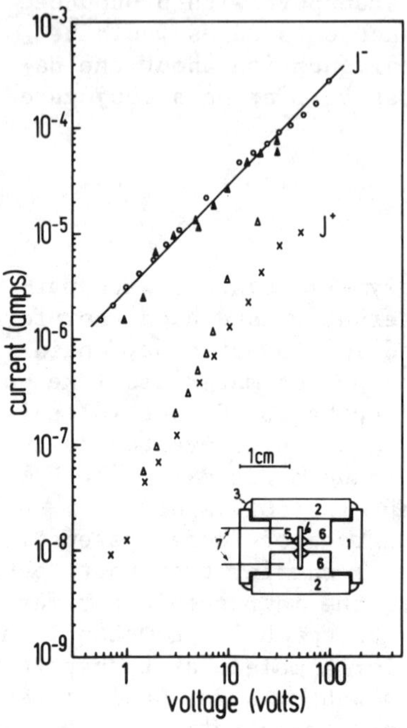

Fig.9

Current-voltage plots measured with DCH crystals contacted with a pure H_2O contact on one side and a J_2 saturated 0.5 n NaJ contact on the other side at different bias of the iodine containing contact. The crystal length was 0.7 cm, its cross-section was 3×10^{-3} cm^2. Triangles indicate data obtained with the same sample at different polarity. To illustrate data reproducibility data points for different samples (crosses and circles) are also included. The inset shows the measuring cell: (1) PMMA sample holder; (2) glass discs; (3) paraffin; (4) crystal; (5) epoxy resin; (6) electrolyte chambers and (7) platinum wires (from ref.(38)).

hysteresis effect upon reversal of the bias (fig.10). Tracer studies with radioactively labelled $J_2/^{123}NaJ$ solutions proved that iodine, probably in the form of J_3^- ions, enters the crystal by field-assisted diffusion and finally accumulates in the exit contact. Upon reversing the field direction the doping effect can partially be reversed. Doping levels of order 10^{18} cm^{-3} were estimated. Although surface diffusion could be eliminated as the essential source of the effect, homogeneous doping cannot be anticipated either, for reasons of crystallographic packing. Instead, ion migration along crystals defects running in line with the polymer backbone must play the dominant role. Nevertheless, these results demonstrate (i) that the cation of PDA chain can be produced electrochemically and is stable at room temperature and (ii) that a search for PDA/dopant systems offering more favorable structural conditions may be rewarding. However, in view of the fact that most likely the effective carrier mobility in PDA in the long time limit is not much larger than it is in $(CH)_x$ and related systems, there is little hope for finding a doped crystalline system that combines high carrier concentration with high transport velocity.

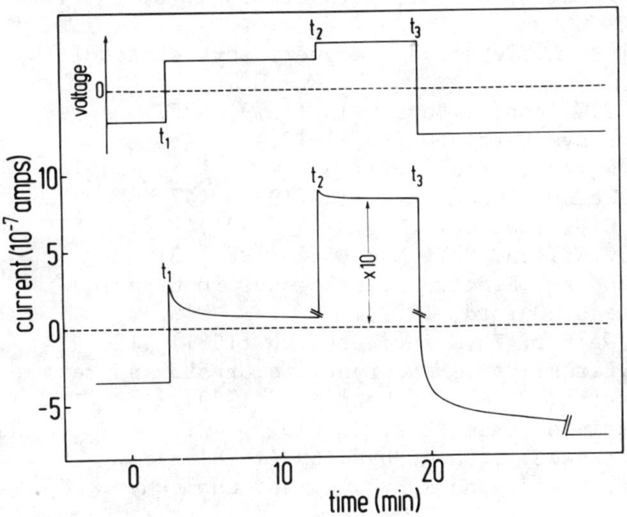

Fig.10

Time dependence of currents measured upon stepwise variation of the applied voltage. The sign of the voltage refers to the iodine containing contact (from ref.(38)).

ACKNOWLEDGEMENT

I am indebted to the members of the Queen Mary and Allied Chemical Groups and to B.Movaghar for valuable discussion and for sending me preprints of their work. Part of this work was supported by the Deutsche Forschungsgemeinschaft and the Fonds der Chemischen Industrie.

REFERENCES

1 Less,K.J. and E.G.Wilson. J.Phys.C 6 (1973) 3110.
2 Wegner,G.. Z.Naturforsch. 24B (1969) 824.
3 Murashov,A.A., E.A.Silinsh and H.Bässler. Chem.Phys.Lett. 93 (1982) 148
4 Suhai,S.. Phys.Rev.B 29 (1984) 4570.
5 Brédas,J.L., R.R.Chance, R.Silbey, G.Nicholas and Ph.Durand. J.Chem.Phys. 75 (1981) 255.
6 Stevens,G.C., D.Bloor and P.M.Williams. Chem.Phys.28 (1978) 399.
7 Knecht,J. and H.Bässler. Chem.Phys. 33 (1978) 179.
8 Spannring,W. and H.Bässler. Chem.Phys.Lett. 84 (1981) 54.
9 Siddiqui,A.S. and E.G.Wilson. J.Phys.C 12 (1979) 4237.
10 Siddiqui,A.S.. J.Phys.C 13 (1980) L1079.
11 Bloor,D., D.J.Ando, F.H.Preston and G.C.Stevens. Chem.Phys.Lett. 24 (1974) 407.
12 Reimer,B., H.Bässler, J.Hesse and G.Weiser. phys.stat.sol.(b) 73 (1976) 709.
13 Reimer,B. and H.Bässler. phys.stat.sol.(a) 32 (1975) 435.
14 Chance,R.R. and R.H.Baughman. J.Chem.Phys. 64 (1976) 3889.
15 Lochner,K., B.Reimer and H.Bässler. phys.stat.sol.(b) 76 (1976) 533.
16 Lochner,K., H.Bässler, B.Tieke and G.Wegner. phys.stat.sol.(b) 88 (1978) 653.
17 Donovan,K.J. and E.G.Wilson. J.Phys.C 12 (1979) 4857.
18 Philpott,M.R.. Chem.Phys.Lett. 50 (1977) 18.
19 Yarkony,D.R.. Chem.Phys. 33 (1978) 171.
20 Sebastian,L. and G.Weiser. Chem.Phys. 62 (1981) 447.
21 Donovan,K.J. and E.G.Wilson. Phil.Mag.B 44 (1981) 9.
22 Donovan,K.J. and E.G.Wilson. Phil.Mag.B 44 (1981) 31.
23 Pope,M. and C.E.Swenberg. Electronic Processes in Organic Crystals (Clarendon Press, Oxford, 1982).
24 Sebastian,L. and G.Weiser. Phys.Rev.Lett. 46 (1981) 1156.
25 Poole,N.J. and D.N.Batchelder. Mol.Cryst.Liq.Cryst. in press.
26 Reimer,B. and H.Bässler. Chem.Phys.Lett. 55 (1978) 315.
27 Lequime,M. and J.Hermann. Chem.Phys. 26 (1977) 431.
28 Shand,M.L. and R.R.Chance. J.Chem.Phys. 69 (1978) 4482.
29 Tokura,Y., Y.Dowaki, T.Koda and R.H.Baughman. Chem.Phys. 88 (1984) 437.
30 Wilson,E.G.. J.Phys.C 8 (1975) 727.
31 Kertesz,M., J.Koller and A.Azman. Chem.Phys. 27 (1978) 273.
32 Lochner,K.,B.Reimer and H.Bässler.Chem.Phys.Lett.41 (1976) 388.
33 Wenz,G. and G.Wegner. Mol.Cryst.Liq.Cryst. 96 (1983) 98.
34 Reimer,B. and H.Bässler. Chem.Phys.Lett. 43 (1976) 81.
35 Reimer,B. and H.Bässler. phys.stat.sol.(b) 85 (1978) 145.
36 Seiferheld,U. and H.Bässler. J.Phys.C 16 (1983) 5189.
37 Wilson,E.G.. Chem.Phys.Lett. 90 (1982) 221.
38 Seiferheld,U. and H.Bässler. Sol.State Comm. 47 (1983) 391.
39 Movaghar,B., D.W.Murray, K.J.Donovan and E.G.Wilson, J.Phys.C 17 (1984) 1247.

40 Scher,H. and E.W.Montroll. Phys.Rev.B 12 (1975) 2455.
41 Hunt,I.G., D.Bloor and B.Movaghar. J.Phys.C 16 (1983) L621.
42 Seiferheld,U.,H.Bässler and B.Movaghar.Phys.Rev.Lett.51 (1983) 813.
43 Baumann,T.,K.J.Donovan,E.Goebel and S.Roth.Materials Science, vol.X (1984) 23 (Proc.4th Int.Conf.on Electrical and Related Properties of Organic Solids, Zamek Ksiaz, Poland).
44 Onsager,L.. Phys.Rev. 54 (1938) 554.
45 Blossey,D.F.. Phys.Rev.B 9 (1974) 5183.
46 Ries,B.,G.Schönherr,H.Bässler and M.Silver.Phil.Mag.B 48 (1983) 87.
47 Rehberg,U.. phys.stat.sol.(a) 51 (1979) 453.
48 Haberkorn,R. and M.E.Michel-Beyerle.Chem.Phys.Lett.23 (1973) 128.
49 Lochner,K., H.Bässler, L.Sebastian, G.Weiser, G.Wegner and V.Enkelmann. Chem.Phys.Lett. 78 (1981) 366.
50 Cade,N.A. and B.Movaghar. J.Phys.C 16 (1983) 539.
51 Brédas,J.L.,R.R.Chance and R.Silbey.Phys.Rev.B 26 (1982) 5843.
52 Burland,D.M. and V.Konzelmann. J.Chem.Phys. 67 (1977) 319.
53 Schein,L.B. and D.W.Brown.Mol.Cryst.Liq.Cryst.37 (1982) 1.
54 Wilson,E.G..J.Phys.C 15 (1982) 3733, 16 (1983) 1039 and 6739.
55 Alexander,S., J.Bernasconi and W.R.Schneider. Rev.Mod.Phys. 53 (1981) 175.
56 Movaghar,B., D.W.Murray, B.Pohlmann and D.Würtz. J.Phys.C 17 (1984) 1677.
57 Spannring,W. and H.Bässler. Ber.Bunsenges.Phys.Chem. 83 (1979) 433; see also the contribution of L.Sebastian and G.Weiser to this volume.
58 Niederwald,H., H.Seidel, W.Güttler and M.Schwoerer. J.Phys.Chem. 88 (1984) 1933.
59 Donovan,K.J., P.D.Freeman and E.G.Wilson. Mol.Cryst.Liq.Cryst. in press.
60 Baughman,R.H., J.L.Brédas, R.R.Chance, R.L.Elsenbaumer and L.W. Shacklette. Chem.Rev. 82 (1982) 209.
61 Shacklette,L.W., R.R.Chance, D.M.Ivory, G.G.Miller and R.H. Baughman. Synthetic Metals 1 (1979) 307.
62 Bloor,D., C.L.Hubble and D.J.Ando. Molecular Metals, ed. W.E. Hatfield (Plenum Press, N.Y., 1979) p.243; see also D.Bloor, NATO ASI Quantum Chemistry of Polymers; Solid State Aspects, eds.J.Ladik and D.M.André.

THE ULTRA HIGH ELECTRON MOBILITY ON POLYDIACETYLENE CHAINS - THE THEORY

E.G. Wilson

Department of Physics, Queen Mary College, London E1 4NS

ABSTRACT

The low field mobility of a charge carrier on a polydiacetylene chain is ultra high, and yet the drift velocity saturates at a low value comparable to the sound velocity. Conventional ideas applicable to conventional semiconductors cannot explain these phenomena. The motion is that of a Solitary Wave Acoustic Polaron (SWAP). The SWAP is characteristic of a one dimensional system. The properties of the SWAP are described.

1. INTRODUCTION

The extraordinary phenomena of the motion of carriers on the polydiacetylene (PDA) chain are summarised in Fig. 1. The carrier velocity V, $V = 2.2 \times 10^3$ m s^{-1}, is essentially independent of the applied field E and close (within a factor of 2) to the sound velocity S, $S = 3.6 \times 10^3$ m s^{-1}. This is so over 4 decades of field. It is also so down to very low fields, such that the low field mobility is ultra high, $\mu > 20$ m^2 s^{-1} V^{-1}. This is greater than the mobility of any conventional semiconductor. The motion is intrinsic to the pure chain and independent of defects which are eventually encountered by the carrier and which either temporarily stop the carrier or lead to its recombination.

The experimental proof of these phenomena is discussed in the previous article by Donovan, Freeman and Wilson (1). The purpose of this article is to present a theory for these phenomena. They cannot be accounted for by conventional theories of carrier mobility

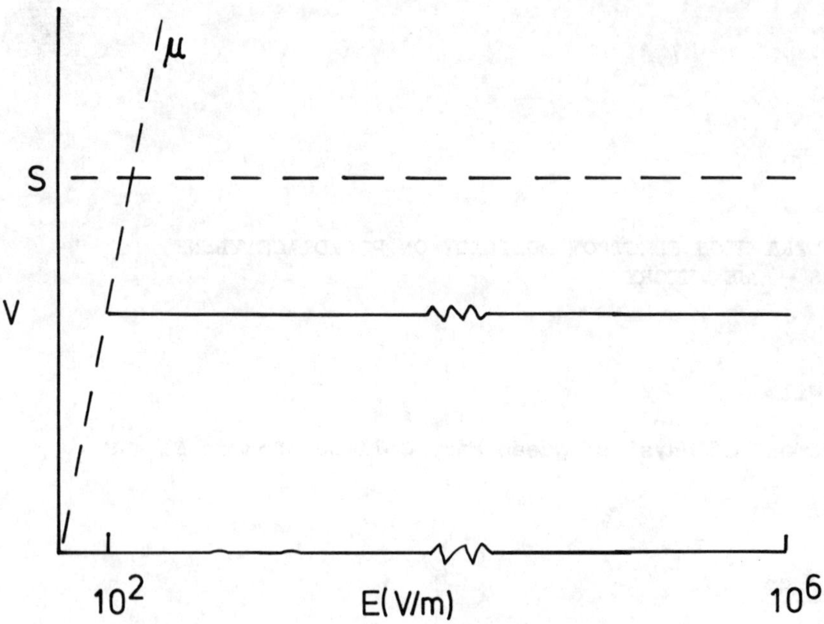

Fig. 1 Schematic illustration of carrier drift velocity V against field E in PDA. S is the sound velocity. The dashed line displays the low field mobility $V = \mu E$.

which are successful in the the conventional semiconductors such as Silicon. Such theories give an ultra large mobility if the electron-phonon interaction is small; however a small electron phonon interaction does not give saturation of V at around S. It is remarkable that the conventional theories are drastically transformed on restricting the electron motion to be in one dimension (1D) characteristic of a polymer. Nevertheless they do not account for the phenomena. In all these theories carriers of effective mass m^* have a mean kinetic velocity v which at temperature T is given by

$$(1/2)m^*v^2 = k_BT \ . \qquad (1)$$

Usually v is 10 to 100 times greater than S. Carriers interact with acoustic phonons by means of a deformation potential D, and a particular carrier experiences all values and directions of kinetic velocity. In 1D this is both forward and backward velocity along the chain. The drift velocity V is then a small bias on this kinetic motion due to the applied field.

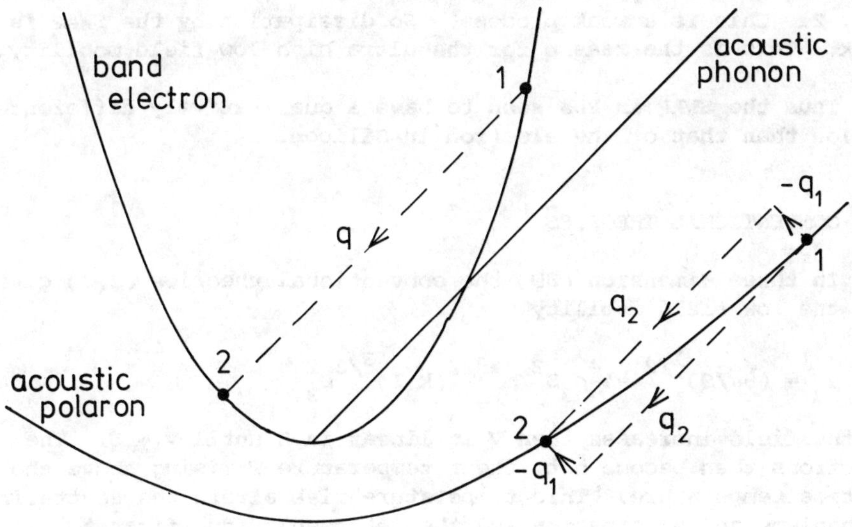

Fig. 2 Schematic illustration of dispersion curves of an acoustic phonon, a band electron and an acoustic polaron. The band electron is undergoing Cerenkov scattering by emission of a phonon q. The acoustic polaron is undergoing Doppler shifted phonon bouncing, the incident phonon $-q_1$ scattering as $+q_2$.

In 1D the lattice deforms around the electron and localises the electron to form a polaron. The important deformation is acoustic, in which the density of the lattice changes; optic deformations occur also but are not important in the polaron motion. The mathematical description of the polaron is like that of solitary waves. The Solitary Wave Acoustic Polaron (SWAP) has a very large effective mass; its kinetic velocity is small. Moreover it cannot move at a velocity greater than S; as its kinetic velocity approaches S the acoustic deformation increases and the SWAP mass increases. In addition back scattering is very rare. So the SWAP kinetic and drift velocities are equal.

A carrier drifting in an applied field acquires energy from the field. In steady state all the energy acquired is dissipated to the lattice. In the conventional theories the dissipation is by Ceenkov emission and absorption of acoustic phonons, illustrated in Fig. 2. This is an efficient process. The SWAP, however, always moving at less than the sound velocity, cannot emit or absorp

acoustic phonons; for to do so violates energy - momentum conservation. So the dissipation is by existing ambient acoustic phonons bouncing with a Doppler shift off the SWAP, as illustrated in Fig. 2. This is a weak process. So dissipation by the SWAP is weak. This is the reason for the ultra high low field mobility.

Thus the SWAP in PDA seen to have a qualitatively different motion than that of the electron in Silicon.

2. CONVENTIONAL THEORIES

In three dimension (3D) the conventional theories (2,3) give for the low field mobility

$$\mu_3 = (8\pi/9)^{1/2}\ e\hbar^4\rho_3 S^2/m^{*5/2}(k_B T)^{3/2} D_3^2 \qquad (2)$$

As the field increases then V is linear in E until V = S. The electrons then become hot, their temperature T rising above the lattice temperature. This temperature rise alters the scattering mechanisms and dissipation and the net result, due first to Shockley, is that V then increases as E to the power one half. This result is shown in Fig. 3, curve 3D. The author, using the same model, but with successive dimensional restrictions, obtained three further solutions, denoted 13D//, 13D⊥ and 1D in Fig. 3 (4).

In all of these the electrons are confined to move in 1D on a chain. In the 13D solution the phonons are nevertheless 3D. This case of hybrid dimensionality is surely the reality in PDA, where the acoustic phonons are 3D. Finally, when the phonons are also 1D, the 1D solution is obtained; this corresponds to an isolated PDA straight chain gas, surely a fiction. There are two deformation potentials to consider in 13D. The first is the change of electron energy due to a stretching of the chains, D//; and the second due to an increase in distance apart of the chains, D⊥. It is possible for an acoustic phonon of wavevector Q travelling at an angle θ to the chain to scatter an electron from $|k_1\rangle$ to $|k_2\rangle$ provided the energy and quasi-momentum selection rules are obeyed, i.e.

$$k_1 = k_2 \pm Q\cos\theta$$
$$\hbar^2 k_1^2/2m^* = \hbar^2 k_2^2/2m^* \pm \hbar SQ \qquad (3)$$

Either of D// or D⊥ can provide the coupling. For simplicity the results are given for either D// or D⊥ existing (if both exist then D⊥ dominates). A characteristic parameter F and angle θ_m exist; θ_m is the maximum angle of Cerenkov phonon emission, by a thermal electron, allowed by the selection rules Eq. (3), and

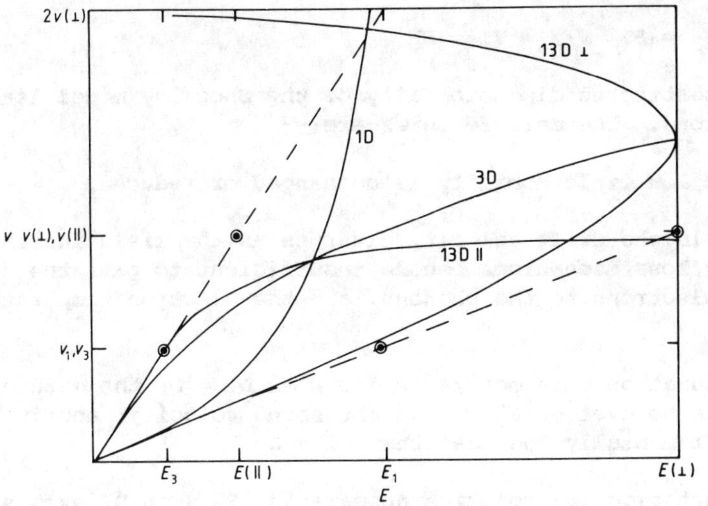

Fig. 3 The drift velocity V against applied field E for electrons undergoing Cerenkov scattering by acoustic phonons. The dashed lines show the low field mobilities. The curve 3D is the usual Shockley 3D solution. The curve 1D is the Shockley solution in 1D. The two curves 13D apply when the electron is 1D and the phonons are 3D. // and ⊥ refer to the type of deformation potential. V_1, V_3, V//, V⊥ are velocities at which departures from the low field behaviour occur. Drawn for convenience with F = 2.

$$F = \tan \theta_m = \hbar^2/2m^* A_c k_B T \ . \quad (4)$$

Here A_c is the area of cross section per single chain in the crystal. In PDA then F is about 4.4. The low field mobilities in the reduced dimensionalities, and for the same deformation potentials and effective mass are

$$\mu(13//) = \mu_3, \ \mu_1 = \mu(13\perp) = \mu_3/F^2 \ . \quad (5)$$

This behaviour of the low field mobilities with dimensionality is illustrated in Fig. 3.

The hot electron effects of Shockley are also drastically affected on reducing the dimensionality. The parameters V_1, V_3, $V//$, and $V\perp$ denote the velocities at which departure from a linear drift velocity with field occurs. They are shown in Fig. 3 together with the non-linear solution that occurs. It can be shown, and for equal D and m^*, that

$$V_1 = V_3 = S, \quad V// = V\perp = SF . \tag{6}$$

Thus restricted dimensionality of the Shockley model leads to new solutions. The main features are:-

a) the low field mobility is unchanged or reduced,

b) in 1D the drift velocity diverges as the field increases; the energy loss mechanisms become insufficient to pass the energy from the electrons to the phonons, and the electron temperature diverges,

c) saturation only occurs in 13D when $D\perp = 0$; the saturated velocity is however greater than the sound velocity; and in PDA it is experimentally the case that $D\perp \neq 0$,

d) a pathological solution appears in 13D when $D\perp$ exists; hot electrons at low field have a drift velocity of $2V\perp$; on increasing the field the electrons cool and the drift velocity slowly falls, as in Fig. 3; and above a critical field the drift velocity again diverges,

e) and finally the scattering probability of an electron having a kinetic velocity equal to S diverges in restricted dimensionality; this gives the SWAP an essential role.

Thus it is clear from this study that models which successfully dominate conventional 3D semiconductors lead to pathological solutions on restricting the electron dimensionality. At best they can describe the electron before the SWAP forms, or in a duration when the SWAP may temporarily break up. In no way can they explain the facts in PDA.

3. SWAP THEORY

The formation of the SWAP in 1D is illustrated in Fig. 4. The electron localises in a region, and by means of the electron lattice interaction exerts a force on the lattice sites. The force causes a lattice displacement. The displacement in turn changes the potential seen by the electron and causes the localisation in the first place. The SWAP at rest is the lowest energy state and inevitably exists

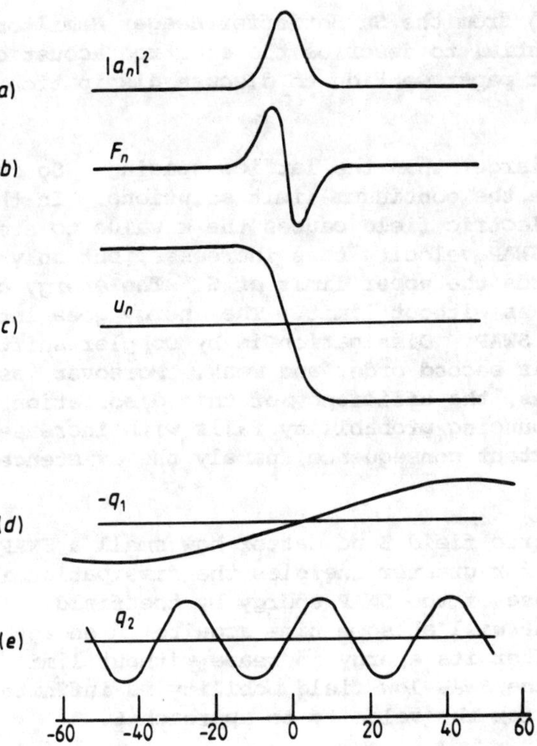

Fig. 4 The structure of the SWAP. The numerical example is appropriate to PDA (6). The abcissa is number of unit cells, each of 0.5 nm length. a) is electron density, b) is force on lattice by electron, c) is lattice displacement. d) and e) are a pair of Doppler shifted bouncing phonons from a SWAP of thermal energy. They are the pair with half the maximum scattering probability. Scattering of short wave phonons giving back scattering of the SWAP is very rare.

in 1D; in contrast there are conditions to be satisfied in 3D. When the SWAP moves then the force on the lattice moves. As the velocity approaches S the lattice displacement increases, for the lattice is being forced at almost its resonance condition. In consequence the electron becomes more tightly bound in the SWAP, the spatial size diminishes and the effective mass and energy of the SWAP increase.

The moving SWAP solution for interaction with acoustic polarons was first produced, in the continuum limit, in strict 1D, by Whitfield and Shaw (5). The dispersion relation is parabolic and particles - like, with a large mass, at small k; it becomes linear and sound - like, with an even larger mass, at larger k. The thermal SWAP in PDA is already sound - like. The SWAP solution was rederived by the author (6) from the Su Shrieffer Heeger Hamiltonian, using a deformation potential to describe the electron acoustic phonon interaction; that paper went on to discuss dissipation and the mobility.

The SWAP is larger than the lattice spacing. So initially it is reasonable to use the continuum limit solutions. In the absence of dissipation an electric field causes the k value to steadily increase with time. The SWAP velocity thus increases, but only assymptotically, and towards the upper limit of S. The energy of the SWAP meanwhile increases without limit. The energy goes into changing the shape of the SWAP. Dissipation is by Doppler shifted phonon bouncing, which is second order and weak. Moreover, as the mass of the SWAP increases, the efficiency of this dissipation falls. This is because the bouncing probability falls with increase of SWAP mass. This has an important consequence, namely the existence of a Runaway phenomenon.

For any electric field E no matter how small a SWAP energy $\epsilon(E)$ exists such that for greater energies the dissipation is too weak to prevent increase of the SWAP energy by the field. Thus by thermal fluctuation a SWAP will at some time acquire an energy greater than $\epsilon(E)$, and thereafter its energy increase without limit, i.e. Runaway. Runaway implies the SWAP low field mobility is infinite; although it may take time for the velocity to approach S.

Thus the SWAP offers the beginnings of a radically new theory of motion to describe a radically new result. Many questions and loose ends however remain. Some are:-

a) the SWAP stability. It is possible for the SWAP to have a positive energy and be unstable. However, the greater the energy then the larger is the binding energy of the electron within the SWAP. So the mechanism of decay is not ejection of the electron leaving the large energy lattice displacement behind. Rather the electron is ejected and the lattice displacement disappear together. This would appear to be a rare process, and needs investigation. A parallel problem is the mechanism of SWAP formation,

b) the survival of the SWAP main charcteristics in hybrid 13D dimensionality. Then the lattice displacement extends to chains adjacent to the chain holding the electron,

c) the properties of small energetic SWAP's when the continuum limit is abandoned. Then the size can approach the lattice repeat distance,

d) the effect of optic phonons on structure and on dissipation,

e) the role of impurities. Particularly for the 13D SWAP it is of great interest to know the chance of a SWAP travelling through, trapping or reflecting at a chain break or impurity.

1. Donovan, K.J., Freeman, P.D. and Wilson, E.G., The Ultra High Electron Mobility on Polydiacetylene Chains - the Facts, previous article.

2. Shockley, W., Electrons and Holes in Semiconductors (Princeton, New Jersey: Van Nostrand, 1950).

3. Conwell, E.M., High Field Effects in Semiconductors, Solid State Physics Supplement 9, 1967.

4. Wilson, E.G., Electron Motion in One Dimensional Semiconductors, J.Phys.C. 15, 3733 (1982).

5. Whitfield, G. and Shaw, P.B., Phys.Rev.B. 14, 3346 (1976).

6. Wilson, E.G., A New Theory of Acoustic Solitary Wave Polaron Motion, J.Phys.C. 16, 6739 (1983).

ELECTRON MOTION IN POLYDIACETYLENE: THE FACTS

K.J. Donovan, P.D. Freeman and E.G. Wilson

Department of Physics, Queen Mary College, London E1 4NS

ABSTRACT

Experiments are described that demonstrate that free carriers travel along the polymer backbone of Polydiacetylenes for distances of up to 1 mm at around the velocity of sound independant of electric field down to low fields of the order of 100 V/m.

1. INTRODUCTION

Polydiacetylene, P.D.A., single crystals of macroscopic dimensions have proved useful in elucidating the nature of the motion of an excess electron in the one dimension, 1D, conduction band of a conjugated polymer backbone.

Because all the chains are all aligned parallel for macroscopic distances one is liberated from considerations of complex chain topology found in other polymers which may reduce the analysis of experiments to an intractable problem. For example, the distance a carrier travels along the polymer chain is identical with the distance it has moved in real space.

The experiments described in this article are conceptually simple and lead to the radical conclusion that electrons travel on the polymer backbone with a velocity, $v_d = 2.2 \times 10^3$ m/s, close to the velocity of sound, $S = 3.6 \times 10^3$ m/s. This velocity is independent of electric field, E, down to the lowest fields at which measurements prove possible, $E = 10^2$ V/m. It then follows from this that the low field mobility, μ_o, is ultra high with a lower limit to its value, $\mu_o > 20\ m^2/V/s$ (1,2). Two P.D.A. derivatives that are available in the form of highly perfect macroscopic single crystals have been studied. The most widely studied of these is the Toluene Sulphonate derivative, T.S. The experiments demonstrate in a simple way that carriers travel of the order of 1 mm along a T.S. backbone before trapping (1,2,3), thus demonstrating their high electrical purity. Further it is shown that recombination centres do not exist, all the carriers eventually discharging at the contacts. By contrast in the DiCarbazole derivative D.C.H. the presence of recombination centres is demonstrated and it is shown that carriers travel of the order of 70 μm before recombination (4,5), otherwise its electrical properties are similar to those of T.S.

2. EXPERIMENT

P.D.A. crystals are contacted with silver paste on the end faces such that E is parallel to the chain direction. The (100) face is illuminated by a Nitrogen laser pulse of duration T. N photons ($\hbar w = 3.68$ eV) are absorbed in a skin depth, δ, and create carrier pairs with a quantum efficiency η. The probability that a carrier pair escape geminate recombination, φ(E), is linear in E at low E as shown by experiment (6,7), theory (8,9,10) and computer simulation (11). A mobile charge Q is thus created given by

$$Q = e(\eta\phi)N\ . \tag{1}$$

Provided most carriers have not trapped in the time T the photocurrent induced at the electrodes will have a peak value I_p given by:

$$I_p = Qv_d/d \quad \text{for } T < \tau \tag{2}$$

where d is the interelectrode separation. After a characteristic time τ the carriers have trapped, however due to subsequent trap limited motion induced charge will continue to accumulate at the electrodes. The carriers will move on average a distance <s> before disappearing due to recombination or discharge at the contacts. The charge Q_m measured using an electrometer relates to the initially created charge through the relation:

$$Q_m = Q\langle s\rangle/d = e(\eta\phi)N\langle s\rangle/d. \tag{3}$$

The ratio of the two measurable quantities is then

$$I_p/Q_m = v_d/\langle s\rangle. \tag{4}$$

Using consecutive laser pulses and measuring I_p and Q_m this ratio can be obtained while avoiding systematic error due to varying sample conditions.

In order to proceed further it is necessary to determine $\langle s \rangle$ unambiguously and to this end an ingenious and unique experiment has been devised.

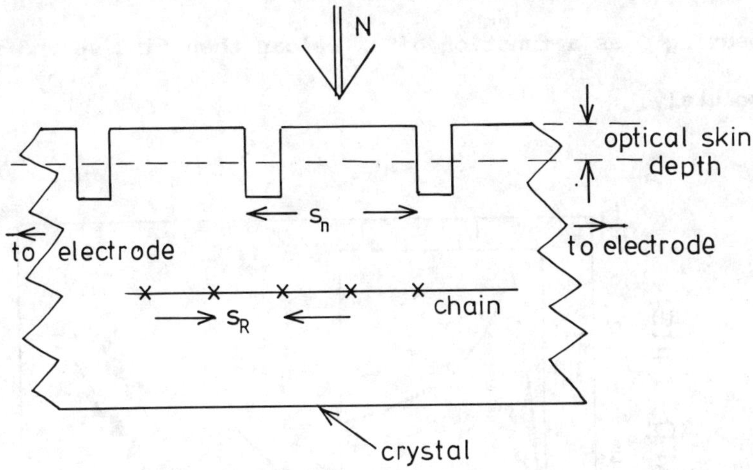

Figure 1. Cuts, of depth greater than the carrier skin depth which stop the flow of carriers along the polymer chains. Carriers of opposite sign accumulate on opposite sides of the cuts and eventually recombine.

In figure 1 is shown a sample whose surface is divided into N equal parts by making (N-1) shallow cuts of separation s_n. The cuts are deeper than the skin depth and act to stop photocarriers while the dark current flows undisturbed in the bulk maintaining the applied electric field. Also shown are intrinsic recombination centres separated on average along the chain by a distance s_R. Thus the average distance carriers travel is given by

$$1/\langle s \rangle = 1/s_n + 1/s_R \tag{5}$$

Thus in the uncut sample

$$\langle s \rangle = s_R \tag{5a}$$

and when $s_n < s_R$

$$\langle s \rangle = s_n \tag{5b}$$

For the former case equation (3) holds. However when one has reduced s_n sufficiently to be in the latter regime then

$$Q_m = Qs_n/d = Q_m(s_n). \tag{6}$$

Denoting Q_m for the uncut sample $Q_m(\infty)$ one finds in general

$$Q_m(\infty)/Q_m(s_n) = 1 + \frac{s_R}{s_n}. \tag{7}$$

By measuring Q as a function of s_n we can then find s_R and therefore v_d absolutely.

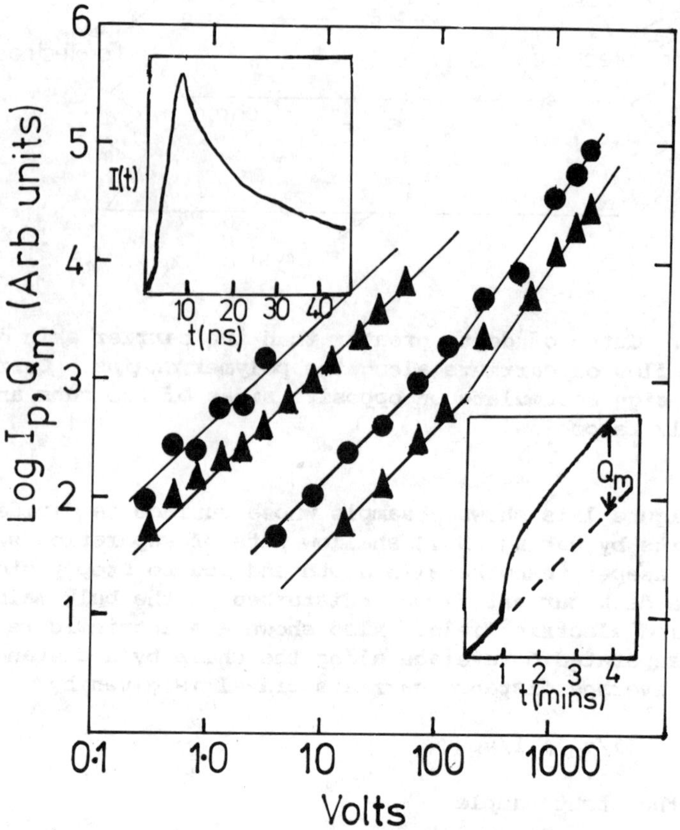

Fig. 2. The field dependence of Q_m ● and I_p ▲ for PDATS. The data that go to lower fields are obtained with focussed light. d=3.6 mm. The insets show a typical pulsed photo current (top left) and a typical charge transfer (bottom right).

3. RESULTS

From figure 2 it is seen that the field dependence of I_p and Q_m are identical and reflect the field dependence of $(\eta\phi)$. Thus s_R is sensibly independent of E and more surprisingly v_d is independent of E. To say more on the absolute values of these quantities further data is needed.

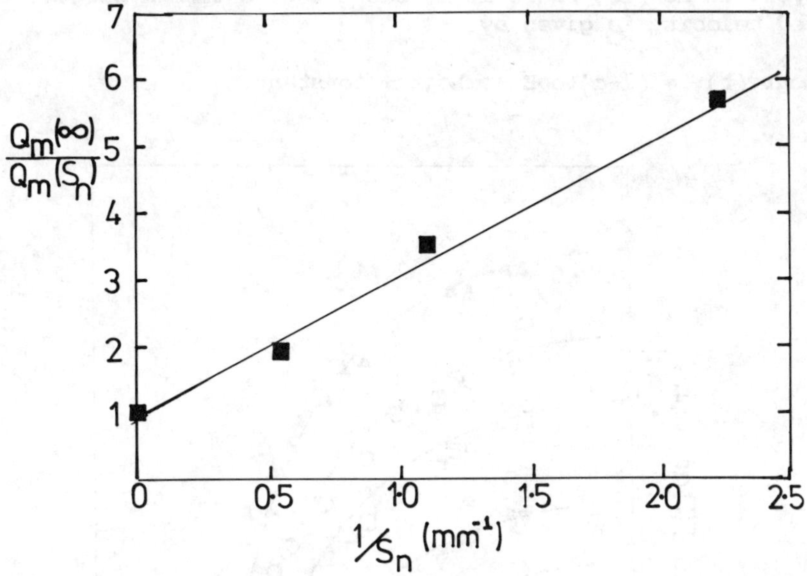

Figure 3 The variation of Q_m with s_n is shown for PDATS the data being presented as suggested in the text in the form $Q_m(\infty)/Q_m(s_n)$ against $1/s_n$.

From the gradient of figure 3 it is possible to extract s_R for T.S. This is found to be 2 mm, which corresponds to half the sample length. In other words the carriers are only recombining at the electrodes. Indeed one cut with a razor has been found sufficient in samples of TS to reduce Q_m by half, though for more cuts the lines are made with a laser focussed to a line. In this circumstance we have for the uncut sample from data such as that in figure 2 that $v_d = 2.2 \times 10^3$ m/s down to fields E = 100 V/m. From figure 2 it appears that the carriers take some hundreds of nanoseconds before trapping. The initial rate of increase of Q is due to the dark current and upon illumination the charge increases initially rapidly before taking some minutes to attain once again the steady rate of increase due to the dark current alone. During these minutes the photocharge is undergoing successive trapping and trap release events. This dispersive transport has been analysed at length elsewhere (12). Through observation of the current decay after a μsec laser pulse over many decades of time the decay is seen to fit the form

$$\text{Log}I(t) = -\alpha \text{Log}t + \text{constant}. \qquad (8)$$

This is in accord with the theory of dispersive trap limited transport in 1D (13) where it is shown that a time averaged trap limited velocity is given by

$$\text{Log}\langle v(t)\rangle = (1-\alpha)\text{Log}E - \alpha \text{Log}t + \text{constant}. \qquad (9)$$

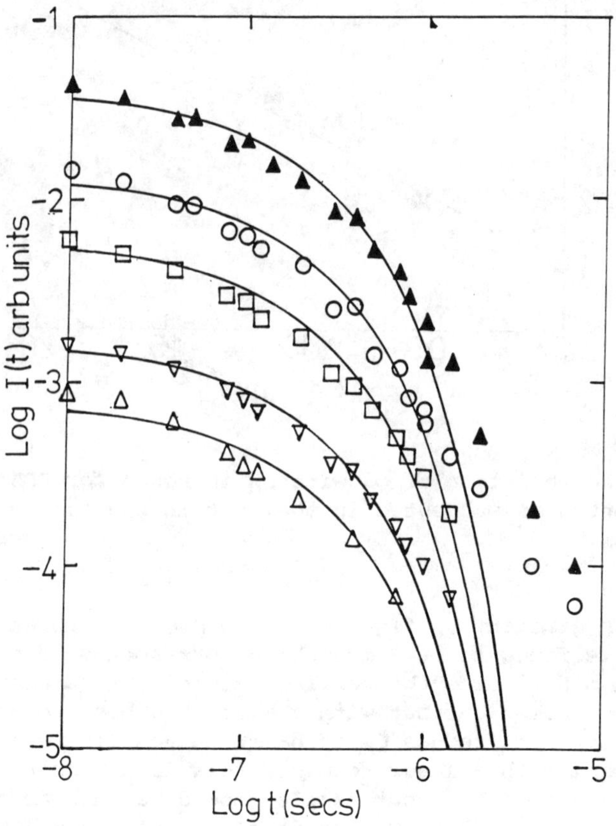

Figure 4 From photocurrent decays in T.S. such as that in the inset of fig. 2 are plotted decay curves at different applied voltages, ▲ 2 kV, ○ 1 kV, □ 500 V, ▽ 200 V, △ 100 V. The solid line represents a theoretical fit of the data to equation (11) as described in the text with τ = 500 nsecs.

From the decays α is found to be $\alpha = 0.85$. In this circumstance the velocity varies with field according to

$$\langle v(t) \rangle \sim E^{1-\alpha} = E^{0.15}. \qquad (10)$$

It is possible to relate this apparent saturation of velocity with E with that observed in the experimental results presented here. To discern whether indeed one may do this it is necessary to measure the decay of the photo current in the nanosecond regime and ascertain whether the current decays here as $t^{-\alpha}$.

From the decay curves shown in fig. 4 it is clear that a $t^{-\alpha}$ decay is not found in the time regime 10 nsecs - 1 μsec though after 1 μsec there is evidence that such a $t^{-\alpha}$ decay is asymptotically approached. Consider carriers created at random positions by the light pulse which subsequently drift freely in 1D with a velocity v_d to randomly positioned traps of mean separation s. It can be shown in these circumstances the current I(t) falls as carriers trap with a decay of the form

$$I(t) = I_o E_2(tv/s) \qquad (11)$$

where I_o is the initial current and E_2 is the exponential integral function (10). The characteristic time τ is

$$\tau = s/v. \qquad (12)$$

From the data in figure 4 τ is 500 nsecs independant of E. This is further evidence that the free drift velocity is saturated. Furthermore from the previously established value for v_d one deduces that s the average intertrap separation is 1.1 mm, a value similar to that found by two independent routes (1,2)

The field dependence of the observables in DCH is shown in figure 5. If one assumes that there is no recombination in DCH as is the case for TS then one can use equation (4) to find v_d from this data. It is found following this route that the drift velocity in DCH is about a factor of fifty greater than in TS. With the expectation that v is a property of the backbone independent of the side group one is led to postulate the existence of recombination centres in DCH with about fifty on a chain. This implies a separation of about 100 μm.

The line cutting experiment as applied to DCH proves to be difficult as the value of s_n must be less than s_R, ~ 100 μm, before its effect on Q_m is at all measurable. In contrast to TS where one line suffices to reduce Q_m by half, in DCH up to 20 lines fabricated with the same techniques as used for TS were not enough to produce any meaninfgul effect and new techniques were used.

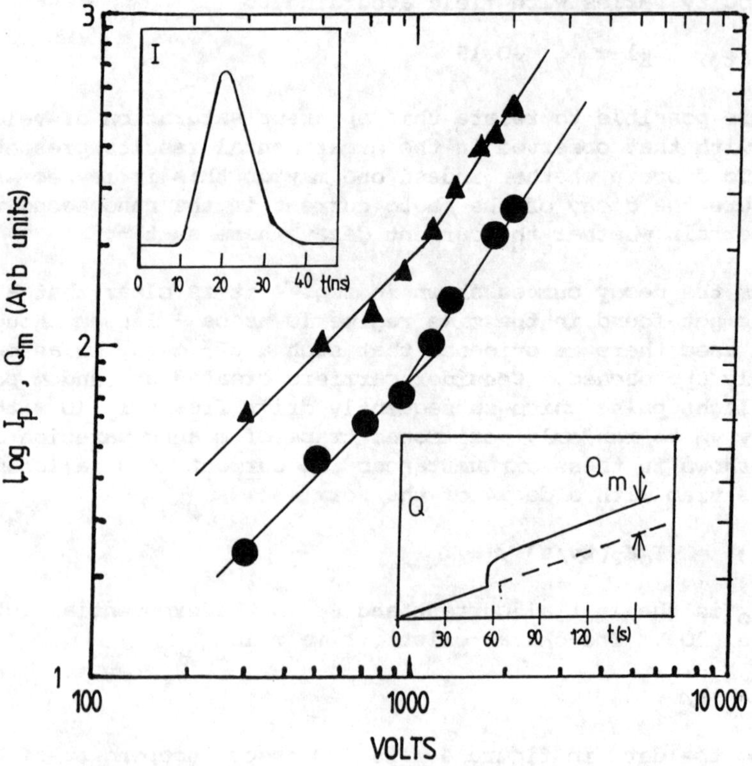

Figure 5 Data similar to that shown in figure 2 for TS is given for DCH▲ = I_p and ● = Q_m. The insets show a typical I_p and a typical Q_m for DCH.

The first of these, that of laser ablation with a ArF excimer laser produced promising results. The sample surface was illuminated by a very high energy 10 nsec pulse at 193 nm through a photolithographically produced Nickel mask. The by products of the photodegraded polymer are ejected from the sample carrying the pulse energy as kinetic energy. This is thus a non-heat evolving process which avoids damaging the sample adjacent to the cuts. With a beam energy of 500 mJ/cm^2/pulse the ablative process excavated 0.1 μm/pulse. It was possible to burn lines of separation, s_n = 100 μm and of width 25 μm using this technique. The resulting change in Q_m gave values for v_d in DCH similar to those in TS although analysis of the experiment was complicated as $s_R \sim s_n$.

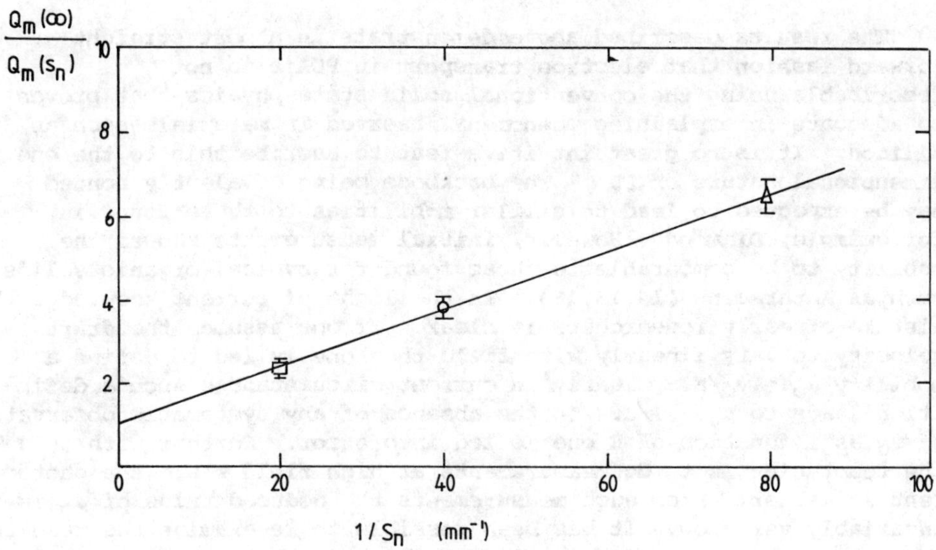

Figure 6 The above data for DCH, shows how Q_m varies with s_n. The data is presented as is that for TS in figure (3), however the data is taken from three separate samples with differing values of s_n.

To attain the regime $s_n < s_R$ another technique was employed. By using an electron beam used in state-of-the-art high definition electron beam lithography it was possible to produce cuts of down to ~ 10 μm separation and of negligible thickness. These walls are thought to be the result of crosslinking. They are shown to stop carriers effectively (4,5) and their construction is described in these references. In figure 6, is shown the effect on Q_m of such walls on three samples with s_n = 12.5 μm, 25 μm and 50 μm. From the gradient of this graph s_R is found to be 70 μm. Using this value and data such as that in figure 5 along with equation (3) the drift velocity is found to be 1.2 x 10^3 m/s comparable to that found for TS. The decay of the photocurrent in DCH shown in the figure inset is much faster than that in TS and is indeed so fast as to call into question the validity of the criteria, $T < \tau$. If this condition is violated then the value of v_d for DCH as given above will be an under-estimate. It should be noted, however, that qualitatively the velocity is still saturated and taking this value of v_d as a lower limit one finds a lower limit to the mobility at the lowest fields at which measurements have been made to be μ_o > 10 m^2/V/s.

4. DISCUSSION

The results described above demonstrate in a most straightforward fashion that electron transport in PDA's is not describable using the conventional solid state physics that proves so adequate in explaining phenomena observed in materials such as silicon. It is no great intuitive feat to ascribe this to the one dimensional nature of PDA. The backbone being covalently bonded may be expected to lead to similar mobilities to those found in, for example, Diamond. However, initial measurements showed the mobility to be comparable to those found for typical organic solids such as Anthracene (14,15,16). In the light of current knowledge the mistake of early researchers is clear. If one assumes the drift velocity to vary linearly with field then one is led to define a mobility $\mu_o = v_d/E$. Clearly in current circumstances such a definition leads to $\mu_o \sim 1/E$. In the absence of any systematic observation of v_d as a function of E one is led into error. Further with $\phi(E) \sim E$ the temptation is to do measurements at high field where the photocurrent is larger. From such measurements the deduced value of μ_o is invariably very low. It has been possible to re-examine the results of other workers in order to find v_d which is the more meaningful quantity. In the cases where this is possible the value is in broad agreement with that reported here. In these results measurements have been done down to fields of 100 V/m. With interelectrode separations of 3 mm - 10 mm such fields entail applied voltages of 300 mV - 1.0 V. Measurements with applied voltages less than this run into problems due to contact potentials. Workers who traditionally use a surface electrode arrangement with electrodes spaced of the order of 100 μm's cannot use such a low field as this would entail voltages of the order 10 mV.

Recently the concept of trap limited drift velocity providing the explanation for the observed saturation have gained much currency (12,13). This has been shown to be inapplicable in explaining these results (3) and would any way require trap limited mobilities to be > 20 $m^2/V/s$. This ultra high value finds no explanation in such trap limited transport theories and would indicate that the intrinsic mobility was even higher!

That one cannot explain an ultra high mobility and a low saturated drift velocity with conventional theory cannot be over emphasised. The latter requires a large electron phonon interaction while the former requires low scattering and a concomitant low electron phonon interaction according to conventional wisdom. The resolution of this apparent paradox is found in the motion of the acoustic solitary wave polaron described in the following article.

1. Donovan, K.J. and Wilson, E.G., Demonstration of high mobility one dimensional semiconducting polymer, J.Phys.C. (1979), 12, 48 57.
2. Donovan, K.J. and Wilson, E.G., Demonstrations of an ultra high mobility organic polymer, Phil.Mag.B. (1981), 44, 9.
3. Donovan, K.J. and Wilson, E.G., Carrier trapping on one dimensional systems. Example: Organic PDATS, J.Phys.C.Letts. (1984), in press.
4. Donovan, K.J., Freeman, P.D. and Wilson, E.G., A novel determination of the number of recombination centres in the one dimensional semiconductor PDA DCH, J.Phys.C.Letts. Submitted for publication.
5. Donovan, K.J., Freeman, P.D. and Wilson, E.G., Micro engineering of a novel organic molecular electronic structure, J.Phys.C.Letts. Submitted for publication.
6. Donovan, K.J. and Wilson, E.G., Photocarrier creation in one dimension, Phil.Mag.B. (1981), 44, 31.
7. Seiferfeld, U., Ries, B. and Bassler, H., Field dependant photocarrier generation in crystalline polydiacetylenes, J.Phys.C. (1983), 16, 5189.
8. Haberkorn, R. and Michel Beyerle, M.E., Onsagers ion recombination model in one dimension, Chem.Phys.Letts. (1973), 23, 128.
9. Blossey, D.F., One dimensional Onsager theory for carrier injection in metal-insulator systems, Phys.Rev.B. (1974), 9, 5183.
10. Wilson, E.G., Electronic conduction in one dimension, J.Phys.C. (1980), 13, 2885.
11. Ries, B., Schonherr, G., Bassler, H. and Silver M., Monte Carlo simulation of the time dependance of geminate recombination, Phil.Mag.B. (1984), 49, 1, 27.
12. Movaghar, B., Murray, D.W., Donovan, K.J. and Wilson, E.G., Anomolous electric field dependance of carrier drift in one dimensional systems. Example: Organic PDATS, J.Phys.C. (1984), 17, 1247.
13. Movaghar, B., Murray, D.W., Pholman, B. and Wurtz, D., Breakdown of linear response theory in one dimensional classical diffusion problems, J.Phys.C. (1984), 17, 1677.
14. Reimer, B. and Bassler, H., Motion and recombination of charge carriers in a polydiacetylene single crystal. Chem.Phys.Letts. (1976), 43, 81.
15. Reimer, B. and Bassler, H., Transient photoconduction in a polydiacetylene single crystal. Determination of transit time, free lifetime and recombination time of charge carriers. Phys.stat.solidi. (b) (1978), 85(1), 145.
16. Chance, R.R., Baughman, R.H., Reucroft, P.J. and Takahashi, K. Transient photoconduction of a polydiacetylene single crystal, Chem.Phys. (1976), 13. 181.

ELECTRIC FIELD DEPENDENCE OF DIFFUSION AND TRAPPING IN POLYDIACETYLENE

B Movaghar*, B Pohlmann** and D Wurtz***

*   GEC Research Laboratories, East Lane, Wembley, UK
** Fachbereich Physik, University of Marburg, FRG
*** Institut fur Theoretische Physik, Heidelberg, FRG

ABSTRACT

Transport in low-dimensional systems exhibits a rich variety of novel features. This manifests itself in unusual electric field and time dependences of drift velocities, Onsager generation rates, and survival fractions in the presence of deep traps. These features are indeed observed in the polydiacetylenes which are expected to be ideal quasi-one-dimensional substances.

INTRODUCTION

One of the most interesting aspects of the polydiacetylenes is the high degree of anisotropy associated with the lattice structure. Assuming a perfect crystal, the minimum interchain, separation $\ell_\perp$ in PTS or DCH is $\ell_\perp \sim 7$ Å whereas the π-electron overlap distance along the chain $\ell_\parallel \sim 1.2$ Å. In terms of a simple one electron tight-binding picture, this gives a ratio of overlap integrals $t_\parallel/t_\perp$ of at least 10^3 making this a rather unique quasi-one-dimensional material. The strong anisotropy leads to a whole series of new and unusual physical effects: single electrons added or subtracted from the polymer backbone will form polarons [1] rather than free electron like states. Coulomb interactions are most important as evidenced in the absorption and photoconductivity action spectra [2]. Photo or thermal excitations lead to the formation of what is most probably a bound exciton-polaron the dissociation product being a polaron (-) anti-polaron (+) pair. Exciton-polarons can exist as interchain or intra-chain excitations. Orenstein and Etemad [3] have suggested the latter as a possible explanatation of the photo-induced defect state observed by them and Hayes and Bloor [4].

Once a 'free' carrier has been created, we then encounter a further class of fascinating transport phenomena associated with the low dimensionality of the system and this is where we want now to focus our attention upon. Let us first consider the diffusion of carriers along a one-dimensional chain. A realistic view of the polymer backbone in PDA is of course not one of perfect order but every now and again we must expect a defect which leads to the formation of a trap (or anti-trap) with probability s [5]. In addition, carriers will also eventually reach recombination centres which lead to the total annihilation of the charge [6]. Depending on the time temperature domain, an ordinary trap may be indistinguishable from a recombination centre. We therefore classify the defects encountered by the particle into two categories: (a) traps with particle conservation (release), (b) with particle annihilation (infinitely deep traps) occurring with concentration x.

The photocurrent

The photocurrent measured experimentally j(t) can be more conveniently analysed in the space of the Laplace variable p

$$j(p) = \int_0^\infty j(t)\, e^{-pt}\, dt \qquad (1)$$

and can be written

$$j(p) = eG(\eta)\; n(p,\eta)\; V_d(p,\eta) \qquad (2)$$

where e is the electronic charge, G, n and V_d are generation rate of free carriers, particle density and drift velocity respectively. The variable η denotes the dimensionless external bias yield E defined by the relation

$$\frac{1-\eta}{1+\eta} = e^{-eaE/kT} \qquad (3)$$

where a is the 'lattice' spacing. Equation (2) expresses the fact that in general all three quantities G, n and V_d are field dependent. The electric-field dependence of the generation rate $G(\eta)$ is a subject of considerable importance in organic materials and in the framework of the Onsager description, is strongly dependent on the dimensionality of the system. It has been calculated and discussed in great detail by several authors [7] and more recently using Monte-Carlo techniques by Ries and Bässler [8]. These authors have also considered anisotropic and finite systems.

The field and frequency dependence of the drift velocity $V_d(p,\eta)$ in a 1-d system has been analysed in great detail in references [5,9] using mean field theory and scaling theory. The disorder was described by allowing a subset of the diffusion or jump rates W to differ and follow a distribution function $\rho(W)$. We shall not repeat all the details here but only recall the main conclusions [9]: the linear response relationship between V_d and the diffusivity D in the low field limit $\eta \ll 1$ given by

$$Vd(p,\eta) = 2\eta a D(p)/p \tag{4}$$

is in general only valid for $p > pc$ a critical frequency determined by $\rho(W)$ and η. In the limit $p \to 0$ Eqn. (4) is only satisfied if the $(\lim_{p\to 0} V_d(p,\eta) \neq 0)$ i.e. if the d.c. conductivity is finite. This leads to a very interesting behaviour if one adopts for PDA the rather realistic distribution function $\rho(W) = (1-\alpha)W^{-\alpha}$ $0<W<1$, $0<\alpha<1$. Here we find that (5)

$$V_d(p,\eta) \underset{p\to 0}{\sim} (p/\eta\overline{W})^{\alpha} 2a\overline{W}\eta/p \tag{5}$$

or

$$V_d(t) \underset{t\to\infty}{\sim} \frac{2a\overline{W}}{(\overline{W}t)^{\alpha}} \eta^{(1-\alpha)} \tag{6}$$

which gives us a sublinear drift velocity with electric field and a vanishing d.c. conductivity. For $\alpha=0.85$ $V_d(t,\eta)$ is almost field independent as found experimentally by Donovan and Wilson [10] in the long time photocurrent decays of PDA-TS ($t>1$ μs).

The survival fraction

Let us now turn our attention to the particle density $n(p,\eta)$. Take a 1-d chain with no disorder in the jump rates W and containing a concentration x of deep recombination centres. For this model we have obtained an exact solution for $n(\eta,t)$ for all times and values of η when $x \ll 1$. The solution must be divided into two regions and can be written (6).

$$n_1(t,\eta) = \frac{4}{\pi^2} e^{-\eta^2 Wt} \int_0^\infty dS \frac{S}{[1+(\eta S/\pi x)^2]^2} e^{-\pi^2 x^2 Wt/S^2}$$

$$\left[\frac{e^{-S}}{1-e^{-S}} + \frac{e^{-S(1-\eta/x)}}{2(1+e^{-S(1/\eta/x)})} + \frac{e^{-S(1+\eta/x)}}{2(1+e^{-S(1+\eta/x)})}\right], \quad \eta<x \tag{7}$$

$$n_2(t,\eta) = n_1(t,\eta) + L(t,\eta) \tag{8a}$$

$$L(t,\eta) = 8x^2(\eta-x)^2 e^{-\eta^2 Wt} \sum_{n=0}^{\infty} \frac{\left(n+\frac{1}{2}\right)\exp\{Wt[(\eta-x)/(2n+1)]^2\}}{(2n\eta+x)^2[2(n+1)\eta-x]^2}, \quad \eta>x \tag{8b}$$

The asymptotic long time forms are simple to evaluate and given by:

$$\underset{t\to\infty}{n(t,\eta)} \sim \exp\left[-3\left(\frac{\pi^2x^2C^2Wt}{4}\right)^{1/3}\right] e^{-\eta^2 Wt} \quad (\eta<x) \tag{9}$$

where $C = (1-\eta/x)$

$$\underset{t\to\infty}{n(t,\eta)} \sim \frac{4(\eta-x)^2}{(2\eta-x)^2} \exp\left[-(2\eta x-x^2)Wt\right] \quad (\eta>x) \tag{10}$$

The crossover field for diffusion to drift limited recombination is given by $\eta=x$ in d=1. Note that $V_d = 2\eta Wa$ is the drift velocity in this case.

Let us now consider dimensionalities d>1. For $\eta=0$ a scaling argument [11] gives

$$\underset{t\to\infty}{n(t)} \sim n(o)\exp\left[-\left(\frac{t}{to}\right)^{\frac{\tilde{d}}{\tilde{d}+2}}\right] \tag{11}$$

where $\tilde{d}$ is the fracton or spectral dimensionality of the system. We can now estimate the crossover field to an exponential (drift limited) behaviour.

Following Rammal and Toulouse [12] we first write for the mean squared distance moved by the particle ($\eta=0$) $\langle R^2(t)\rangle$ and number of new sites visited in a time t, S(t) on a fractal as

$$\langle R^2(t)\rangle \underset{t\to\infty}{\sim} Dt^{2\nu} \tag{12}$$

$$S(t) \underset{t\to\infty}{\sim} St^{\gamma} \tag{13}$$

Introducing an electric field $\eta>0$ changes $S(t)$ to $S(t,\eta)$ which we write as

$$S(t,\eta) \underset{t\to\infty}{\sim} S_1 t^{\beta} \tag{14}$$

Crossover occurs when $S(t,\eta) > S(t)$. Using the length scale determined by the average trapping distance (a/x), we obtain $1<d<2$

$$\eta > c(\tilde{d})\, x^{(\beta-\gamma)/\nu} \tag{15}$$

Rammal and Toulouse have shown that $2\alpha = \overline{d}/\overline{a} = \tilde{d}$ where $\overline{d}$ is the fractal or Hausdorff dimensionality of the system, and $\nu = \tilde{d}/2\overline{d} = 1/2\overline{a}$.

Assuming $\beta=1$ then gives $(c\sim 1)$

$$\eta > x^{[2\overline{a}-\overline{d}]} = x^{2-\overline{d}} \quad \text{if } (\overline{a}=1)\ ,\ 1<\overline{d}<2 \tag{16}$$

The crossover field to drift limited behaviour increases with increasing dimensionality of the system and becomes independent of trap concentration when $\overline{d}>2$.

Equation(11) for $\tilde{d}>2$ is of little value in practice because by the time the asymptotic form has set in $n(t)/n(0)<10^{-6}$ and therefore not observable experimentally [13].

The First Passage Time method [14] is more appropriate in this case and for most of the experimentally relevant time domain, one has to a good approximation.

$$n(t,\eta) = n(0)\exp\left[-xS(t,\eta)\right]\ ,\ \overline{d}>2 \tag{17}$$

Finally let us consider a disordered 1-dimensional chain where $V_d(p,\eta)$ obeys for example (5). In general the procedure is as follows: we first solve for $V_d(p,\eta)$ using the methods discussed in Reference [9], for a given $\rho(W)$.

For distribution functions of interest here, this can be done by introducing effective jump frequency and field dependent jump rates $W_c^{+,-}(p,\eta)$ so that

$$V_d(p,\eta) = \frac{ea[W_c^+(p,\eta) - W_c^-(p,\eta)]}{p} \tag{18}$$

In terms of $W_c^{+,-}$, $n(p,\eta)$ now becomes (15)

$$n(p,\eta) = \frac{1}{p} - \frac{x^2}{p}\left[\frac{1}{a^+} + \frac{1}{a^-}\right] \int_0^\infty \zeta\, e^{-\zeta x}\, F(\zeta)\, d\zeta \tag{19}$$

$$F(\zeta) = \left[\frac{\mathrm{Sinh}(a^+\zeta/2)\ \mathrm{Sinh}(a^-\zeta/2)}{\mathrm{Sinh}(a^+ + a^-)\zeta/2}\right] \tag{20}$$

where $a^{+,-} = \dfrac{p}{p+g_0^\pm}$ (21)

and

$$2g_0^\pm = -p + (W_c^\pm - W_c^\mp) + \left[p^2 + 2p(W_c^+ + W_c^-) + (W_c^+ - W_c^-)^2\right]^{\frac{1}{2}} \tag{22}$$

When $W_c^\pm(p,\eta) = W(1\pm\eta)$ i.e. no disorder on the chain, (19) reduces to the Laplace transform of (7) and (8). Unfortunately (19) is no longer analytically invertible into time space when $W(p)$ is frequency dependent as in (5) for example.

The inversion has to be done numerically, but the asymptotic $t\to\infty$ behaviour of $n(t,\eta)$ is always given by (9,10) as long as $\lim_{p\to 0} \{V_d(p,\eta)p\} \neq 0$. in which case we simply replace $W \to D(0)/a^2$ in (7,8) where $D(0)$ is the long time diffusivity on the chain.

The absence of diffusion in the long time limit as expressed by (6) leads to a qualitatively new asymptotic decay law. To see this let us consider the region $\eta >> x$ satisfied in most good crystals of PDA. Here $j(p,\eta)$ can be well approximated by

$$j(p,\eta) \simeq eG(\eta)\ \frac{pV_d(p,\eta)}{p+2xp\frac{V_d}{a}(p,\eta)} \tag{23}$$

with $V_d(p,\eta)$ given by (5).

$$j(t,\eta) \sim \eta e G(\eta) \begin{cases} \dfrac{1}{(\eta t)^{1-\mu}} & \text{(a)}\quad t<t_c \\[2ex] \dfrac{1}{(\eta t)^{1+\mu}} & \text{(b)}\quad t>t_c \end{cases} \qquad [1-\mu = \alpha] \tag{24}$$

where t_c is the average time taken to reach a recombination centre

$$t_c \sim \left(\frac{1}{2x}\right)^{\frac{1}{1-\alpha}} (\eta\overline{W})^{-1} \qquad (25)$$

Note that (25) looks exactly like the famous dispersive transport behaviour discovered by Scher and Montroll [16]. Here the deep traps act like boundaries and produce a knee in the photocurrent decay curve, the 'transit time' t_c scales however normally with with field, but the total photocurrent has a most interesting and unusual field dependence. Note that in 1-d [7] $G(\eta)\alpha\eta$.

The qualitative change in the $t\to\infty$ behaviour of $j(t,\eta)$ is evidently related to the vanishing of V_d as $t^{-\alpha}$ ($t\to\infty$). If $V_d(t)$, $t\to\infty$ is finite, we recover the long time forms given by equations (9,10). In general we can therefore expect a typical $j(t,\eta)$ decay curve to exhibit all three regions (24) and (9,10). Experimentally (24) has already been observed by Donovan and Wilson. These authors are working in the region $\eta >> x$, they did not observe (24b) consistent with their conclusions regarding the absence of recombination centres ($x<<1$) and $\alpha=0.85$ in the good crystals. In helium bombarded materials, Seiferheld and Bässler [17] observed (9,10) in the long time limit. Hunt and Bloor [18] observed (9,10) and verified both field and temperature scaling in PDA-10H. From their data one must conclude that the long time diffusivity is temperature independent and extremely small $D(0) \sim 10^{-14}$ cm^2/s.

This suggests that the rate determining steps are associated with tunnellling across barriers (chain ends or to another chain) rather than phonon assisted hopping processes.

CONCLUSION

Transport in the region of dimensionality $\overline{d}\sim 1$ evidently exhibits a rich variety of unusual features. These features have been observed experimentally in the photocurrents of PDA which can therefore be regarded as the closest material to d=1. In reality there are of course always three time domains when $0<t<t_1$ the motion of changes is 1-dimensional, when $t_1<t<t_2$, the carriers will move on a fractal lattice with $1<\overline{d}<2$, the reason is that crossing to another chain will take place at preferred points in the lattice, where defects, impurities, bond-distortions are present. Finally, when $t>t_2$, motion is on a 3-d anisotropic lattice with defects. Experimentally we only appear to observe the first and possibly some part of the second (fractal) region. This suggests that interchain jump rate $W_\perp$ is insignificant except at special points in the lattice in a large class of PDA's. An experimental estimate of $W_\perp$ is as yet not available and would be highly desirable. The nature of the transport process in the short and long time limit remains a puzzle and is a a matter of controversy. In the short time limit Wilson [14] suggest that carriers build polarons with low field saturated drift velocities. In the long time limit, the data of Hunt et al [18] and Bässler et al [20] suggest that in $t^{-\alpha}$ α is T independent, the same applies to the long time diffusivity D(0). This is indeed novel and very interesting, more data on temperature dependent transport and in particular very low temperature data is required to clarify the situation in the polydiacetylenes.

ACKNOWLEDGEMENT

The authors are grateful to Professor Bässler, Professor Bloor and Dr B Ries for helpful dicussions.

REFERENCES

1 Cade, N. and Movaghar B. J. Phys. C. 16 (1983) 539

2 Bloor, D., Hubble, C.L. and Ando, D.J. in Molecular Metals Ed. Hatfield, W.E. (New York Plenum) (1979)

3 Orenstein, J. and Etemad, S. J. Phys. C., 17 (1984) L297

4 Hayes, W. and Bloor, D. (To be published)

5 Movaghar, B., Murray, D.W., Donovan, K.J. and Wilson, E.G. J. Phys. C., 17 (1984) 1247

6 Movaghar, B., Pohlmann, B., and Wurtz, D. Phys. Rev. A., 29 (1984) 1568

7 Haberkorn, R., and Michel-Beyerle, M.E. Chem. Phys. Lett. 23 (1973) 128

8 Ries, B., Schönherr, H., Bässler, H., and Silver, M. Phil. Mag. B., 49 (1984) 259

9 Movaghar, B., Murray, D.W., Pohlmann, B., and Wurtz, D. J. Phys. C. (1984) 1677

10 Donovan, K.J., and Wilson, E.G. Phil. Mag. B., 44 (1981) 9

11 Webman, I. Phys. Rev. Lett., 52 (1984) 220

12 Rammal and Toulouse, G., J. de Physique Lett., 44 (1983) L13

13 Zumofen, G., Blumen, A., and Klafter, S. J. Phys. Lett., 45 (1984) L49

14 Scher, H., Alexander, S. and Montroll, E.W. Proc. Nat. Acad. Sci. USA, 77 (1980) 3758

15 Movaghar, B., Pohlmann, B. and Wurtz, D. (To be published)

16 Scher, H., and Montroll, E.W. Phys. Rev. B., 12 (1975) 2455

17 Seiferheld, U., Bässler, H., and Movaghar, B. Phys. Rev. Lett. 51 (1983) 813

18 Hunt, I., Bloor, D., and Movaghar, B. J. Phys. C. 16 (1983) L623

19 Wilson, E.G. J. Phys. C. 16 (1983) 6739

20 Bässler, H., and Seiferheld, U. (Private Communication)

THE STUDY OF ELECTRONIC EXCITATIONS OF POLYDIACETYLENES AND RESONANCE RAMAN SPECTROSCOPY

David N. Batchelder

Department of Physics, Queen Mary College, London E1 4NS

1. INTRODUCTION

Much of what is known about the electronic excitations of PDAs has been learned from optical and resonance Raman (RR) spectroscopy (1, 2). In addition to measurements on single crystals the polymer has been studied in solutions, precipitates and thin films. This paper will attempt to review the experimental evidence for the nature of the electronic excited states in PDAs which are accessible by these spectroscopic techniques. Starting from the optical spectra for single crystals the effect of changing environment and conformation on the dominant electronic transition will be explored. Additional information, particularly about vibrational coupling, which can be brought out by RR spectroscopy will be included. Finally, electronic transitions which have been detected by RR spectroscopy when small molecules such as oxygen are adsorbed by the PDA backbone will be discussed. Selection of the spectroscopic data has been made with the aim of giving a representative rather than a complete picture of the experimental situation.

2. OPTICAL SPECTROSCOPY OF SINGLE CRYSTALS

Over the visible region of the electromagnetic spectrum the optical properties of PDAs are dominated by electronic transitions of the π-electrons on the conjugated backbone. This is in spite of the fact that in many PDAs the backbone carbon atoms on which the π-electrons are located may account for less than 10% of the atoms on the chain. The changes in the reflection spectrum which occur upon polymerisation of a TS monomer crystal vividly illustrate this anomaly (3). In Fig. 1 the reflectivity of the monomer crystal is never greater than 10% and there are no spectral features which stand out. By contrast the polymer has a very sharp peak at 2 eV in which the reflectivity, R, reaches approximately 70% for light with the $\underline{E}$-vector parallel to the direction, $\underline{b}$, of the polymer chain. This is sufficiently high to give the crystals their characteristic metallic lustre.

The reflectivity of a molecular crystal is only indirectly related to the transition dipole matrix elements $\underline{M}$ which determine the electronic structure (4). The Kramers-Kronig relations may be used, however, to derive the real and imaginary parts of the complex index of refraction, $\underline{n} = n + ik$, from the spectral dependence of R. For normal incidence these are given as a function of the angular frequency, ω, of the incident light by

$$n = (1 - R)/(1 + R - 2R^{1/2}\cos\phi)$$

$$\text{and} \quad k = 2R\sin\phi/(1 + R - 2R^{1/2}\cos\phi) \qquad (1)$$

$$\text{where} \quad \phi(\omega) = \frac{2\omega}{\pi}\int_o^\infty \frac{\ln R^{1/2}(u)\ du}{\omega^2 - u^2}$$

For samples with surface defects affecting R it is clear that the values of n and k will be subject to error. An additional practical problem is that R can only be measured over a finite frequency interval so correction factors must be included to complete the range of integration. The complex dielectric constant which relates the response of the medium to the applied $\underline{E}$-field is given by $\underline{\epsilon} = \epsilon_1 + i\epsilon_2$, where $\epsilon_1 = n^2 - k^2$ and $\epsilon_2 = 2nk$. For $k > n$ it follows that the real part of the dielectric constant will be negative. The absorption coefficient which can be measured directly for sufficiently thin crystals is given by $\alpha = 2\omega k/c$.

Application of Eqn. 1 to the data in Fig. 1 over the 1 to 10 eV range gave the absorption spectra for light polarised parallel and perpendicular to the $\underline{b}$-direction for both the monomer and polymer crystals (3). The results in Fig. 2 show clearly that peaks A, A', and B must be associated with electronic transitions of the polymer backbone as they do not appear in the monomer spectra. In addition the direction of $\underline{M}$ for the transitions must lie approximately along the polymer chains as the absorption bands only appear for the $\underline{E} \parallel \underline{b}$ spectrum. Peak C which appears in the absorption spectrum of the monomer but not the polymer has been identified with the diacetylene moiety of the monomer molecule. The rest of the features were attributed to electronic transitions on the sidegroups.

The reflectivity of PDA-TS crystals has been studied in greater detail over the visible spectral region (5 - 8). Figure 3 shows

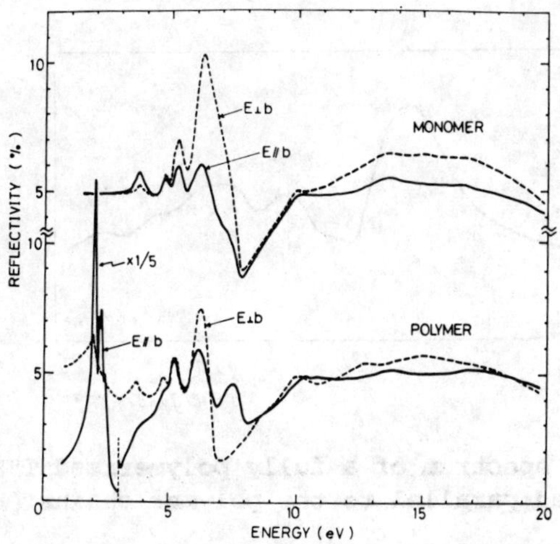

Fig. 1 Polarized reflectivity spectra on the (100) surface of monomer (upper) and polymer (lower) single crystals of TS diacetylene at room temperature.

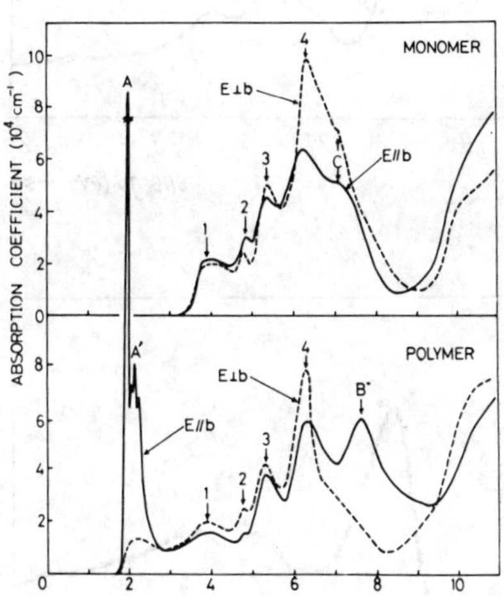

Fig. 2 Absorption spectra of monomeric (upper) and polymeric (lower) TS crystals obtained from the Kramers-Kronig transformation of the reflectivity spectra.

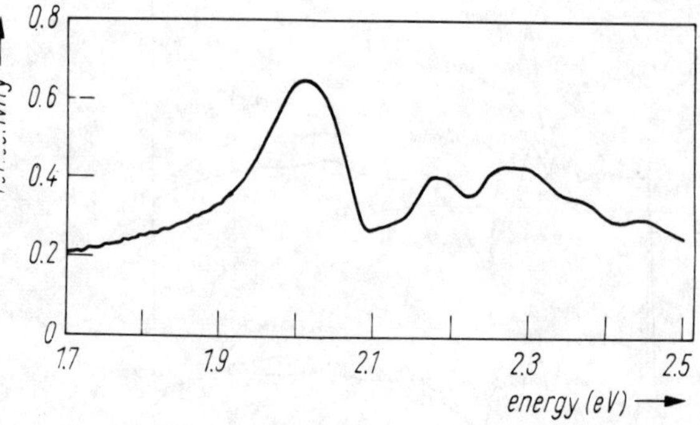

Fig. 3 Reflection spectrum of a fully polymerized TSHD crystal with light polarised parallel to the polymer chains (b-axis).

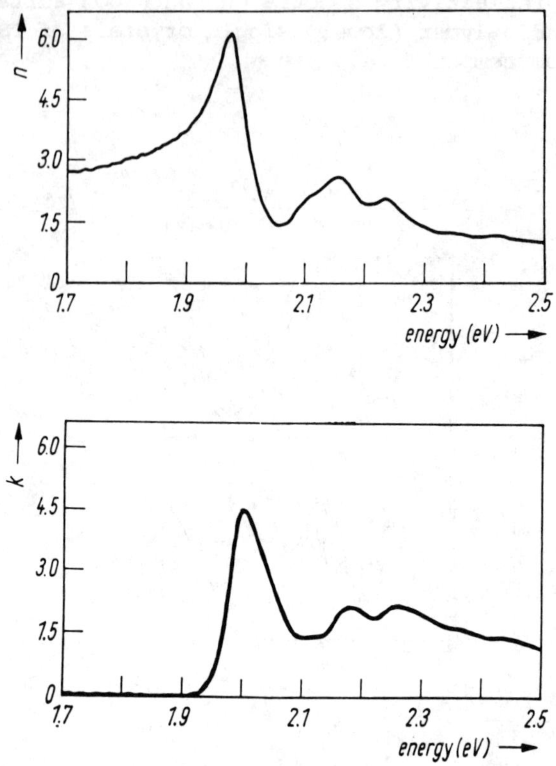

Fig. 4 Refractive index of a TSHD polymer crystal for light polarized parallel to the polymer chains (b-axis). a) The real part, n, and b) the imaginary part, k.

typical data for this range which includes peaks A and A', while Fig. 4 illustrates the n and k values derived from that data (5). Data of a very similar nature have been recorded for a number of other PDA crystals, e.g. PDA-DCH (9, 10), -TCDU (11), -ETCD (12), -IPUDO (13), -PFBS (14) and -4BCMU (15). It is generally accepted that peaks A and A' represent respectively the 0-0 transition of the π-electron system and its vibrational sidebands and, since no photoconduction is associated with the absorption, that the excited state is excitonic in character. The C_{2h} symmetry of the backbone suggests that peak A is due to the allowed transition from the $1'A_g^-$ ground state to the $1'B_u^+$ excited state. The transition to the $2'A_g^-$ state which is expected to lie at a higher energy (16) is optically forbidden while that to the $2'B_u$ is allowed; the latter has been tentatively identified with peak B in Fig. 2 (3). The vibrational modes which couple strongly with the $1'B_u^+$ excited state to produce peak A' should have A_g symmetry and also dominate the resonance Raman (RR) spectrum.

The absorption spectrum for $\underline{E} \parallel \underline{b}$ derived from the data of Fig. 3 is shown in Fig. 5. Also included is that measured directly by transmission through a thin crystal (< 10 μm) with $\underline{E} \perp \underline{b}$ (17). The location of the 0-0 transition and associated vibronic structure is in good agreement but the relative heights are considerably different. This discrepancy most likely arises from the greater difficulty in performing accurate reflectivity experiments (none of the crystals used has been optically flat) and the approximations made in the K-K analysis. The reported values of the principle absorption coefficients for PDA-TS have been found to be 9×10^5, 2×10^3 and < 5 cm^{-1} (5), showing that $\underline{M}$ lies at a small angle to the $\underline{b}$-direction. The values of ϵ_1 and ϵ_2, for a different set of data (6), are shown in Fig. 6, both at 295 K and at 9 K. As expected from the 295 K n and k values in Fig. 5 the sign of ϵ_1 goes negative over a spectral range of 0.4 eV. This condition was exploited by Philpott to demonstrate the propagation of polariton waves over the surface of PDA-TS crystals (18). Such modes arise from the coupling of the electromagnetic radiation to the polarization created by electronically excited molecules in the surface region. The splittings of the peaks A and A' at 9 K shown in Fig. 6 for both ϵ_1 and ϵ_2 arise from the phase transition which will be discussed below.

3. RESONANCE RAMAN SPECTROSCOPY OF SINGLE CRYSTALS

The Raman spectra obtained from PDA single crystals are at least as striking as the optical spectra (2). When the laser beam has a photon energy in the region of peaks A and A' of the absorption spectrum the intensity of the Raman scattering is so great that it is possible to obtain good spectra with less than a

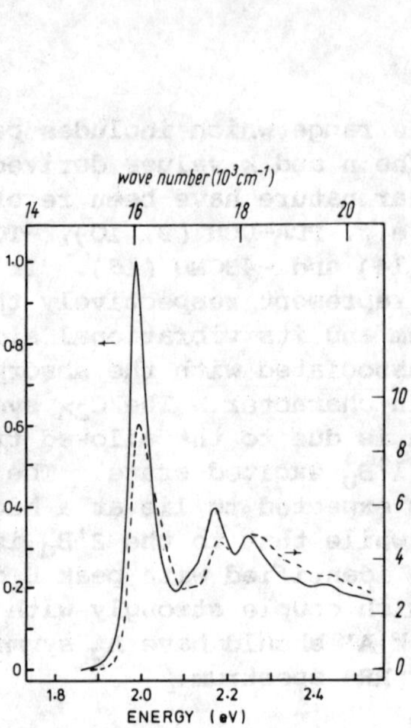

Fig. 5 Comparison of the relative optical absorption of a PDA-TS crystal obtained by transmission with E⊥b (solid line) and the absorption coefficient determined by Kramers-Kronig analysis of reflectivity data with E∥b (dashed line).

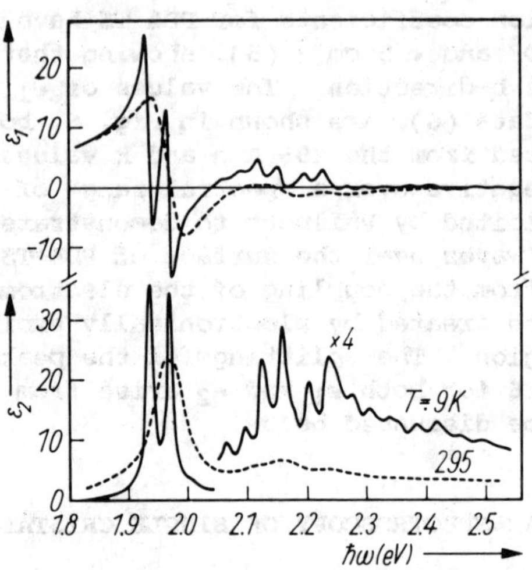

Fig. 6 Real and imaginary part of the dielectric constant $\epsilon_1 + i\epsilon_2$ at 9 K (full curve) and at room temperature (dashed curve).

milliwatt of incident power. In addition there are many fewer lines present in the spectra than would be expected from group theoretical considerations. These features are common to resonance Raman (RR) scattering in which the intensities of totally symmetric vibrational modes which are strongly coupled to the electronic excited state are greatly enhanced. For PDAs this means that the vibrational modes of the polymer backbone can be distinguished from those of the side-group in a way that is impossible in infrared spectroscopy. Thus RR scattering has been a very useful probe in studying the structure of PDAs in various morphologies. For the purposes of this paper it is more important that RR spectroscopy can also provide detailed information about the electronic excited states. The excitation profile (EP) is obtained by measuring the Raman cross section for a vibrational mode as a function of the incident laser photon energy. By application of a theoretical model involving the Frank-Condon overlap integrals it is possible to obtain from the EP of a vibrational mode an estimate of the strength of the interaction with the electronic excited state, z.

The Raman spectrum of a single crystal of PDA-TS at 295 K is shown in Fig. 7 (17). Although the unit cell of C_{2h}^5 symmetry contains 92 atoms and 66 totally symmetric A_g modes, only four lines and overtones and combinations of those modes are present in the spectrum. This is in agreement with the predictions of a planar model for the vibrational modes of the backbone for which the eigenvectors are shown in Fig. 8 (19). Mode 1 (2086 cm^{-1}) involves considerable stretching of the triple bond and mode 2 (1485 cm^{-1}) stretching of the double bond. Modes 3 (1203 cm^{-1}) and 4 (952 cm^{-1}) approximately describe rotation about those bonds. The EP for mode 1 is shown in Fig. 9 together with the theoretical curve; the EPs of the other three modes were similar in character. The relative heights of the two peaks gave the value of z while their separation gave the vibrational frequency of mode 1 in the electronic excited state. One difficulty in interpreting the measurements of the EP was the very large correction factors for absorption and reflectivity which had to be applied to the raw intensity data (20). The procedures which were followed have been recently verified, however, by the very similar results which have been obtained for Langmuir-Blodgett multilayer films of the cadmium salt of PDA-10/8AFA in which the correction factors were very small (19). A summary of the properties obtained for the four modes is given in Table 1. A calculation of the absorption coefficient using the values of z determined from RR spectroscopy was in very good agreement with that obtained by transmission through a thin crystal in Fig. 5.

The data from Table 1 were combined with the vibrational model illustrated in Fig. 8 to obtain information about the change in structure of the polymer backbone which occurs in the transition

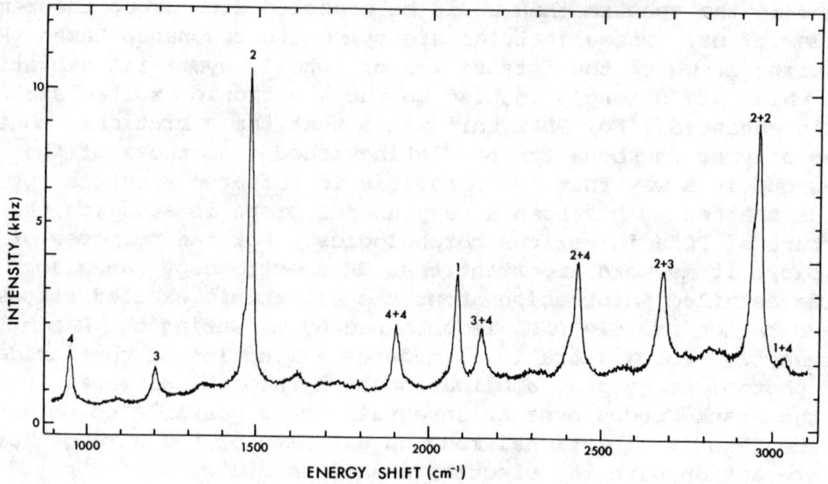

Fig. 7 Raman spectrum of a PDA-TS single crystal for an incident laser photon energy of 2.15 eV. The Raman bands corresponding to vibrational modes 1 to 4 and their overtones and combinations are labelled.

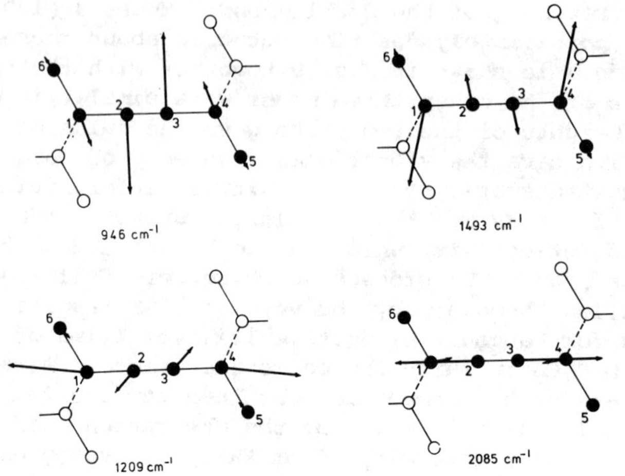

Fig. 8 Eigenvectors for the normal modes 1-4 calculated for PDA-TS.

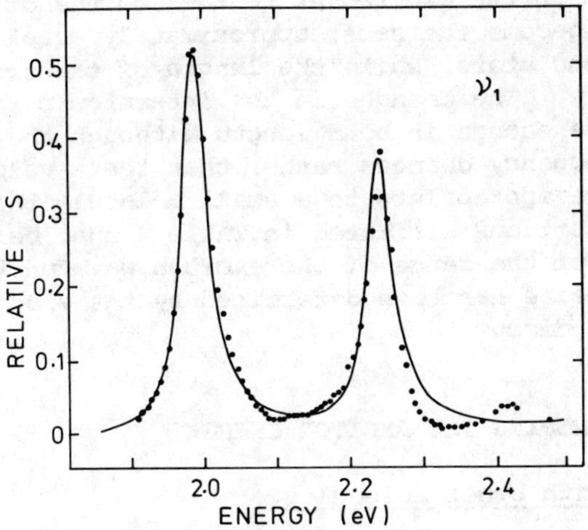

Fig. 9 Resonance Raman excitation profile for mode 1. The points represent the experimental data while the solid lines have been evaluated using the theoretical model.

Table 1 Electronic and vibrational properties of PDA-TS[a]

Mode	f/cm^{-1}	F/cm^{-1}	z	D/nm	(dE/dQ)/eV nm^{-1}
1	2086	2040	0.40	0.0021	40.0
2	1485	1420	0.55	0.0034	33.0
3	1203	1170	0.21	0.0014	9.2
4	952	940	0.31	0.0024	9.6

[a] Vibrational mode wavenumbers for the ground (f) and excited (F) states and the parameters (z, D and dE/dQ) describing the interaction between the 2.0 eV exciton and those modes are listed for each of the four strongly Raman-active modes. The reduced mass for all the vibrational modes has been taken as that for a carbon atom so that the displacement of the normal mode equilibrium position, D, is given in nm and the exciton-vibration coupling constant dE/dQ in eV/nm.

Table 2 Interatomic separations, r, and effective force constants, K, in the ground and excited states for PDA-TS[a]

Parameter	Ground state	Excited state
r_{12}/nm	0.143	0.140
r_{23}/nm	0.119	0.121
r_{14}/nm	0.136	0.140
K_1/N m^{-1}	450	474
K_2/N m^{-1}	550	481
K_3/N m^{-1}	1260	1172
K_4/N m^{-1}	23	19
K_5/N m^{-1}	55	56

[a] The subscripts refer to the atom positions in Fig. 1. The five independent force constants have been taken to be: for bond stretching, $K_1 = K_{12} = K_{34} = K_{16} = K_{45}$, $K_2 = K_{14}$ and $K_3 = K_{23}$, and for bond bending, $K_4 = K_{416} = K_{412} = K_{612} = K_{345} = K_{341} = K_{145}$ and $K_5 = K_{123} = K_{234}$.

from the ground to the excited state. The results summarised in Table 2 show that in the excited state the lengths of the single and double bonds become the same, approximately equal to their mean value in the ground state, while the length of the triple bond increases slightly. The changes in the interatomic force constants simply reflect the change in bond length although they were estimated from the frequency changes rather than the z values. The model used is more appropriate to a small molecule than an infinite chain so the distortions estimated in Table 2 must be taken as the average values over the range of the exciton wavefunction. An estimate of that size has been determined by the electroreflectance measurements discussed below.

4. FACTORS DETERMINING THE EXCITON ENERGY

4.1 Interaction with other molecules

The energy E of an exciton in a molecular crystal with two molecules per unit cell, as is the case for most PDAs for which x-ray structures exist, is given by

$$E = E_o + D + W + \pm L \qquad (2)$$

where E_o is the excitation energy of the free molecule and D is the difference in the van der Waals' interaction energy of a molecule in the ground state and the excited state (4, 21). W arises from dispersion forces due to two electron excitations; this is usually negative since it essentially gives the lowering of E due to the polarization of the medium. S and L arise from exchange of excitations between translationally equivalent and inequivalent molecules respectively. The sum D + W is known as the solvent shift as it is also present when the molecule is in solution. The term S is known as the exciton shift and 2L is the Davydov splitting.

For PDAs the interpretation of Eqn. 2 must be somewhat different than for the usual molecular crystal since in this case the molecules are effectively of infinite length. Thus E_o must include the effects of the translational symmetry along the polymer chain and describe an energy band in which E_o is a function of the exciton wavevector (4). The contribution of the W term could in principle be broken up into two parts, one arising from neigbouring polymer chains but with another from its own sidegroups as the polarisabilities of PDA sidegroups have a very wide variation. The sidegroups also directly affect E_o if their packing is such as to make a mechanical contribution to the equilibrium length of the repeat unit along the polymer chain. Thus different PDAs will not have identical values for E due to the differing contributions of the sidegroups to the polarisability and the equilibrium value of $\underline{b}$.

4.2 Effects of hydrostatic pressure

By far the largest term in Eqn. 2 for PDAs is W. As the backbone is usually at some distance from the atoms involved in the contact forces between neighbouring chains then D must be considerably smaller than in molecules like anthracene where it has been estimated to be less than 12 meV (21). The difference of the energies of the maximum in the absorption coefficients for $\underline{E} \parallel \underline{b}$ and $\underline{E} \perp \underline{b}$ should be just equal to 2L; the data in Fig. 5 and other results (5) suggest that the Davydov splitting in PDA-TS must be less than 5 meV. S is usually about the same size as L. The magnitude of W has been estimated by measuring the change in E with hydrostatic pressure, P, (22 - 25) and assuming that W is proportional to R^{-6}, where R is the mean separation between polymer chains. In this case

$$\frac{dE}{dP} = \frac{dE}{dP} - \frac{6W}{R}\frac{dR}{dP} \qquad (3)$$

where $\frac{1}{R}\frac{dR}{dP}$ is the mean linear compressibility between neighbouring chains. The chain repeat unit, $\underline{b}$, is independent of pressure (24, 26) so W will only depend upon the interchain separation, R. The magnitude of the first term was estimated to be 12 meV/GPa using a simple molecular orbital model for the electronic excited state (27). From the experimental result for dE/dP of -70 ± 2 meV/GPa (22, 25), the value of W was estimated to be -320 ± 40 meV. This is similar to the value of -230 meV found for anthracene (21). The magnitude of W is thus about 15% of the energy of the exciton in the single crystal; it will be a contributing factor in most processes which perturb the polymer backbone but aside from the effect of hydrostatic pressure it must usually play a minor role. For a PDA-TS chain a significant fraction of W must be contributed by the polarisabiity of its own sidegroups.

4.3 Effects of uniaxial strain

Studies of the optical reflection spectra of a PDA-TS single crystal fibre subjected to uniaxial stress parallel to the $\underline{b}$-direction showed that peak A in Fig. 1 shifted to higher energy (28). The energy of the peak increased linearly with strain by 37 meV/% up to the maximum 4% strain at which the fibre broke. This change in dimension is expected to affect W but the primary cause of the shift must have been the increase in the bond lengths along the backbone. The resonance integrals which determine the energy vary exponentially with distance while W only goes as the sixth power. This expectation was confirmed by the good agreement obtained with the prediciton for the shift made using a molecular orbital model (27) for the electronic excitation.

The strain dependence of the energy of peak A has also been observed for PDA chains in partially converted monomer crystals by optical absorption measurements done as a function of degree of polymerisation or of changing temperature. The repeat unit of the polymer chain is in each case determined primarily by that of the monomer unit cell since most chains will be embedded in a monomer matrix. The energy of peak A in the initial PDA-TS polymer formed in the monomer is higher by just the amount which would be predicted for the 3% increase in b̲ if the 37 meV/% value for the strain dependence is applied. Figure 10 shows the temperature dependence of the energy of peak A as observed by optical absorption in nine different monomer crystals where the conversion to polymer is less than 1% (29). Most of the observed data can be interpreted by assuming that the energy shifts are directly related to the thermal expansion of the monomer. Thus the 600 meV shift observed on cooling the lithium salt of 10/12-AFA from 380 K to 4 K is the result of a very large coefficient of thermal expansion which is typical for paraffinic materials. The fully polymerised crystals have a very small coefficient of thermal expansion in the chain direction which is reflected in a correspondingly small change in the exciton energy.

4.4 Modulation spectroscopy

The discussion above illustrates how the application of hydrostatic or uniaxial stress can shift the exciton energy and features such as peaks A and A' in the reflection spectrum. If such perturbations are applied periodically and the reflected light analysed using a phase sensitive detector than various spectral features which are not obvious in the direct reflection spectrum can be enhanced (30). The reflectance modulation is given by

$$\Delta R/R = \alpha(\epsilon_1, \epsilon_2)\Delta\epsilon_1 + \beta(\epsilon_1, \epsilon_2)\Delta\epsilon_2 \qquad (4)$$

where the coefficients α and β are known functions of the real and imaginary parts of the dielectric constant. The theories of the various effects, as distinguished by the choice of the modulation parameter - electric field, stress or temperature - predict specific line shapes for $\Delta\epsilon_1$ and $\Delta\epsilon_2$.

The piezomodulated spectrum of a single crystal of PDA-TS was measured with the applied strain in the polymer chain direction (8). As discussed above the primary effect of such a uniaxial strain is to shift the reflection spectrum linearly to higher energy (28). Thus it might be expected that the piezomodulated spectrum should approximate a simple derivative of the reflection spectrum. Figure 11 shows that this is the case. Similar results were found for PDA-TCDU (11).

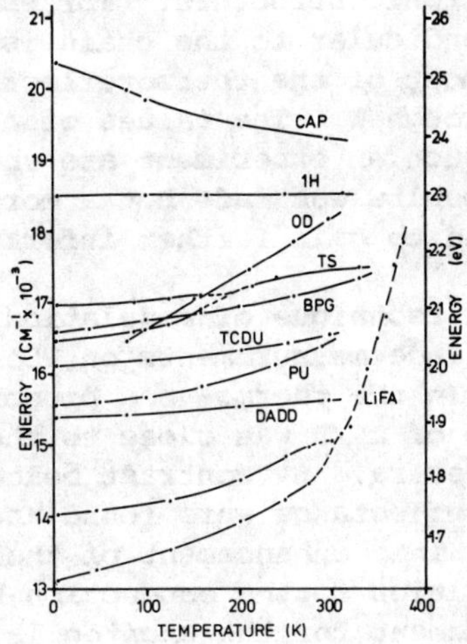

Fig. 10 Polymer chain absorption energy for a number of partially polymerized diacetylenes over the temperature range 2 to 400 K.

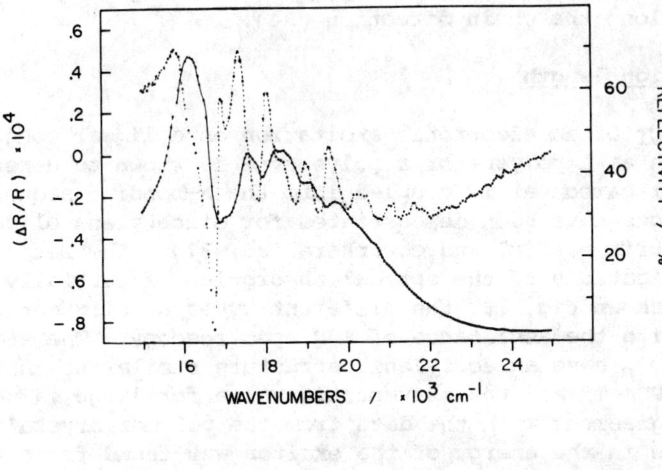

Fig. 11 Direct (solid line) and piezomodulated (dashed line) reflection spectra of PTS at 300 K. Bandpass is 30 cm^{-1}

Thermoreflectance is more complex to analyse as the primary effect of the heat pulse is to cause thermal expansion which in turn perturbs the electronic structure. For PDA-TS crystals the thermal expansion perpendicular to the chain is much greater than that parallel (31) so most of the thermoreflectance signal would be expected to come through W. The values of $\Delta\epsilon_1$ and $\Delta\epsilon_2$ which have been obtained in such an experiment are shown in Fig. 12 (32). Some analysis of the results was made but a more detailed microscopic model is required to gain further information.

The most successful technique of modulation spectroscopy has been the electroreflectance measurements on PDA-DCH and -TS single crystals (9, 33, 34). In the thermo- and piezoreflectance measurements the maximum value of $\Delta R/R$ was close to the location of peak A in the reflectance spectra. By contrast below 10 K the strongest signals in the electroreflectance were found about 0.4 eV higher in energy than peak A. This enhancement of the signal has been identified with a transition to the conduction band of the polymer. The transition dipole moment for the exciton is much greater than that for the band-to-band transition and no evidence for the latter has been found in optical or RR spectroscopy. At very low temperature, however, the perturbation of the applied electric field can exceed the lifetime broadening and produce a very strong electroreflectance signal at the band edge (34). Previously it had only been possible to identify the band edge by photoconduction measurements. The electroreflectance signal for the exciton is believed to arise from its charge transfer character; it costs more energy to send an electron in the direction of the field than the opposite. Application of such a model to the results yielded evidence that the electron in the excited state is spread out through two or three unit cells along the chain direction (34).

4.5 Conjugation length

The energy of an electronic excitation on a linear conjugated molecule such as a polyene or a polyeneyne is known to depend upon the number of carbon atoms coupled into the π-bonding sequence (35). Similar effects have been demonstrated for diacetylene oligomers in a monomer matrix by Sixl and coworkers (36, 37). The results of their investigations of the optical absorption of partially polymerised TS are shown Fig. 13; the different types of oligomer were identified with the assistance of ESR spectroscopy. The stable oligomers $(SO)_n$ have an acetylenic structure similar to that of the backbone of PDA-TS and the assymptotic value for large values of n is in good agreement with the data from the polymer crystals. Little change in the energy of the exciton was found for n values greater than 10. The diradical oligomers $(DR)_n$ have the butatriene structure in which the exciton has a lower energy than in the acetylenic bonding sequence; these were not found to be stable beyond

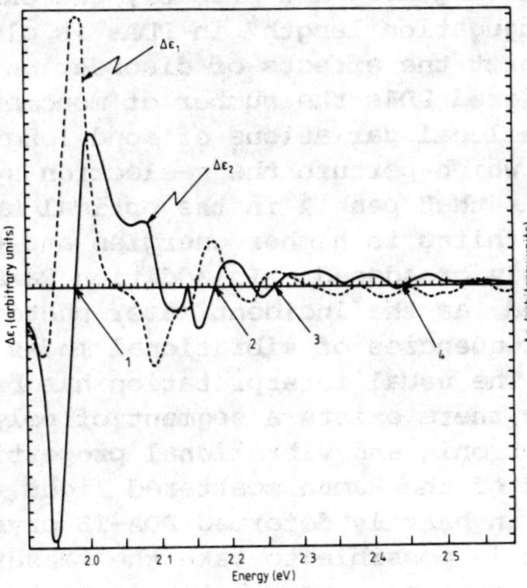

Fig. 12 Temperature modulation of ϵ_1 and ϵ_2 which corresponds to to the thermoreflectance data for PDA-TS at room temperature.

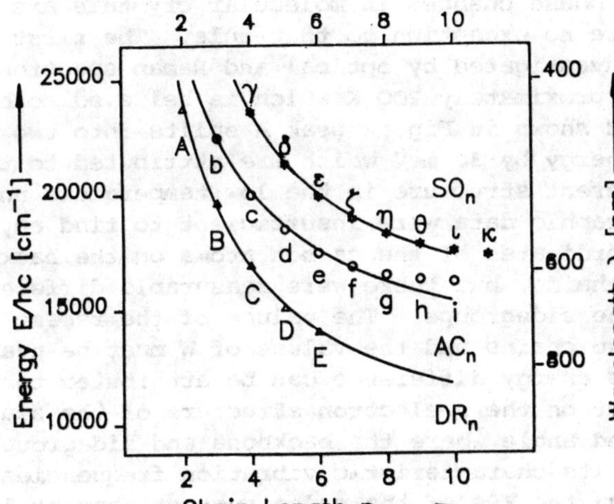

Fig. 13 Absorption energy of diradicals (DR), asymmetric carbenes (AC) and stable oligomers (SO) as a function of chain length in partially polymerized PDA-TS.

n = 6.

The short chain oligomers are probably the only case for which the concept of "conjugation length" in PDAs is clear but it has also been used to interpret the effects of disorder on the optical and RR spectra. In disordered PDAs the number of monomer units is very large but there are local variations of bond length and angle or backbone planarity which perturb the π-electron system on the backbone. The result is that peak A in the optical absorption and reflection spectra shifts to higher energies and the spectral features are considerably broadened. In addition Raman photoselection is commonly observed; as the incident laser photon energy is increased so the frequencies of vibrational modes measured in the RR spectra increase. The usual interpretation has been that for each laser photon energy there exists a segment of polymer chain with its characteristic electronic and vibrational properties which will give maximum enhancement of the Raman scattered light. These effects have been observed in heavily deformed PDA-TS crystals (38) and PDA solutions (39). It is possible to take the measured energy for the peak of optical absorption, compare it to a plot of the exciton energy as a function of n for PDA oligomers such as Fig. 13, and define an effective conjugation length. This can be a qualitative assessment at best since it neglects the other factors such as local variations in stress or density which could also affect the energy.

4.6 Crystallographic phase changes

Structural phase changes in molecular crystals are very common and the PDAs are no exception to that rule. The first phase change to be investigated by optical and Raman spectroscopy was that of PDA-TS at approximately 200 K which is believed to be second order (40). As shown in Fig. 6 peak A splits into two components separated in energy by 32 meV which are attributed to the two chains of different structure in the low temperature unit cell (41). The crystallographic data were insufficient to find any differences between the coordinates of the carbon atoms on the backbone of the two different chains, but there were measurable differences in the positions of the sidegroups. The values of the repeat unit are the same for the two chains and the values of W must be nearly identical. Thus this energy difference can be attributed to the peturbing effect on the π-electron structure of the small difference in the bond angle where the backbone and sidegroup are joined. Each chain has its characteristic vibration frequencies as shown by RR spectroscopy; the EPs of the modes suggest that at least in partially polymerised crystals there is very little sign of any interaction between the two types (42).

There are numerous first order phase transitions in PDAs which have been described as thermochromic since the colour of the

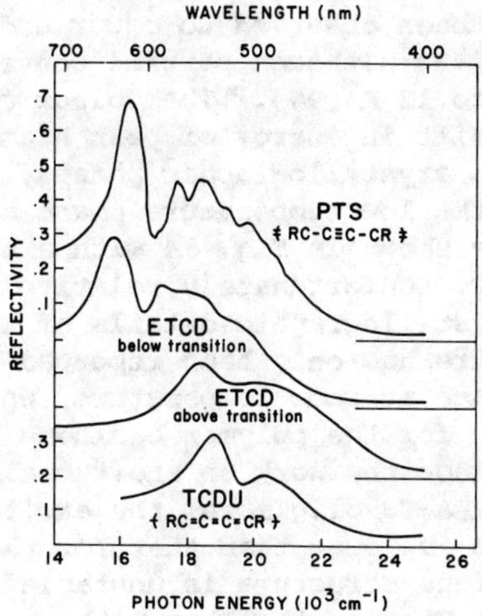

Fig. 14 Normal incidence reflection spectra for PTS, TCDU, ETCD below the phase transition, and ETCD above the phase transition.

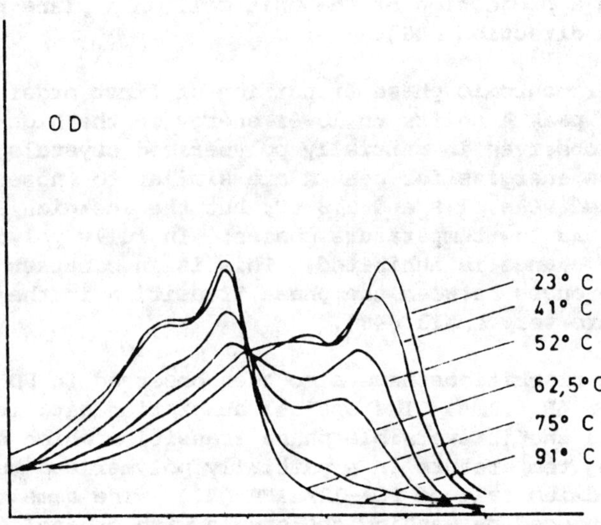

Fig. 15 Transmission spectra of a partially polymerized multilayer of the cadmium salt of the 12-8 acid at various temperatures.

crystal changes at the transition. Single crystals with urethane substitutents in the sidegroup often appear in this class of PDAs, e.g. PDA-ETCD (12), -IUPDO (13) and -4BCMU (15). The PDA-TCDU transition has only been observed to occur under the influence of compressive stress (43) although partial conversion was observed in crystals cooled to 12 K (44). The colour change in turn has been related to a shift in energy of peak A in the reflection spectrum between the two crystallographic phases. Typically peak A is at about 1.9 eV in the low temperature phase and 2.3 eV in the high temperature phase as shown in Fig. 14 with a transition temperature of about 110 - 120 C. Unfortunately relatively little information exists about the crystallographic details of the transitions. A full crystal structure has only been reported for PDA-TCDU, in its high temperature phase at room temperature, which reported a near butatriene structure for the polymer backbone (45). This is an unexpected result since the work on short chain oligimers shows that, at least for PDA-TS oligimers, the exciton energy for the butatriene structure is lower than that for the acetylenic structure and the butatriene structure is unstable for more than six monomer units (36). Most of the PDAs with urethane substituents have maximum reflectivities in the 40 - 50% range as compared to PDA-TS and -DCH near 70%. This lower value of the reflectivity could be due to poorer crystal quality. Disorder in the chain direction has prevented the determination of a complete crystal structure for another urethane derivative, PDA-PUDD; it was only possible to obtain a projection of the unit cell on a plane perpendicular to the chain direction (46).

A thermochromic phase transition of first order in which the energy of peak A shifts to lower energy in the high temperature phase is observed in partially polymerised crystals of PDA-DCH (47). The energies for peak A are similar to those for the urethane substituted PDAs, 1.8 and 2.6 eV, but the energies are reversed in the high and low temperature phases. In fully polymerised PDA-DCH the phase change is inhibited. This is in contrast to PDA-TS where the differences between the phase transition in the polymer and monomer are very small (48).

Phase transitions have also been observed in PDA mono- and multilayer LB films. The optical absorption data in Fig. 15 illustrate the irreversible phase transition which occurs with increasing temperature in a partially polymerised multilayer sample of the cadmium salt of PDA-12/BAFA (49). The same transition can also be induced by washing the sample with ethanol (50), or exposing the sample for longer times in the UV source used to effect polymerisation (51). The initial polymer forms with peak A at approximately 1.9 eV while in the final phase the peak has shifted to 2.3 eV. The spectra for both phases are characteristic of partially disordered crystalline material. Sufficient electron diffraction data exist to show that the structure in the sidegroups is different in the two phases. Pure paraffin crystals often have several phase transitions prior to melting so such behaviour is not unexpected in a PDA with paraffinic sidegroups. It is interesting to note that the initial polymer formed in multilayers of the pure acid of PDA-16/2AFA has peak A at 2.3 eV while a gradual shift to the phase with the lower energy value occurs with increasing polymerisation (52). Heating then returns the material to the original phase, however.

The structure of a PDA is determined by the balance between the strong interatomic forces along the backbone and the weaker but considerably more numerous interactions between atoms in adjacent sidegroups. In the crystal the energy of the exciton is a function of the chain repeat unit, the planarity of the backbone, the bond angles where the sidegroups are attached and the magnitude of W which is largely determined by the polarisability of the sidegroups and their packing. The optical absorption spectra of partially polymerised crystals in Fig. 10 suggest that there is a continuous range of energies available to the exciton on a PDA chain from nearly 1.6 to 2.6 eV. The exciton energy for any particular PDA is just determined by the structure which in turn is a result of the interplay between the interatomic forces of the backbone and sidegroups. It is possible that an electronic phase transition of the backbone could trigger a structural phase transition in a crystal where these forces are delicately balanced. Most if not all phase transitions in PDA crystals, however, do not require such an explanation.

4.7 Spectroscopy of PDA solutions

The analysis of the structure of PDA molecules in solution has presented interesting problems (53). Optical and RR spectroscopy have contributed to progress in the area but by themselves are unable to provide any definitive answers. Typical absorption spectra for PDA-TCDU in DMF/methanol solution are shown in Fig. 16 (54). For volume fractions of DMF below 70% the shape of the absorption spectrum is remarkably similar to that of the single crystal except that peak A is 110 meV higher in energy. This suggests that the structure of the polymer in the methanol-rich solution must be very similar to that in the crystal. The shift in the exciton energy is so small that it could arise from changes in W or minor rearrangements of the packing of the sidegroups. It is clear that the backbone must be straight and essentially planar over at least 10 repeat units. The optical spectra are unable to determine if the polymer exists as an isolated chain, a cluster of chains or as a chain folded lamella. It would be somewhat surprising, however, if the vibrational sidebands of the exciton could be retained in the detail shown in Fig. 16 for a completely isolated

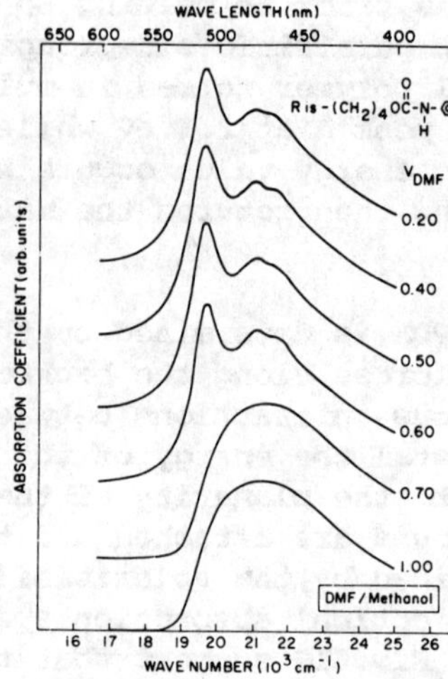

Fig. 16 Optical absorption spectra for polyTCDU solutions. The volume fraction of dimethylformamide (DMF) in methanol is indicated. Polymer concentrations (~10^{-5} mol/L) are the same for all spectra. The spectra are arbitrarily offset.

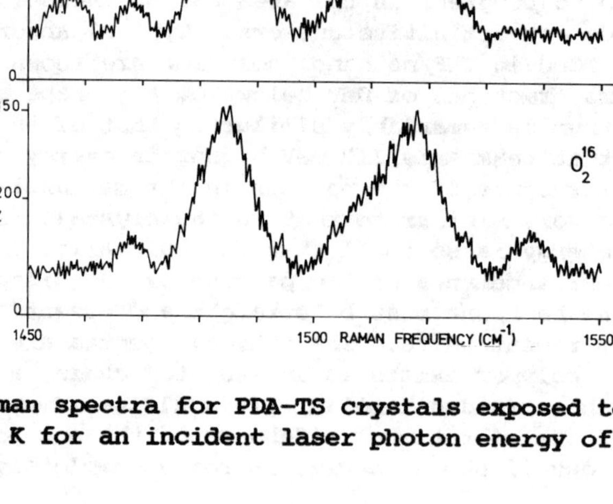

Fig. 17 Raman spectra for PDA-TS crystals exposed to $^{16}O_2$ and $^{18}O_2$ at 295 K for an incident laser photon energy of 2.60 eV.

chain. In the DMF-rich solution the vibrational sideband structure in the absorption spectrum is lost and replaced by a relatively featureless peak. This broadening could be either homogeneous or inhomogeneous, that is it could arise from shortened exciton lifetimes or perturbations of the shape of the polymer backbone; it is most likely to be a combination of both (55). Raman photoselection has provided evidence of the inhomogeneous contribution (39). In either case it is clear that the structure of the polymer in the DMF rich solution has considerably greater disorder. The absorption spectrum is not inconsistent with that expected for a wormlike coil (53).

5. RESONANCE RAMAN SPECTROSCOPY OF POLYDIACETYELENES WITH SURFACE-ADSORBED MOLECULES

Resonance Raman spectroscopy of PDAs with surface adsorbed molecules has provided evidence for electronic transitions which thus far have not been observed in optical spectra (56, 57). Figure 17 shows the RR spectra in the vicinity of 1500 cm^{-1} for PDA-TS crystals with $^{16}O_2$ and $^{18}O_2$ adsorbed on the surface. For crystals prepared in an O_2-free environment it is possible to adsorb and desorb the gases reversibly. In each spectrum in Fig. 17 there are two Raman lines which are not present in the spectra of O_2-free crystals such as shown in Fig. 7. One line at 1517 cm^{-1} is present for both molecular isotopes and so must be associated with a vibrational mode of the polymer backbone. The other has a value of 1455 cm^{-1} for adsorbed $^{18}O_2$ and 1538 cm^{-1} for $^{16}O_2$. As these frequencies are approximately in the ratio of the square root of the isotopic mass ratio it has been suggested that this Raman line must correspond to the vibrations of the O_2 molecule. In vacuum the vibrational frequency of $^{16}O_2$ is 1555 cm^{-1} and a small frequency decrease on adsorption would be expected (58).

The EP for the 1517 cm^{-1} line is shown in Fig. 18; there is a dramatic increase in the Raman cross section of this line relative to that of the intrinsic vibrational modes of the polymer backbone with decreasing temperature. The EP has been interpreted as arising from an electronic transition localised on the backbone which has an energy of 2.39 eV. It is possible that this is the normally forbidden $1'A_g^- \rightarrow 2'A_g^-$ transition (16) which has been made allowed by the symmetry breaking properties of the adsorbed O_2. Additional evidence for this assignment comes from two photon absorption measurements at 2.34 eV (58). In this case the 1517 cm^{-1} Raman line is likely to have B_u symmetry as opposed to the usual A_g symmetry imposed by RR spectroscopy. This interpretation is consistent with the predictions of the vibrational model (19). Alternatively the 2.39 eV electronic excitation and the 1517 cm^{-1} vibrational mode may be intrinsic properties but characteristic of

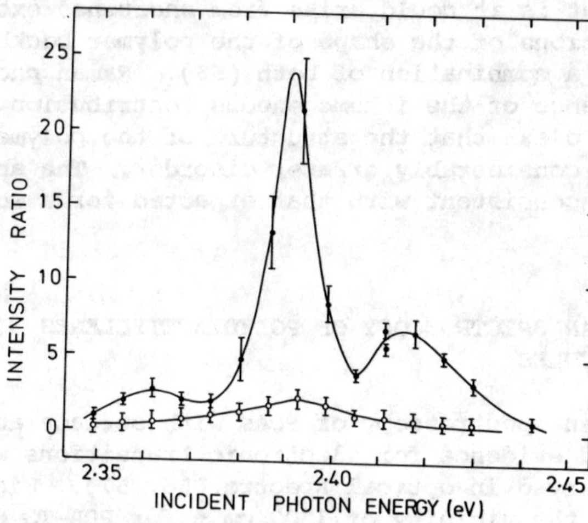

Fig. 18 Excitation profiles of the 1517 cm^{-1} Raman line at 295 K (open circles) and 4 K (full circles).

a polymer chain in which the structure has been perturbed by the adsorbed O_2. A similar Raman line has been observed for O_2 adsorbed on PDA-PUDD (56) but none has been observed on PDA-DCH. This is not unexpected as the sidegroups pack very tightly about the backbone in PDA-DCH and may exclude O_2 from it. It is significant that the surfaces of PDA-TS and -PUDD crystal experience photooxidation if simultaneously exposed to air and light while those of PDA-DCH do not.

The EPs for the 1455 cm^{-1} line of adsorbed $^{18}O_2$ and the 1538 cm^{-1} line of adsorbed $^{16}O_2$ were found to be essentially the same with a peak at 2.60 eV. Observation of the Raman line of O_2 suggests that a charge transfer excitation must occur with an electron moving from the PDA-TS backbone acting as donor and the O_2 acting as acceptor. This result is consistent with the results of optical absorption spectroscopy of O_2 adsorbed on a wide range of hydrocarbons in solution (59, 60). Similar effects were observed for I_2 adsorbed on PDA-TS crystal surfaces; the EP suggested that there was a charge transfer excitation at 2.57 eV. This again would be consistent with the results found in studies of I_2 adsorbed on a variety of hydrocarbons in solution (59).

6. CONCLUSIONS

Optical and RR spectroscopy have assisted in the identification and characterisation of the electronic excited states of the PDA backbone. The same techniques have been shown to be an invaluable tool in the probing of the structure of the PDA backbone in a wide variety of morphologies.

1. Bloor, D, in "Photon, Electron, and Ion Probes of Polymer Structure and Properties" ed D.W. Dwight, T.J. Fabish and H.R. Thomas, ACS Symposium 162 (American Chemical Society, Washington DC, 1981) p.81.

2. Batchelder, D.N. and D. Bloor, in "Advances in Infrared and Raman Spectroscopy Vol 11" ed R.J.H. Clark and R.E. Hester (Wiley Heydon, Chichester, 1983) p.133.

3. Tokura, Y., T. Mitani and T. Koda, Chem.Phys.Lett. 75, 324 (1980).

4. Philpott, M.R., Ann.Rev.Phys.Chem. 31, 97 (1980).

5. Bloor, D. and F.H. Preston, Phys.Stat.Sol. (a) 37, 427 (1976).

6. Reimer, B., H. Bässler, J. Hess and G. Weiser, Phys.Stat.Sol. (b) 73, 709 (1976).

7. Philpott, M.R., in "Topics in Surface Chemistry" ed. E. Kay and P.S. Bagus (Plenum Press, New York, 1978) p.329.

8. Eckhardt, C.J., H. Muller, J. Tylicki and R.R. Chance, J.Chem.Phys. 65, 4311 (1976).

9. Sebastian, L. and Weiser, G, Phys.Rev.Lett. 46, 1156 (1981).

10. Hood, R.J., Muller, H., C.J. Eckhardt, R.R. Chance and K.C. Yee, Chem.Phys.Lett. 54, 295 (1978).

11. Muller, H., C.J. Eckhardt, R.R. Chance and R.H. Baughman, Chem.Phys.Lett. 50, 22 (1977).

12. Chance, R.R., R.H. Baughman, H. Muller and C.J. Eckhardt, J.Chem.Phys. 67, 3616 (1977).

13. Eckhardt, H., C.J. Eckhardt and K.C. Yee, J.Chem.Phys. 70, 5498, (1979).

14. Chance, R.R., K.C. Yee, R.H. Baughman, H. Eckhardt, C.J. Eckhardt, J.Polymer Sci. Polymer Phys.Ed 18, 1651 (1980).

15. Chance, R.R., G.N. Patel and J.D. Witt, J.Chem.Phys. 71, 206 (1982).

16. Dinur, U. and M. Karplus, Chem.Phys.Lett. 88, 171 (1982).

17. Batchelder, D.N. and D. Bloor, J.Phys.C. 15, 3005 (1982).

18. Brillante, A., I. Pockrand, M.R. Philpott and J.D. Swalen, Chem.Phys.Lett. 57, 395 (1978).

19. Lyall, I.R.J. and D.N. Batchelder, to be published.

20. Lewis, W.F. and D.N. Batchelder, Chem.Phys.Lett. 60, 232 (1979).

21. Rice, S.A. and J. Jortner, in "Physics of Solids at High Pressure" ed. C.T. Tomizuki and R.M. Emrich (Academic Press, New York, 1965) p.63.

22. Cottle, A.C., W.F. Lewis and D.N. Batchelder, J.Phys.C. 11, 605 (1978).

23. Jankowiak, R., J. Kalinowski, B. Reimer and H. Bassler, Chem. Phys.Lett. 54, 483 (1978).

24. Lochner, K., H. Bässler, H. Sowa and H. Ahsbahs, Chem.Phys. 52, 179 (1980).

25. Lacey, R.J., D.N. Batchelder and G.D. Pitt, J.Phys.C. 17, 4529 (1984).

26. Batchelder, D.N., J.Polymer Sci. Polymer Phys.Ed. 14, 1235 (1976).

27. Cojan C., G.P. Agrawal and C. Flytzanis, Phys.Rev.B. 15, 909 (1977).

28. Batchelder, D.N. and D. Bloor, J.Phys.C. 11, L629 (1978).

29. Bloor, D. and C.L. Hubble, Chem.Phys.Lett. 56, 89 (1978).

30. Seraphin, B.O., in "Optical Properties of Solids" ed. F. Abeles (North-Holland, Amsterdam, 1972) p.163.

31. Reimer, B., H. Bässler and T. Debaerdemaeker, Chem.Phys.Lett. 43, 85 (1976).

32. Stizza, S., I. Davoli, V. Prantera, A. Bianconi, J. Glinski and J. Kalinowski, J.Phys.C. 16, 2165 (1983).

33. Sebastian, L. and G. Weiser, Chem.Phys.Lett. 64, 396 (1979).

34. Sebastian, L. and G. Weiser, Chem.Phys. 62, 447 (1981).

35. Baughman, R.H. and R.R. Chance, J.Polymer Sci.Polymer Phys.Ed. 14, 2037 (1976).

36. Gross, H. and H. Sixl, Chem.Phys.Lett. 91, 262 (1982).

37. Siegel, D., H. Sixl, V. Enkelmann and G. Wenz, Chem.Phys. 72, 201 (1982).

38. Batchelder, D.N., R.J. Kennedy, D. Bloor and R.J. Young, J.Polymer Sci.Polymer Phys.Ed. 19, 677 (1981).

39. Shand, M.L., R.R. Chance, M. LePostollec and M. Schott, Phys.Rev.B. 25, 4431 (1982).

40. Bloor, D., D.A. Fisher, D.N. Batchelder, R. Kennedy, A.C. Cottle and W.F. Lewis, Mol.Cryst.Liq.Cryst. 52, 83 (1979).

41. Enkelmann, V., Acta.Cryst. B33, 2842 (1977).

42. Batchelder, D.N. and D. Bloor, Chem.Phys.Lett. 38, 37 (1976).

43. Müller, H. and C.J. Eckhardt, Mol.Cryst.Liq.Cryst. 45, 313 (1978).

44. Iqbal, Z., R.R. Chance and R.H. Baughman, J.Chem.Phys. 66, 5520 (1977).

45. Enkelman, V. and J.B. Lando, Acta.Cryst. B34, 2352 (1978).

46. Fisher, D.A., Ph.D. Thesis, University of London (1978).

47. Kennedy, R.J., I.F. Chalmers and D. Bloor, Makromol.Chem. Rapid Commun. 1, 357 (1980).

48. Patillon, J.N., P. Robin, P.A. Albouy, J.P. Pouget and R. Commes, Mol.Cryst.Liq.Cryst. 76, 297 (1981).

49. Lieser, G., B. Tieke and G. Wegner, Thin Solid Films 68, 77 (1980).

50. Tieke, B., G. Lieser and G. Wegner, J.Polymer Sci.Polymer Chem. Ed. 17, 1631 (1979).

51. Tieke, B., V. Enkelmann, H. Kapp, G. Lieser and G. Wegner, J.Makromol.Sci. (Chem) A15, 1045 (1981).

52. Tieke, B. and G. Lieser, J.Coll.Interf.Sci. 83, 230 (1982).

53. Wenz, G., M.A. Muller, M. Schmidt and G. Wegner, Macromolecules 17, 837 (1984).

54. Chance, R.R., Macromolecules 13, 396 (1980).

55. Bloor, D., D.N. Batchelder, D.J. Ando, R.T. Read and R.J. Young, J.Polymer Sci. Polymer Phys. Ed. 19, 321 (1981).

56. Batchelder, D.N., N.J. Poole and D. Bloor, Chem.Phys.Lett. 81, 560 (1981).

57. Poole, N.J. and D.N. Batchelder, Mol.Cryst.Liq.Cryst. 105, 55 (1984).

58. Reimer, B. and H. Bässler, Chem.Phys.Lett. 55, 315 (1978).

59. Mulliken, R.S., "Molecular Complexes" (Wiley, New York, 1969).

60. Chien, J.C.W., J.Phys.Chem. 69, 4317 (1965).

ELECTRIC-FIELD SENSITIVE DEFECT STATES IN FULLY AND PARTIALLY POLYMERIZED PTS

G. Weiser and L. Sebastian

Fachbereich Physik, University of Marburg, F.R. Germany

Polydiacetylene crystals with toluene sulfonate side groups (PTS) show a defect absorption which contributes to photoconductivity and gives rise to a strong and narrow electroabsorption peak near 1.75 eV for electric fields and polarization of light parallel b. This defect absorption is not related to oligomers, and only part of it responds to an electric field. It is assumed that the field sensitive defect absorption arises from optical excitation of electrons, localized on the defect, into free electron states.

1 INTRODUCTION

Electroabsorption and electroreflectance have provided a sensitive and selective tool to study delocalized electron states in semiconductors (1). Singularities of the band structure, like band extrema respond particularly sensitive to electric fields and the corresponding optical transitions are clearly resolved even in case of low oscillator strength. In case of localized states, however, the external fields in the range of 10 kV/cm cannot compete with the much larger local fields and the field induced perturbation of optical transitions becomes unresolvable. This limitation qualifies electromodulated spectroscopy as a selective tool to study delocalized states in organic crystals.

In polydiacetylene single crystals electroreflectance revealed the weak interband transitions, hidden under the stronger vibronic excitons (2). The data yield the band gap, the exciton binding energy and derive in case of DCHD a surprisingly small effective mass $m^* = 0.05\ m_o$ (3). Charge transfer excitons too respond to an external field. Owing to their large dipole moment these transitions shift in an electric field which gives rise to field induced changes of the optical spectra. Such study has been performed on the CT exciton in PMDA:anthracene (4). The method revealed considerable charge transfer contributions to the excitons in polydiacetylenes (5) and resolved the weak charge transfer transitions in anthracene (6).

PTS single crystals show below the π-π^* excitations of the polymer backbone weak absorption which is probably related to defects (7). Different from absorption tails in other PDA crystals excitation of PTS in this range contributes to photoconductivity (8) and gives rise to an electroabsorption signal, whose evolution during the polymerization process has been studied.

2 Experimental Results

2.1 Photopolymerization at Low Temperature

Photoproducts like long oligomers (9) seem a possible source for defect absorption. They were prepared at T = 10 K in monomer crystals by irradiation for 1 h with UV light, λ = 300 nm and followed by illumination with λ = 367 nm (10). Fig. 1 shows the resulting absorption spectra in the region below the π-π^* excitations of the polymer. The transitions are polarized parallel to the polymer backbone and they respond to an electric field parallel to this direction. Field F|b did not cause any measurable changes. The electroabsorption signal $\Delta\alpha$ rises quadratically with the field-strength and

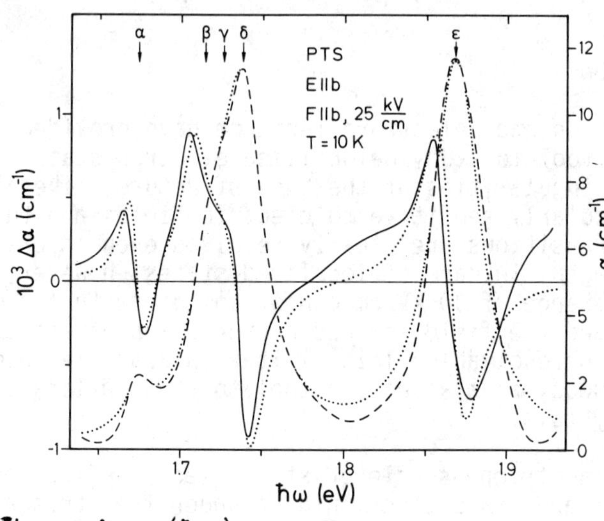

FIGURE 1: Absorption $\alpha(\hbar\omega)$ and electroabsorption $\Delta\alpha$ of photoproducts in PTS-monomer. Full curve: experimental spectra, dashed curves: fit by Lorentzians and a charge transfer model for $\Delta\alpha$.

Transition	$\hbar\omega_o$/eV	γ/meV	α_{max}/cm^{-1}	$m/10^{-9}$ecm	δ/%
α	1.675	17	1.3	3.1	4.8
β	1.715	25	2.6	3.1	3.8
γ	1.726	25	2.6	3.1	3.8
δ	1.739	30	9.2	3.7	2.0
ε	1.868	35	11.5	3.8	2.0

TABLE 1: Transition energy $\hbar\omega_o$, linewidth γ and peak amplitude α_{max} of the photoproducts in PTS as obtained from an approximation by Lorentzians. The charge transfer dipoles m and δ the preference for charge transfer along the field are obtained from the electroabsorption spectra.

the relative changes $\Delta\alpha/\alpha$ are of the same order as observed for the excitons in PTS. Five transitions α- ε are observed, the strongest related to oligomers of 7 (β), 6 (δ) and 5 (ε) monomer units (11). Transition α may correspond to an oligomer of 8 units. Transition energies, oscillator-strength and line width are summarized in Tab. 1. The field induced changes $\Delta\alpha$ are analyzed with the CT model applied before the excitons of the polymer (5):

$$\Delta\alpha = \frac{d\alpha}{d(\hbar\omega)} \delta(\Delta E) + \frac{1}{2} \frac{d^2\alpha}{d(\hbar\omega)^2} (\Delta E)^2 \qquad (1)$$

where $\Delta E = qrF$ refers to the energy shift of the charge transfer dipole qr in the electric field F and δ to the preference for charge transfer parallel to the field. Whereas for the excitons in the polymer the second derivative of α prevails, the spectra of the oligomers are close to a first derivative partly because of the largerlinewidth. The charge transfer moments are about 2/3 of the value obtained for the excitons in PTS. The first derivative shape of may suggest a quadratic Stark shift of the transitions by

$$\Delta E = \frac{1}{2} p F^2 \qquad (2)$$

This interpretation leads to a huge polarizability $p \approx 1.5\ 10^{-21} cm^3$ along the axis of the molecules. It is likely that the oligomers show large polarizabilities and that this contributes to the first derivative part of $\Delta\alpha$.

2.2 Defects in Thermally Polymerized Samples

Thermally polymerized samples have a defect absorption for polarization $E \parallel b$ which is quite different from the absorption of the photo-

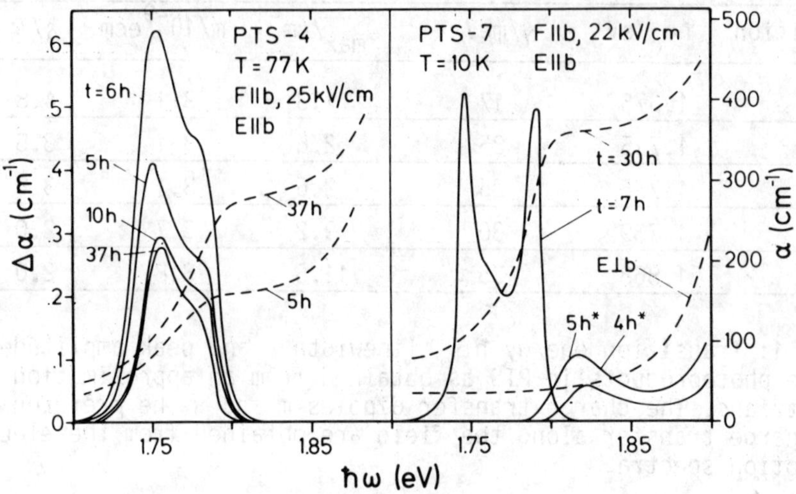

FIGURE 2: Defect absorption (dashed curves) and electroabsorption spectra (full curves) at different annealing times in two different samples. Annealing temperature T = 80°C.

products. A weak absorption band rises near 1.75 eV to a bump at 1.8 eV, about 150 meV below the π-π^* excitons. The threshold of this absorption band is quite sensitive to an electric field F ∥ b, giving rise to the electroabsorption spectra $\Delta\alpha$(Fig.2). No response was found for field F or polarization of light E perpendicular to b. Whereas the defect absorption band is similar in all samples, the electroabsorption spectrum $\Delta\alpha$ shows some pecularities. A few samples like PTS-7 show like the excitons in PTS a doublet structure with a splitting of 42 meV at 10 K slightly larger than that of the excitons (13). In most samples the doublet is reduced to a peak at 1.75 eV and a shoulder on the high energy side. The electric field enhances the absorption. There is no sign reversal of $\Delta\alpha$ like in the spectra of the excitons or of the photoproducts which excludes energy shifts or density of states effects as explanation of these spectra. It is obvious that $\Delta\alpha$ is not a derivative of the absorption band.

Both defect absorption and electroabsorption grow during the thermal polymerization process though differently. The absorption band saturates after about 6 h at 80°C. The electroabsorption peak increases rapidly when the autocatalytic range is reached, together with a red shift by about 70 meV similar as the excitons (Fig. 2, PTS-7). As shown in Fig. 3 the growth of $\Delta\alpha$ follows closely the polymerization curve (14). When most of the material is polymerized, however, $\Delta\alpha$ decreases again to about half its peak value quite in contrast to the absorption spectrum.

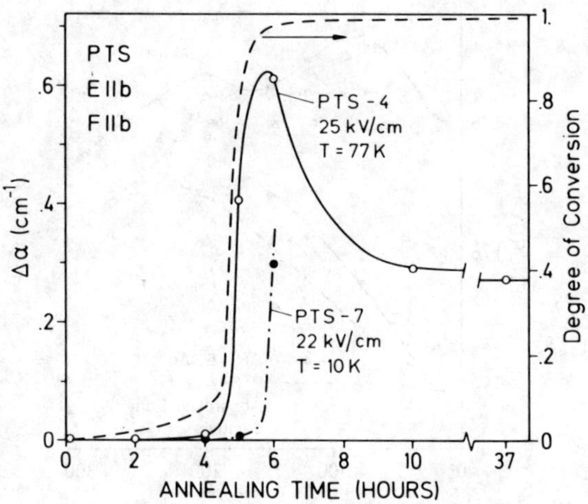

FIGURE 3: Growth of the electroabsorption peak at 1.75 eV with annealing time in two different samples in comparison with the polymer conversion curve (ref. 14). Annealing temperature T = 80°C.

A phase transition in PTS at low temperatures (15) causes splitting of the π-π^* transitions. Similar splitting occurs for the defect electroabsorption peak as shown in Fig. 4. At higher temperature down to 150 K the electroabsorption spectrum shows a broad, slightly asymmetric peak which shifts to higher energy as the temperature decrea-

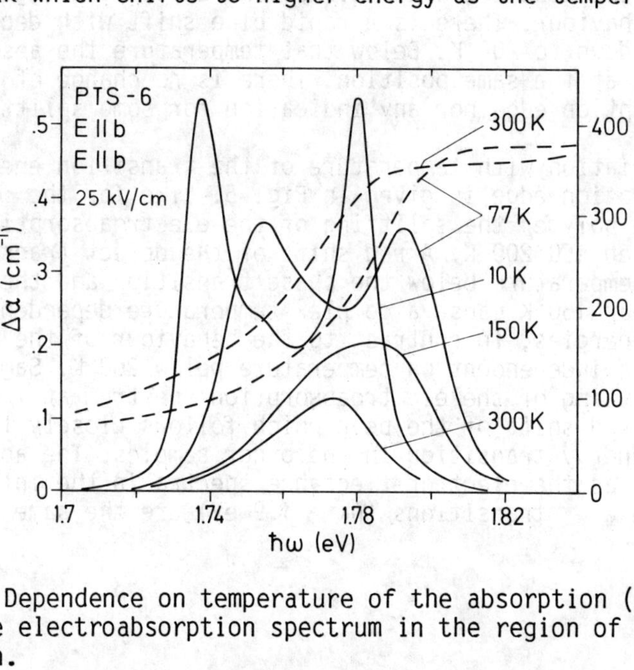

FIGURE 4: Dependence on temperature of the absorption (dashed curve) and of the electroabsorption spectrum in the region of defect absorption.

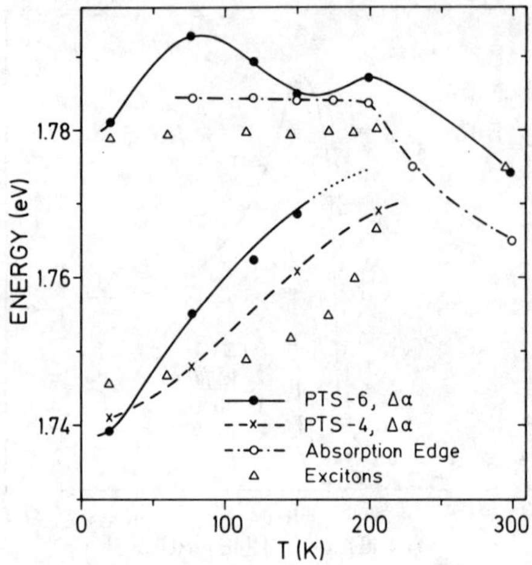

FIGURE 5: Temperature shift of the electroabsorption peaks of two samples (full curve: PTS 6, dashed curve: PTS 4 where no splitting is observed) in comparison with the shift of the absorption edge (represented by $\alpha = 250\ cm^{-1}$) and with the position of the π-π^* exci- tons (open triangles). The transition energies of the excitons have been reduced by 190 meV to fit into the scale.

ses. With further decreasing temperature the peak grows, narrows and splits into two distinct transitions. The absorption edge shows a different behaviour. There is a rapid blue shift with decreasing temperature down to 200 K. Below that temperature the absorption edge remains at the same position. There is no change of the slope of the absorption edge nor any indication for some splitting.

The variation with temperature of the transition energies and of the absorption edge is given in Fig. 5. Like for the π-π^* transitions of the polymer the splitting of the electroabsorption peak occurs between 150-200 K. A red shift of the defect transition with decreasing temperature below the phase transition and the rapid splitting near 150 K cause a complex temperature dependence of the transition energies, in contrast to the behaviour of the absorption edge which is independent on temperature below 200 K. Samples without the splitting of the electroabsorption spectrum at low temperature show a red shift of the peak which follows closely the shift of the low energy transition in the other samples. The absorption edge as well as the electroreflectance spectra in the intrinsic range of the π-π^* transitions ($\hbar\omega$ > 1.9 eV) are the same for all samples.

3 DISCUSSION

3.1 Summary of Experimental Results

The electroabsorption signal of the defect absorption in thermally polymerized PTS is very different from that of the photoproducts which is consistent with a red shift of the transitions in an electric field, either by some charge transfer contribution to the π-π^* transitions or by an extremely large polarizability. Such explanation does not apply to the response of the defect in thermally polymerized samples. The main observations are summarized in this section.

The defect absorption is observed for polarization $E||b$ and responds to fields $F||b$ only. This clearly shows a correlation of the π-π^* transitions of the polymer. It is present in all samples and develops with the thermal conversion to the polymer, very similar to the electroreflectance signal of the band transition near 2.4 eV which is small in partly polymerized samples but growth strongly in the autocatalytic range of conversion.

Only a small part, the threshold of the defect absorption is sensitive to the field. Most of the broad absorption band which overlaps with the absorption spectrum of the π-π^* excitons of the polymer arises from transitions into states which are not sensitive to electric fields.

The electroabsorption signal at 1.75 eV is very large. Fields of 20 kV/cm cause changes $\Delta\alpha/\alpha \simeq 10^{-3}$, much larger than the signal obtained near 2 eV from the excitons.

The electroabsorption signal $\Delta\alpha$ is always positive and shows no sign reversal in the spectrum. Such shape is not consistent with a field induced energy shift of electron states nor with a Franz-Keldysh effect of free carriers which must lead to a sign reversal of $\Delta\alpha$.

Absorption near 1.75 eV leads to photoconductivity. The creation of free carriers may be the result of competing rates of ionization and relaxation in the excited state (16,17) or of direct excitation into delocalized band states.

Only some samples show splitting of the electroabsorption signal at low temperature. The splitting is very similar to that of the excitons near 2 eV.

3.2 Model for the Field-Sensitive Defect Absorption

From the experimental observations we conjecture that the electroabsorption spectrum arises from a transition of an electron locali-

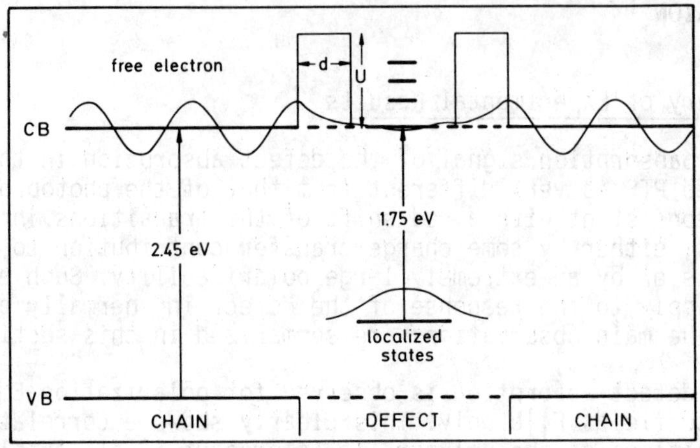

FIGURE 6: Model for the field-sensitive defect absorption in PTS. The free electron states of the undistorted chain are separated from the defect region with its localized states by abarrier of height U and width d. Tunneling accounts for some spatial overlap of free electron states and localized defect ground state which is modulated as an external field alters the barrier height.

zed on a defect into a delocalized band state. The model is sketched in Fig. 6. In the polymer chain optical excitation ($E \| b$) at 2.45 eV leads to free electrons moving along the b-axis and represented as a plane wave. A defect of the chain creates a barrier of height U and width d which reflects the free electron wave. On the defect the energy levels are different and localized. The free electron wave, however, can tunnel through the barrier and its finite amplitude gives rise to a transition of an electron localized at the defect to the free electron state. The absorption α_o depends on the tunnel factor of the free electron state

$$\alpha_o \propto \exp(-2(2m\,U)^{1/2}\,d/\hbar) \tag{3}$$

Electric fields $F \| b$ increase or decrease the barrier height depending on the orientation of the field which for $U \gg eFd$ yields a field dependent absorption

$$\alpha(F) = \alpha_o \cosh((2m/U)^{1/2} e\,Fd^2/2\hbar) \tag{4}$$

The field-induced change of the absorption

$$\Delta\alpha = \alpha(F) - \alpha_o = \alpha_o \cdot (e^2 d^4 m/2\hbar^2 U)\,F^2 \tag{5}$$

which is positive and depends quadratically on the external field in accordance with the experiment. Absorption and dependence on the field decrease with rising barrier height U. Inserting a barrier width of 25 Å and a height U = 0.5 eV yields $\Delta\alpha/\alpha \simeq 10^{-3}$ as observed in fields of 20 kV/cm. Increasing barrier height rapidly reduces $\Delta\alpha$ and the field sensitive absorption may be quite unique to PTS. Photon energies of 1.75 eV are sufficient to reach the field sensitive edge of the conduction band. Transitions into localized excited states are not sensitive to the field. They are located above of the conduction band and the excited electrons therefore may escape to the free electron states giving rise to photoconductivity. If the localized excited states are below the conduction band recombination is likely to dominate which seems the case for other polydiacetylenes.

The origin of the defect absorption is not known. Chain ends which are present in large concentration may contribute to the defect absorption band but they take not part in the field sensitive absorption. Studies on soluble PDA's indicate that chain ends do not annihilate (18) during polymerization whereas the decrease of the electroabsorption signal at 1.75 eV with prolonged annealing reveals that the number of defects is reduced by a factor of 2 in the final stage of conversion. Studies on PTS point to a large density of edge dislocations with Burgers vector parallel to b (19). These dislocations do not create chain ends but cause large local strain on the polymer backbone over several repeat units. Similar strain may results from screw dislocation with Burgers vector $\perp$b which in one-dimensional systems are generated by chain ends (20). The density of dislocations should decrease when the conversion is saturated. We suggest therefore that absorption of light in those strained regions of the polymer backbone where the bonds are not yet broken is responsible for the field sensitive absorption at 1.75 eV in PTS. It appears also possible that from the two configurations of the polymer side groups in the low temperature phase one is prefered in strained regions, accounting for the absence of splitting in some samples.

This work has been supported by the Deutsche Forschungsgemeinschaft.

REFERENCES

1. for a review see: Semiconductors and Semimetals, Vol. 9. eds. R.K. Willardson and A. Beer (Academic Press, New York, 1971)
2. L. Sebastian and G. Weiser, Chem.Phys.Lett. 64 (1979) 396
3. L. Sebastian and G. Weiser, Phys.Rev.Lett. 46 (1981) 1156

4. D. Haarer, M.R. Philpott and H. Morawitz, J.Chem.Phys. 63 (1975) 5238
5. L. Sebastian and G. Weiser, Chem.Phys. 62 (1981) 447
6. L. Sebastian, G. Weiser, G. Peter and H. Bässler, Chem.Phys. 75 (1983) 103
7. R. Reimer, H. Bässler, phys.stat.sol. (a) 32 (1975) 435
8. K. Lochner, B. Reimer and H. Bässler, phys.stat.sol. (b) 76 (1976) 39
9. H. Sixl, W. Hersel and H.C. Wolf, Chem.Phys.Lett. 53 (1978) 39
10. H. Gross and H. Sixl, Chem.Phys.Lett. 91 (1982) 262
11. H. Gross, H. Sixl and E. Knapp, Chem.Phys. 84 (1984) 321
12. D. Bloor and F.H. Preston, phys.stat.sol. (a) 37 (1976) 427
13. B. Reimer, H. Bässler, J. Hesse and G. Weiser, phys.stat. sol. (b) 73 (1976) 709
14. R.R. Chance and G.N. Patel, J.Polym.Sci.Polym.Phys. 16 (1978) 859
15. V. Enkelmann and G. Wegner, Macromol.Chem. 178 (1978) 635
16. W. Spannring and H. Bässler, Ber.Bunsenges.Phys.Chem. 83 (1979) 433
17. A.S. Siddiqui, J.Phys. C: Solid State Phys. 13 (1980) 2147
18. G. Wegner, Macromolekular Chem. Suppl. 6 (1984) 347
19. R.J. Young, J. Polymer Science: Polymer Physics 20 (1982) 961
20. P. Predecki and W.O. Stratton, J.Appl.Phys. 37 (1966) 4053

SPECTROSCOPIC STUDIES OF A DIACETYLENIC SURFACTANT

P.E. Schoen, P. Yager and R.G. Priest

Code 6190, Naval Research Laboratory, Washington, DC 20375, U.S.A.

Spectroscopic studies have been performed on aqueous dispersions of di-(10,12-tricosadiynoyl) phosphatidylcholine before and after polymerization with ultraviolet light. Non-resonance Raman and infrared spectroscopies of both the polymer and monomer reveal the structure of the hydrocarbon chains and head groups of the lipids.

INTRODUCTION

Despite their potential importance, there are few spectroscopic studies of polymerizable surfactants. The only well-characterized polymerized surfactants are the polydiacetylenic fatty acids. The polymerization mechanism (1,4 addition) was only recently discovered as a result of studies on crystals of non-surfactant monomers which become brightly colored on polymerization [1], indicating the formation of polymer concurrent with a solid-solid phase transition. Several groups, particularly that of Ringsdorf, have pioneered the study of polydiacetylenic fatty acids [2-5]. A significant amount of spectroscopic work has been done on non-surfactant polydiacetylenes [6-9]. In addition, Lando and co-workers have used electron diffraction from mono- and bilayers of fatty acids to establish the conformation of the diacetylene backbone chains in this ordered surfactant system [10,11], and others have performed Raman spectroscopy on Langmuir-Blodgett films of similar surfactants [12]. This provides one with a powerful "basis set" of information from which to study other polydiacetylenic systems.

The surfactants of interest in drug encapsulation and other areas requiring liposomal structures have two alkyl chains, and have more complex behavior than simple fatty acids. Diacetylenic surfactants of this class have been synthesized, most notably by Ringsdorf [13], Chapman [14-16] and O'Brien [17]. The polymerization can be followed by eye, as the polymers are intensely colored. Although both positively and negatively charged species have been made, the most thoroughly studied of these two-chain amphiphiles is a phosphatidylcholine which has a naturally-occurring head group attached to two diacetylenic hydrocarbon chains. Diacetylenic phospholipids, as these have been called, have a single chain melting phase transition in aqueous dispersion. The transition temperature is dependent on the length of the hydrocarbon chains and the positioning of the diacetylenic groups [16]. The absorption spectra of dispersions of these types of lipid have been monitored during polymerization and before and after heating, and changes in the absorption spectra are observed in both cases.

We have concentrated our initial studies on di-(10,12 tricosadiynoyl) phosphatidylcholine ($DC_{23}PC$) which has been used in this lab and in others [16-19]. When dry or in aqueous dispersion below the transition temperature of 38 deg. C, $DC_{23}PC$ has been known to polymerize to a red polymer having an absorption spectrum similar to that seen for many diacetylenes. It has not been reported to form a blue polymer under any conditions, and upon heating the red form turns yellow. We have found that dispersions of $DC_{23}PC$, which are liposomal above 38 deg. C, will on cooling spontaneously form unusual tubular structures [19] similar to but not identical to "cochleate cylinders". Such structures have also been observed but not characterized by Chapman [16] for a shorter chain lipid, but not observed for this compound. In an attempt to elucidate the forces which govern the conversion to tubules from the liposomes, and how the monomeric chain packing affects the polymerization process, we have applied various spectroscopic techniques to the $DC_{23}PC$ system. We report here the first detailed vibrational spectroscopy on a polymerizable lipid, and the first analysis of the $DC_{23}PC$ spectroscopic data in light of the unusual tubular form which this lipid assumes in the monomeric state.

MATERIAL AND METHODS

The lipid di-(10, 12 tricosadiynoyl) phosphatidlycholine was synthesized as previously described [17] by Avanti Polar Lipids from fatty acids prepared by P and L Laboratories. The $DC_{23}PC$ gave a single spot by TLC. Dilignoceroyl (24:0) phosphatidylcholine and dinervonoyl (24:1 15-cis) phosphatidylcholine were purchased from Avanti Polar Lipids and used without further purification. Dispersions were prepared by hydrating the lipids in distilled water at temperatures above their melting temperatures with vortexing, and were either used immediately or stored frozen under N_2. The appearance of the dispersions was checked by optical microscopy to check for the formation of liposomes or other structures before spectroscopy. Polymerization was performed by brief exposure to 13 mW/cm^2 254 nm light in a Rayonet model RPR-100 reactor at room temperature.

The samples of lipid dispersion for Raman spectroscopy were pelleted in hematocrit capillaries and maintained in a temperature-controlled block during laser illumination. Above the phase transition temperature over a hundred milliwatts of light could be used, but with the partially polymerized samples powers were reduced to less than 10 mW in most cases. Slit widths used on the monochromators were such that resolution was between 5 and 7 cm^{-1} depending on the wavelength of excitation and scattering, and differences between the alignment of the two monochromators used limits the accuracy of stated band locations to +/- 3 cm^{-1}.

RESULTS

As has been previously mentioned, $DC_{23}PC$ in excess water has a single broad phase transition peaked at about 38 deg. C, above which polymerization does not occur. However, unlike most other phosphatidylcholines, the structures formed by this lipid seem to depend greatly on the history of the sample. Above the phase transition the lipid forms stable liposomes, but as the temperature is dropped the liposomes may either collapse into an amorphous structure or, if the temperature is lowered slowly and the liposomes are larger than 1 micrometer, the liposomes convert quantitatively to long hollow tubular structures [19]. We have tried to differentiate between these two low temperature structures in our spectroscopic studies. The initially formed polymer of $DC_{23}PC$ is bright red if the lipids are dry or in the form of large hydrated structures. The absorption spectrum shown in Figure 1 is typical for a sample of polymerized tubules. When $DC_{23}PC$ dried from chloroform solution is polymerized, in addition to the "red" polymer peaks near 490 and 520 nm, strong absorption peaks appear at 560 and 590 nm, and appreciable portions of the material appear blue under the optical microscope. There are at least two different species of polymer in the dry material, then, although neither absorbs as far to the red as the polymerized 10,12 tricosadiynoic acid which absorbs at 570 and 620, or polymerized achiral diacetylene two-chain amphiphiles, which can absorb as far to the red as 580 and 650 nm [18-20]. As can be seen in Figure 1, we find that in tubule preparations there is a small absorption band present near 600 nm in addition to the main bands near 500 nm which has not been previously discussed. The monomer structure in the tubules, then, is nearly as ordered as in the dry material,

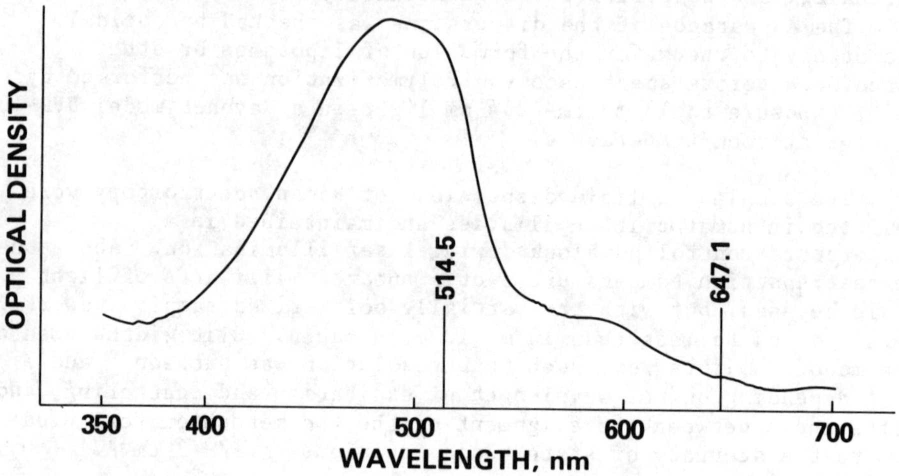

Fig. 1 Visible absorption spectrum of $DC_{23}PC$.

although even dry $DC_{23}PC$ is not so ordered as other polymerizable diacetylenic surfactants.

In order to differentiate the structures of the tubules and the "amorphous" cold lipid we prepared several samples for infrared absorption spectroscopy: both polymerized and un-polymerized tubules and amorphous material. These were then dried *in vacuo* onto AgBr plates for observation. The spectra shown in Figure 2 include an absolute spectrum (a) of unpolymerized dry tubules, and a difference spectrum (b) between this sample and polymerized tubules. The infrared spectrum of the amorphous material (not shown) was found to differ very little from that of unpolymerized tubules. At the resolution of the instrument used, the difference spectrum of Figure 2 displays slight but noticeable features. Note that the C≡C stretching is so weak in the infrared as to be invisible in this figure. Particularly large difference bands are observed in the C=O stretching region at about 1720 cm^{-1}, the CH_2 scissoring region, and the CH_2 rocking mode. These are all similar to the changes seen between the subphase and gel phases of another lecithin, dipalmitoyl phosphatidylcholine [21]. (The subphase of dipalmitoyl PC is known wo be nearly anhydrous, and has very highly ordered hydrocarbon chains. In contrast the gel phase of dipalmitoyl PC is fully hydrated and less ordered.) Small but distinct changes in the $DC_{23}PC$ C-C stretching modes and C-H stretching modes further indicate that the chain packing becomes poorer on polymerization. This is consistent with our observation that the perfectly straight monomeric tubules become twisted and distorted on polymerization, presumably because of changes in unit cell dimensions on formation of the polymer backbone [19].

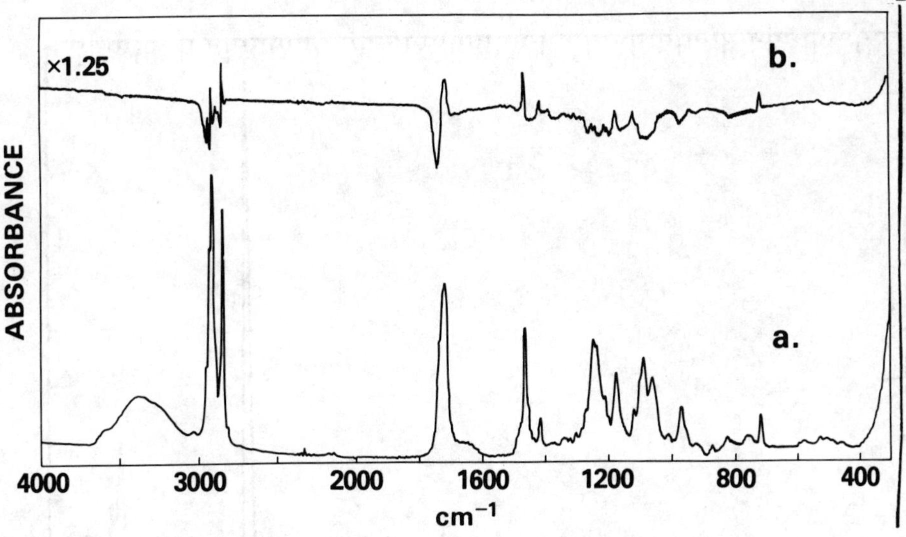

Fig. 2 Infrared absorption of $DC_{23}PC$.

The Raman spectrum of the unpolymerized $DC_{23}PC$ differs from those of other phospholipids in several respects. The spectrum of Fig. 3 was recorded from a (nominally) unpolymerized sample below the transition temperature. The 514.5 nm laser radiation caused a small degree of polymerization as is evidenced by the 2111 cm^{-1} polymer triple bond stretch band. The monomer triple bond stretch occurrs at 2261 cm^{-1}. The degree of polymerization was so slight that no color was visible to the eye. Several major differences from what one might expect from a simpler long-chain lecithin are evident such as the appearance of the CH_2 scissoring (or deformation) modes at approximately 1440 cm^{-1} and the vibrationally-coupled C-H stretching region. At first glance the appearance of the sharp peak at 1420 cm^{-1} in the CH_2 scissoring manifold indicates that the chains are in an orthorhombic packing similar to that seen in dipalmitoyl phosphatidylcholine at very low temperatures [23]. However the C-H stretching region does not resemble the manifold seen in such cases. The 2825 cm^{-1} peak seen for $DC_{23}PC$ is not observed in any lipid or hydrocarbon of which we are aware, and the presence of a small 2852 cm^{-1} peak and a sharp one at 2912 cm^{-1}, and a shoulder on the high side of the 2890 cm^{-1} compound the discrepancy. It appears that either the coupling between the various vibrational modes involved in the complex Fermi resonance-factor group splitting of the hydrocarbons [24] is significantly different in the diacetylenic lipids, or the packing of the hydrocarbon chains is quite different from that observed in other systems.

In Fig. 4 is shown the spectrum of $DC_{23}PC$ with the polymer

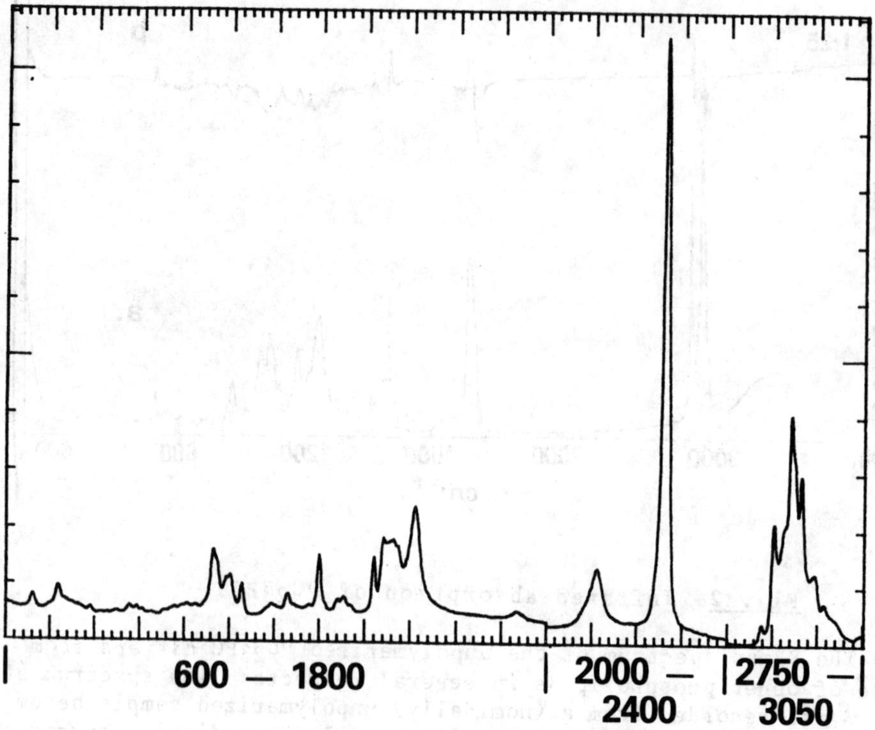

Fig. 3 spectrum of $DC_{23}PC$ showing slight polymerization.

spectrum subtracted out. The skeletal optical region from 1060 to 1130 cm^{-1} shows a triplet of sharp lines, characteristic of vibrational uncoupling of the upper and lower halves of the fatty acyl chains on either side of the diacetylenic group - a phenomenon observed in crystalline fatty acids with cis double bonds [22].

To prove that the unusual vibrational structure of this compound is not caused by the great length of the tricosadiynoyl chains we studied the spectra of two simpler lecithins, dilignoceroyl phosphatidylcholine with two saturated fatty acyl chains of 24 carbon length and dinervonoyl phosphatidylcholine with chains of the same length, but with a single cis double bond at the 15 position. Both lipids form liposomes which are stable above and below their transition temperatures. The spectra differ very little from that seen for any other well-packed phospholipids. None of the differences can account for the unusual hydrocarbon chain bands seen in the $DC_{23}PC$ spectrum.

The perturbation of the vibrational structure of hydrocarbon chains by the diacetylenic unit persists even when the motional restrictions of the crystalline state are removed. The spectrum of the monomeric lipid above its phase transition temperature is shown

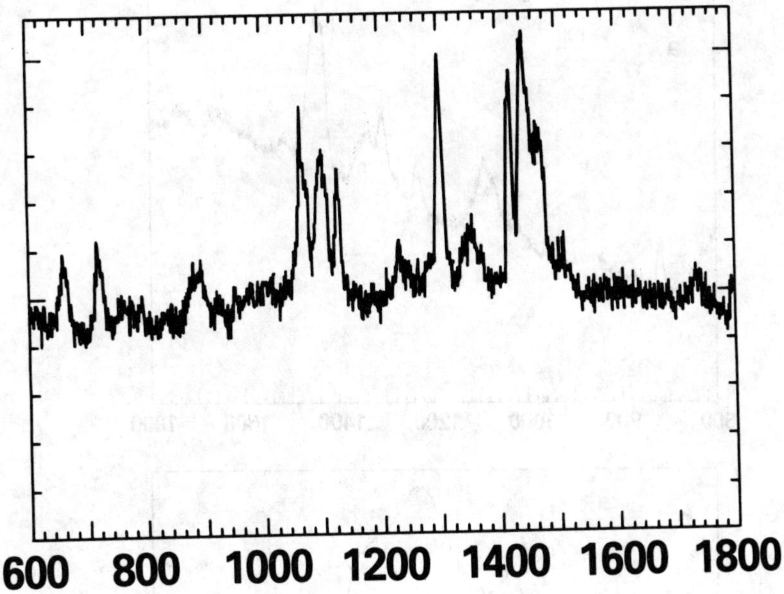

Fig. 4 Raman difference spectrum of $DC_{23}PC$.

in Fig. 5, and in many ways the changes seen on raising the temperature are typical of those seen in other systems. However, the atypical intensities of the band envelope in the C-H stretch region (5b) emphasize the unusual character of the $DC_{23}PC$ spectrum. The diacetylenic group halfway down the chains must strongly perturb the vibrations of the surrounding hydrocarbon regions, altering the spectrum even when whatever factor group splitting imposed by packing in the crystal is released by melting.

It proved impossible to observe the non-resonance enhanced Raman spectrum of the monomer for long periods of time with 514.5 nm light without obscuration by the polymer spectrum. For this reason it is preferable to use the 647.1 nm krypton laser line. When the lipid is deliberately polymerized, the resonance Raman effect produces spectra which depend greatly on the exact conditions of observation, polymerization, and the laser frequency. This complex situation is the subject of another publication [25].

DISCUSSION

While our spectroscopic investigations to date represent more of a survey than an exhaustive study of $DC_{23}PC$, several conclusions may be made about the system. The formation of the highly unusual tubule structure by this phospholipid can only be explained by a

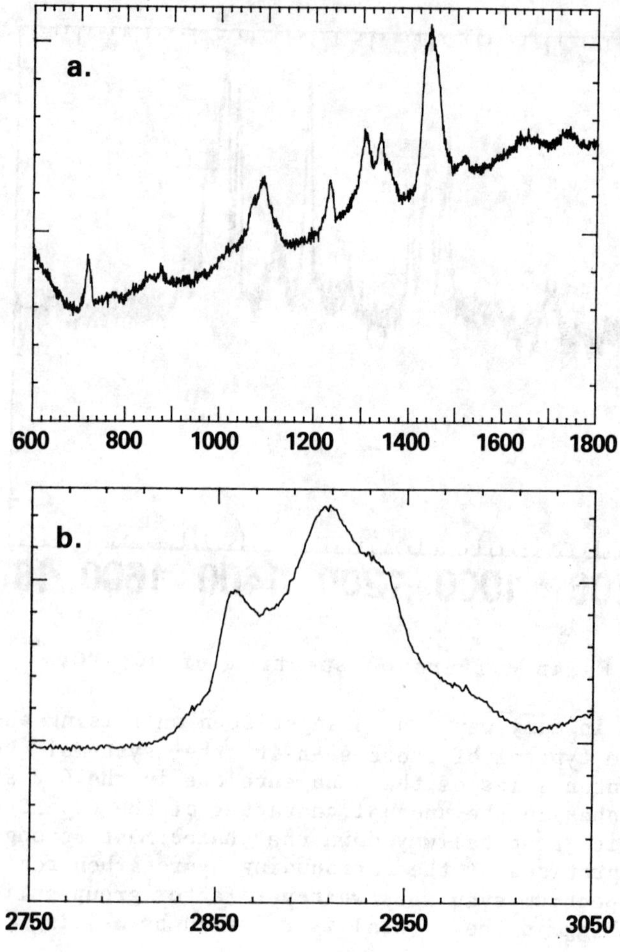

Fig. 5 Raman spectrum of $DC_{23}PC$ at 50° C (unpolymerized).

fundamental difference between the low temperature phase of $DC_{23}PC$ and that of other better-studied phospholipids. As we have found that neither dilignoceroyl nor dinervonoyl PC forms tubules, the key must be in the perturbation of hydrocarbon chain packing by the diacetylene group. As quantitative conversion to tubules is possible under our experimental conditions (see figures in Ref. 15), it is clear that other low temperature states are metastable. The tubules differ from liposomes in that they have two quite different radii of curvature--one narrowly defined near a micrometer, and another near infinity. There may be a specific chain crystal packing which exists unaltered throughout the tubules to support such a regular structure. The helical pattern of polymerized fibers within polymerized tubules [19] supports this hypothesis.

We see only subtle differences between the non-resonance Raman or infrared spectra of $DC_{23}PC$ samples which appear completely different under the optical microscope, e.g. between the infrared spectra of dry tubules and "amorphous" (presumably microcrystalline) lipid. On the molecular scale these structures must be similar.

The polymerization itself seems to disturb the hydrocarbon chain packing and the head group conformation, as seen by changes in the infrared spectrum and some indications of changes in the Raman active C-H stretching vibrations. As yet we cannot identify the crystallographic packing in which the chains fit because of the difficulty of interpretating the $DC_{23}PC$ spectra by analogy with other lecithin spectra. The vibrational perturbation of the diacetylenic group is sufficient to muddy the analysis of the spectra from the rest of the hydrocarbon chains. The spectra do not fit any existing category for alkane chains. However, if the aforementioned similarity between the tubule chain packing and that in fatty acid monolayers is correct, then the packing change on polymerization must be similar to that found by Day and Lando in the fatty acid systems [11]. These include a change from a tilted structure to one in which the hydrocarbon chains are perpendicular to the bilayer, accompanied by a contraction of the packing along the direction of polymerization. The hydrocarbon chain packing of phospholipids is in general looser than that in fatty acids, partially because of the unequal penetration of the bilayer by the two hydrocarbon chains imposed on the lipid molecule by the chiral head group, and because of the mismatch in cross sectional area between headgroup and chains. This loose packing may account for the "relaxed" red polymer forming first in most cases, and for there to be some strained blue polymer formed only in the highly crystalline tubule structures and dry lipids. Ongoing work in this laboratory with this and other model compounds should make possible a better analysis of the chain conformations.

ACKNOWLEDGEMENT

We would like to thank Dr. James P. Sheridan for experimental and interpretive assistance and Dr. Alok Singh for preparation of additional materials. This work was supported in part by the Air Force Office of Scientific Research and the Office of Naval Research.

REFERENCES

1. G. Wegner, Z. Naturforsch, 246, 824 (1969)

2. D. Day and H. Ringsdorf, J. Polym. Sci., Polym. Lett. Ed., 16, 205 (1978)
3. J.P. Fouassier, B. Tieke, and G. Wegner, Israel J. Chem., 8, 227 (1979)
4. C. Bubeck, B. Tieke, and G. Wegner, Ber. Bunsenges. Phys. Chem., 86, 495 (1982)
5. C. Bubeck, B. Tieke and G. Wegner, Ber. Bunsenges. Phys. Chem., 86, 499 (1982)
6. M.L. Shand, R.R. Chance, M. LePostollec and M. Schott, Phys. Rev. B, 25, 4431 (1982)
7. D. Bloor, R.J. Kennedy and D.N. Batchelder, J. Poly. SCi.: Poly. Phys. Ed., 17, 1355 (1979)
8. D.N. Batchelder and D. Bloor, J. Phys. C Solid State Phys., 15, 3005 (1982)
9. D. Bloor, D.N. Batchelder, D.J. Ando, R.T. Read and R.J. Young, J. Poly. Sci.: Poly. Phys. Ed., 19, 321 (1981)
10. D. Day and J.B. Lando, Macromolecules, 13, 1478 (1980)
11. D. Day and J.B. Lando, Macromolecules, 13, 1483 (1980)
12. B. Tieke and D. Bloor, Makromol. Chem., 180, 2275 (1979)
13. R. Buschl, B. Hupfer, and H. Ringsdorf, Makromol. Chem., Rapid Commun., 3, 589 (1982)
14. O. Albrecht, D.S. Johnston, C. Villaverde, and D. Chapman, Biochim. Biophys. Acta, 687, 165 (1982)
15. M. Pons, D.S. Johnston, and D. Chapman, J. Polym. Sci., Polym. Chem. Ed., 20, 513 (1982)
16. J. Leaver, A. Alonzo, A.A. Durrani, and D. Chapman, Biochim. Biophys. Acta, 732, 210 (1983)
17. D.F. O'Brien, T.H. Whitesides and R.T. Klingbiel, J. Polym. Sci., Polym. Lett. Ed., 19, 95 (1981)
18. E. Lopez, D.F. O'Brien, and T.H. Whitesides, J. Am. Chem. Soc., 104, 305 (1982)
19. P. Yager and P.E. Schoen, Mol. Cryst. Liq. Cryst., 106, 371 (1984)
20. B. Hupfer and H. Ringsdorf, in Polymer Science Overview, G.A. Stahl, ed., A.C.S. Symposium Series (Washington DC, American Chemical Society, 1981)
21. D.G. Cameron and H.H. Mantsch, Biophys. J., 38, 175 (1982)
22. J. L. Lippert and W.L. Peticolas, Biochim. Biophys. Acta, 282, 8 (1972)
23. N. Yellin and I.W. Levin, Biochem. Biophys. Acta, 489, 177 (1977)
24. R.G. Snyder, S.L. Hsu, and S. Krimm, Spectrochim. Acta, 34a, 395 (1978)
25. P.E. Schoen and P. Yager, to appear in J. Poly. Sci.

OPTICAL PROPERTIES OF POLYACETYLENE*

E. Mulazzi

Dipartimento di Fisica dell'Università di Milano and GNSM,
Via Celoria 16 - Milano 20123 - Italy

Resonant Raman Scattering properties of cis and trans polyacetylene are studied in the framework of the tight binding approximation. The Raman spectra and the Raman Excitation Profiles are reported for both the isomers and the different results are discussed on the basis of the models.

1. INTRODUCTION

Recent experimental results on Resonant Raman Scattering (RRS) spectra from trans (1,2) and cis $(CH)_x$ (3) and $(CD)_x$ have shown that the two isomers present different properties for the electronic excited states and their couplings to the vibrational modes responsible of the light scattering. The main differences in the RRS spectra consist of the long progression of overtones in the spectra of the cis $(CH)_x$ samples while in the trans $(CH)_x$ spectra only the first order processes have been observed and carefully analyzed at least with the laser frequencies used so far (1.65 eV $\leq \Omega_L \leq 3.7$ eV (4). Broad and weak second order scattering has been observed for $\Omega_L \cong 2.7$ eV). Moreover and more important in the cis $(CH)_x$ RRS spectra it is possible to observe the following properties. a) The line shapes of the first order and higher order processes show very small changes as function of the increasing incident laser frequency. b) The intensities of the different lines in the spectra, as function of the incident laser frequency, follow the usual behavior already observed e.g. in the spectra of β carotene in CS_2 solution. c) No observable changes have been detected on the line shapes and on the intensity law, which can be ascribed to sample dependence and to its history. On the contrary in the trans $(CH)_x$ RRS spectra it is possible to observe that the line shapes change as function of the incident laser frequency and of the history of the samples. These different properties related to the change in line shapes and line widths in the two isomers can be attributed to the fact that the disorder (which is always present and at larger extent in the trans samples since they are obtained by isomerization) influences in a different way the electronic and vibrational properties of cis and trans polyacetylene. While in the cis isomer the disorder can induce only local strain which can change a little the electronic and dynamical properties of long conjugation segments, in the trans the disorder is reflected not only in a small perturbation due to local strain, but mainly in the distribution of different conjugation lengths of the segments which constitue a given chain. This fact in the trans samples brings as a consequence that the selective resonance with electronic transitions of different conjugation length segments, determines the scattering of light due to vibrational modes of different frequencies.

In this paper we show that the different properties of RRS spectra which we have just discussed of both isomers, can be explained in terms of the electronic and dynamical properties of conjugated segments, which constitue a given chain, studied in the tight binding approximation. In particular for the cis $(CH)_x$ we consider only the long conjugation length segment properties, while for the trans $(CH)_x$ samples we take into account the properties of different conjugation length segments. The different contributions to the RRS spectra are weighted in this last case by a bimodal distribution which is supposed to simulate the distribution of long and short conjugation segments (5). In the following we present briefly the models used in order to perform the calculation of the spectra. We refer the reader to (5) and (6) for a more exhaustive presentation of the two models.

2. RRS FROM CIS$(CH)_x$

We have evaluated the electronic states, the dipole moments for the transitions ($\pi \rightarrow \pi^*$) and the electron vibration (e-v) interaction in the excited states of long conjugated segments in the framework of the tight binding approximation (7). The excited e-v interaction matrix elements are calculated in a self-consistent way with those entering the lattice dynamics calculations (8). See for the details (6). The e-v couplings calculated considering these e-v matrix elements are reported in Fig. 1.

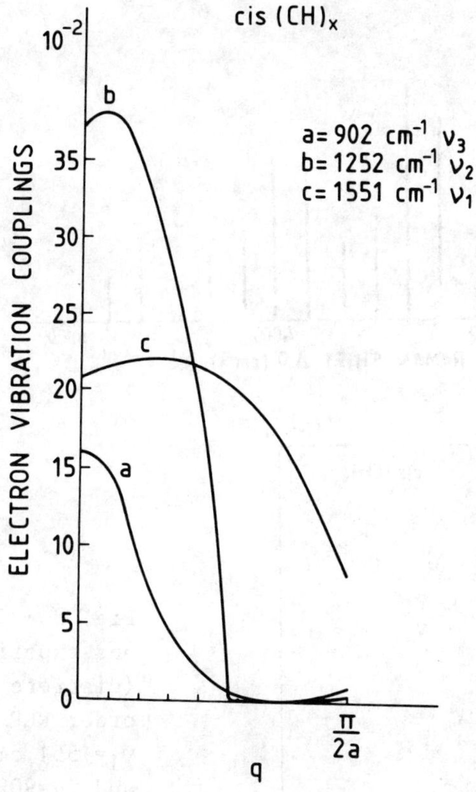

Fig. 1 - Electron vibration couplings for cis $(CH)_x$, as function of q for the three vibrational modes as indicated.

By using the expression of RRS cross sections for long segments given in (5) and the e-v couplings (Fig. 1) in the excited states, the electronic and dynamical properties of cis $(CH)_x$, we have evaluated the RRS spectrum shown in Fig.2 and the Resonant Excitation Profiles (REP) reported in Fig. 3. The details of the theoretical calculations together with the spectra and the REP of cis $(CD)_x$ will be reported in (6). From Fig. 3 it is possible to observe a very good agreement between the theoretical calculations and the experimental data, without any ad hoc fitting parameters

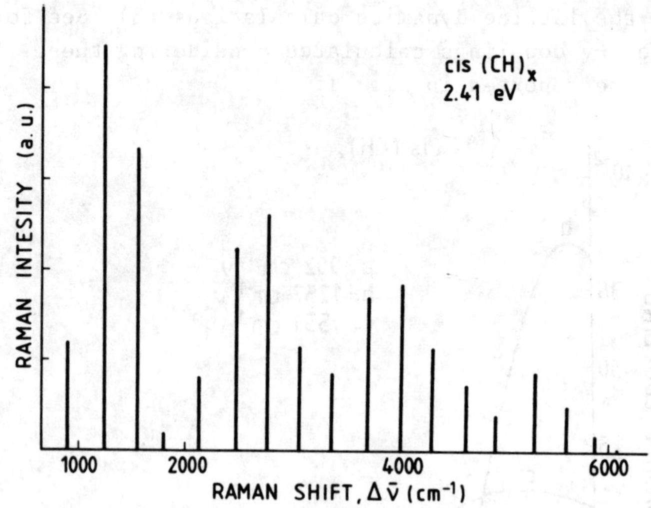

Fig.2 - RRS spectra calculated for Ω_L=2.41 eV

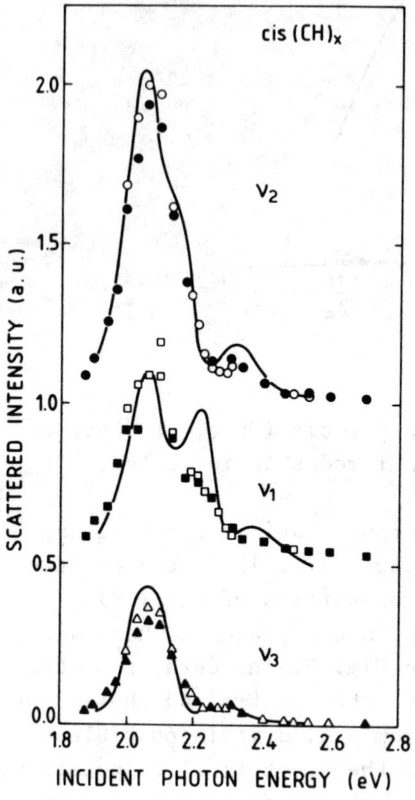

Fig. 3 - Calculated (——) and experimental (3) (discrete points) first order REP for cis $(CH)_x$. ν_1=1551 cm^{-1}, ν_2=1252 cm^{-1} and ν_3=909 cm^{-1}.

3. RRS FROM TRANS $(CH)_x$

As already discussed in the introduction, in the trans $(CH)_x$ case we consider a model in which a distribution of segments of different conjugation lengths is taken into account. From this reason we have to consider the electronic states and the electronic properties for long conjugated segments and for finite segments as described in (5). In the same reference it is possible to find the details for the calculations of the e-v couplings in the excited electronic states and the dispersion law for the vibrational frequencies as function of the double bond number (N) in the conjugated segments considered. We report here in Fig. 4 only the RRS spectra calculated for different Ω_L, by considering the same bimodal distribution with $N_1=100$, $\sigma_1=50$, $N_2=15$, $\sigma_2=7$ and G=0.75 (5). Good agreement is found between the experimental data in (4) and the calculated spectra of Fig. 4 and this proves that the model is able to explain the changes in the line shapes of the spectra as function of the incident laser frequency at least in the frequency region which has been considered.

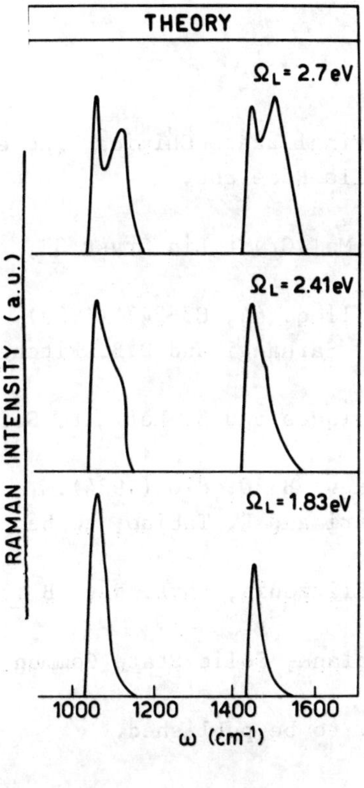

Fig. 4 - Calculated RRS spectra of trans $(CH)_x$ for Ω_L as indicated. Parameters are given in the text.

Recent experimental results (9) show that in the frequency region $\Omega_L \leq 1.5$ eV the first order RRS line widths become very narrow (note that the frequencies of the principal peaks are the same as those of Fig.4) and the higher order processes become more evident. This is in agreement with the model proposed here,because in that frequency region only the contribution of long conjugated segments is important and then the linewidths must become narrow. In fact the mechanisms responsible of the broadening and of the structured line shapes are completely uneffective here, since the electronic transitions of short conjugation length segments are out of resonance with the incident light frequency. On the other hand in that frequency region one has to consider also a new mechanism of scattering from the excitonic states, which are in resonance with the incident light frequency and then the exciton vibration couplings have to be taken into account in the calculations. Preliminary experimental results (9) seem to indicate this fact, since the relative intensities for the first order processes due to ω_{C-C} and $\omega_{C=C}$ are different from those observed in (4) and calculated for the $\pi\to\pi^*$ transitions, considered here and in (5).

REFERENCES

*Work partially supported by Progetto Finalizzato Chimica Fine e Secondaria del Consiglio Nazionale delle Ricerche.

1. See for instance Fitchen D.B., Mol.Cryst.Liq.Cryst.83, 95 (1982).
2. Lefrant S., J. Phys. (Paris) Colloq. 44, C3-247 (1983).
3. Lichtmann L.S., E.A. Imhoff, A. Sarhangi and D.B. Fitchen, J. Chem. Phys. 81, 168 (1984).
4. Mulazzi E., G.P. Brivio, E. Faulques and S. Lefrant, Solid State Commun. 46, 851 (1983).
5. Brivio G.P., E. Mulazzi, Phys.Rev. B 30, 876 (1984).
6. Mulazzi E., R. Tiziani, L. Piseri and R. Tubino, to be published.
7. Cojan C., G.P. Agrawal and C. Flitzanis, Phys. Rev. B 15, 909 (1977).
8. Piseri L., R. Tubino, G. Dellepiane, Solid State Commun. 44, 1589 (1982).
9. Shott M., Mol. Cryst.Liq.Cryst. to be published.

SPECTROSCOPY AND KINETICS OF A PLANAR-NONPLANAR CONFORMATIONAL TRANSITION IN POLYDIACETYLENE SOLUTIONS

R.R. CHANCE

Corporate Technology, Allied Corporation, Morristown, New Jersey 07960

M. W. WASHABAUGH

Biochemistry Department, Johns Hopkins University, Baltimore, Maryland 21205

D. J. HUPE

Merck & Co., Rahway, New Jersey 07065

ABSTRACT

Poly3BCMU, a polydiacetylene with bulky urethane substituents, readily dissolves in common organic solvents such as $CHCl_3$ to yield a yellow polymer solution. On addition of a nonsolvent, such as hexane, a dramatic color change takes place to yield a blue polymer solution. The variation in optical properties during this color transition is reviewed and discussed. New data on the kinetics of the transition are presented. The results strongly support a model which invokes a planar-nonplanar conformational transition involving the individual polydiacetylene chains in solution.

I. INTRODUCTION

Polydiacetylenes are generally insoluble even in exotic organic solvents and cannot be characterized via normal polymer characterization techniques. The first exceptions to this rule were the butoxycarbonylmethylurethane (BCMU) substituted polydiacetylenes.(1) The substituents for these polymers, commonly referred to as poly3BCMU and poly4BCMU, are shown below.

3BCMU 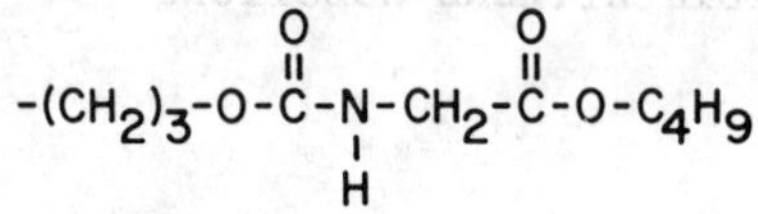

4BCMU 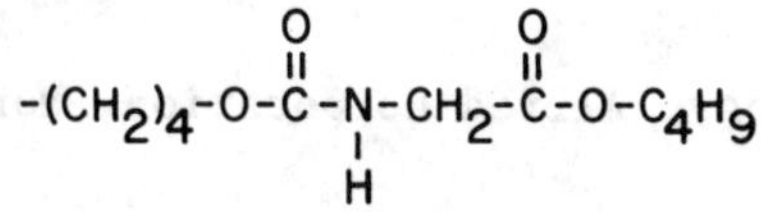

These polydiacetylenes are soluble in common organic solvents, such as $CHCl_3$, due to the high entropy content of the complicated urethane substituent groups. The discovery of these soluble polymers has lead to a number of interesting developments: the first determination of the chain length in a polydiacetylene, [about 1000 repeat units for both polymers(2)], characterization of the chain dynamics for polydiacetylenes in solution(3), discovery of the remarkably strong two photon absorption now known to be characteristic of the polydiacetylene backbone(4), and discovery of a conformational transition in the polymer solutions, referred to as a "visual conformational transi-tion"(1).

The conformational transition in BCMU polymers can be induced by temperature or a change in solvent. Conformational transitions are commonly observed in polymer solutions. The polydiacetylene system is unique because of the sensitivity of the electronic properties to backbone conformation. The result is dramatic color changes during the conformational transition. In poly4BCMU, the transition is accompanied by a yellow to red color change, a $2000 cm^{-1}$ shift in optical absorption. In poly3BCMU the color change is yellow to blue, an absorption shift of more than 5000 cm^{-1}. In the blue solution phase of poly3BCMU the optical properties are very similar to those of its crystalline form.(5)

Additional soluble polydiacetylenes have been discovered which show temperature- or solvent-induced color changes with absorption shifts similar to poly4BCMU(6-8). These solutions are believed to consist of aggregates in the red phase(6,9). Evidence is mixed as to whether polyBCMU solutions in their red or blue phases consist of aggregates or single chains. Lim and Heeger(3), using light scattering measurements, argue for the single chain interpretation based on the lack of variation of the hydrodynamic radius with polymer concentration. Wenz et al(6,9) believe the solutions are composed of aggregates as large as 700 chains. Their analysis is based on an interpretation of the angular dependence of the light scattering intensity of these solutions. These light scattering data will not be dealt with further here since they are to be discussed in another paper(10). In any case, light scattering measurements on the solutions cannot address the important question of the molecular origin of the conformational transition. In this article we will show that the driving force for the transition is purely intramolecular in origin and involves the hydrogen bonding network of the urethane substituent groups. If aggregation takes place, it must be subsequent to the conformational transition. These conclusions are based on the variations of optical properties of the BCMU polymers in solution. The conformational transition is discussed in Section II, the kinetics of the transition in Section III, and the microstructure of polymer chains in solution in Section IV. We concentrate our attention on poly3BCMU for which the most extensive data are available.

II. THE CONFORMATIONAL TRANSITION

A $CHCl_3$ solution of Poly3BCMU is bright yellow in appearance, the absorption maximum being about 21 000 cm^{-1}(470 nm). (See Figure 1.) The spectrum of a poly3BCMU crystal or film(5) has a peak absorption at about 16 000 cm^{-1}, a value which is typical of polydiacetylene (PDA) crystals. We attribute the blue-shift of the solution absorption relative to that of the crystal to the backbone disorder required for maximizing sidegroup flexibility in solution. The electronic properties of conjugated polymers are determined by the conjugation length, which is defined as the length over which backbone planarity is maintained without interruption. Conjugation length, rather than chain length, measures the continuity of the π electron network. The disruption of the backbone required to produce the observed blue-shift in the solution spectra could be a systematic distortion [helical or "wormlike" (9) structure] or discrete structural defects distributed along the chain. As will be demonstrated in Section IV, the discrete defect model is the most plausible, consistent with currently available data. The conjugation length scale (number of polymer repeat units) given in Figure 1 is based on the

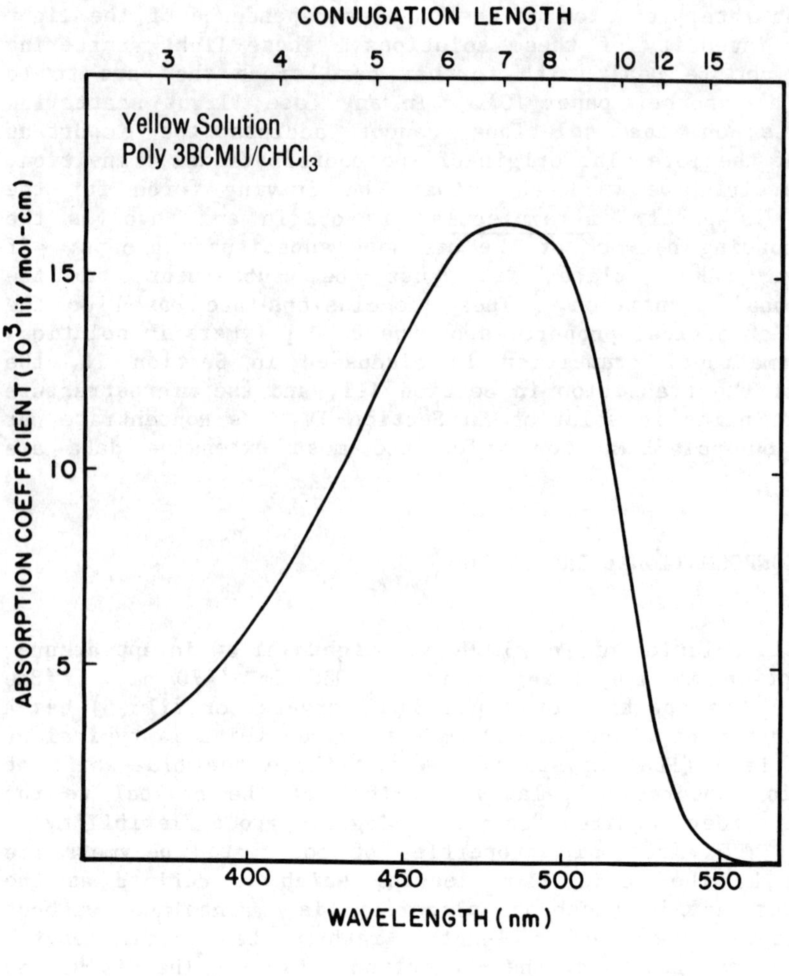

Figure 1. Absorption spectrum for poly3BCMU in $CHCl_3$ solution. The dependence of the absorption maximum on conjugation length given by the scaling at the top of the figure is taken from reference 11.

absorption behavior of model compounds for PDA oligomers.(11) Thus the yellow solution of poly3BCMU is viewed to contain a distribution of relatively short conjugation lengths, the peak absorption corresponding to 7 units (or 14 conjugated muliple bonds).

On addition of a nonsolvent, such as hexane, a dramatic color change takes place to yield a deep blue solution. Addition of excess hexane results in polymer precipation as a blue solid. The color change and the precipitation point can be monitored by titration.(1) The resulting phase diagram for poly3BCMU is shown in Figure 2 with polymer concentration given as moles of monomer repeat units per liter. [The results for poly4BCMU are very similar.(1)] Above 0.003 mol/lit, the color change and precipitation point cannot be distinguished as separate events. At lower concentrations a well-defined color change occurs prior to the precipitation point. The precipitation point varies with polymer concentration as would be expected for an aggregation phenomena. The color transition is completely independent of polymer concentration. This strongly suggests that the color transition is a single chain phenomenum and that chain aggregation cannot be the driving force for the color transition. If the blue solution is actually composed of small aggregates as Wenz *et al*(6,9) suggest, the aggregation must occur subsequent to some other physical change involving the individual polymer chains. Since the color of the solutions typifies much longer conjugation lengths, this physical change probably involves considerable increase in the planarity of the chains in solution. These experimental results have been confirmed by other workers.(3,12)

The spectroscopic changes that take place during the transition are shown in Figure 3 for various values of X_c, the mole fraction of $CHCl_3$ in hexane. The optical absorption spectra are seen to shift to lower energy rather abruptly by ~5000 cm^{-1} at X_c~0.71. The lowest energy optical transition in the $CHCl_3$/hexane solution is located at 16 000 cm^{-1}, a value very close to that of the crystal. Therefore, the conjugation length appears to have increased substantially during the conformational transition. It is also important to point out that spectral changes are discontinuous, i.e., at the first indication of spectral change, the 16 000 cm^{-1} optical transition is already evident (X_c=0.79). In fact intermediate spectra such as X_c=0.79, 0.71, and 0.67 can be accurately represented as weighted mixtures of the X_c=1.0 and X_c=0.41 spectra. In other words there is a fairly well-defined isobestic point with no suggestion of long-lived intermediates in the transformation.

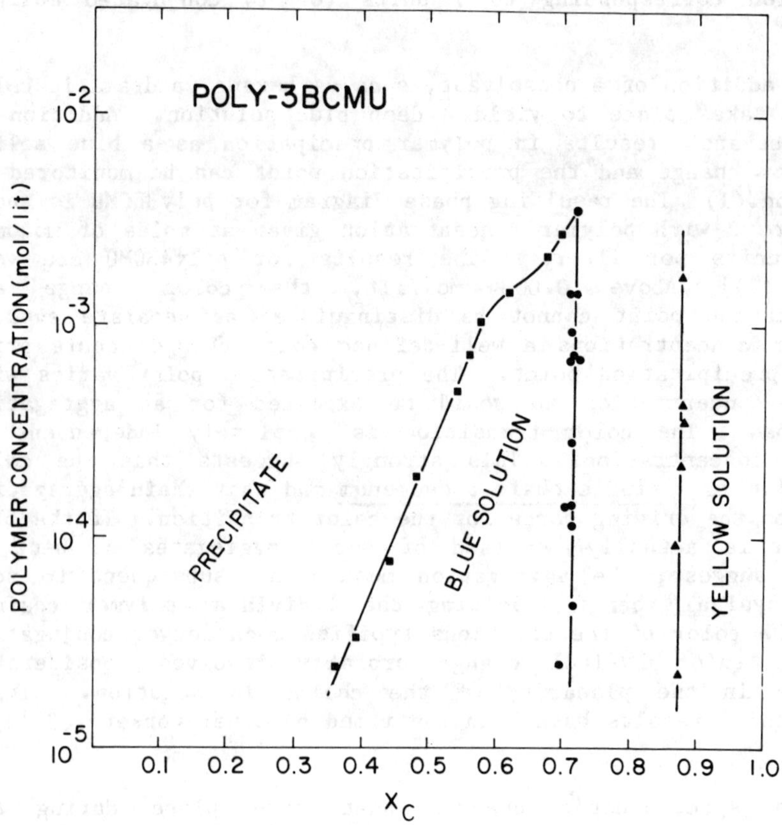

Figure 2. Phase diagram for poly3BCMU in $CHCl_3$/hexane mixtures. X_c is the mole fraction of $CHCl_3$ in hexane. The squares (precipitation or cloud points) and the circles (color transition points) are determined by titration with hexane. A red-purple endpoint was chosen as an indication of the color transition point. The triangles are obtained in a back titration with $CHCl_3$ to a yellow endpoint. Polymer concentration is given as moles of monomer repeat units per liter. At polymer concentrations greater than 0.003 mol/lit, the color transition and precipitation points appartently occur simultaneously. (From reference 1)

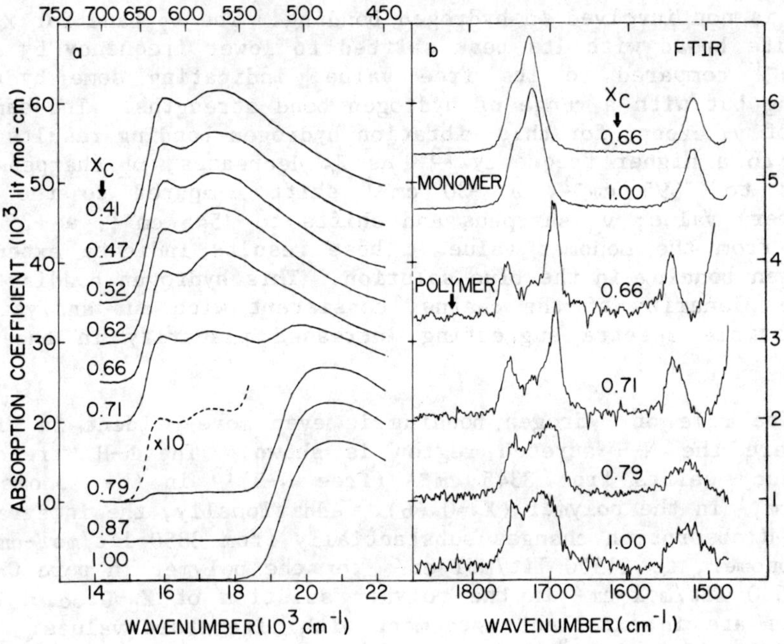

Figure 3. Visible and Fourier transform infrared spectra of poly3BCMU in various mixtures of $CHCl_3$ and hexane. The mole fraction of $CHCl_3$ in hexane is indicated. Polymer concentrations are 1.9×10^{-5} mol/lit for the visible spectra and 1.3×10^{-3} for the infrared spectra. (From reference 1)

The FTIR spectra in Figure 3 represent the primary evidence for the molecular origin of the conformational transition, the formation of N-H•••O=C intramolecular hydrogen bonds between the urethane functionalities on adjacent substituent groups. (The hydrogen-bonded configuration of poly3BCMU is illustrated in Figure 4.) The spectral region of the C=O stretch of the ester group (ν_1), the C=O stretch of the urethane group (ν_2), and the amide II band (ν_3) are shown in Figure 3. Note that the monomer spectra are virtually unaffected by X_c variations; the observed frequencies in the monomer are ν_1=1745 cm^{-1}, ν_2=1725 cm^{-1}, and ν_3=1520 cm^{-1}. These are quite reasonable "free" values, i.e., in the absence of hydrogen bonding.[13] The frequency of the ν_1 band of the polymer is essentially the same as the monomer and shows no significant variation with X_c. This is expected, since the ester group is not involved in hydrogen bonding. The ν_2 band for X_c=1.0 is quite broad with its peak shifted to lower frequency by about 10 cm^{-1} compared to its free value, indicating some hydrogen bonding but with a range of hydrogen bond strengths. The same is true of ν_3 except for this vibration hydrogen bonding results in a shift to a higher frequency.[13] As X_c decreases, ν_2 sharpens and shifts to 1695 cm^{-1}, a -30 cm^{-1} shift compared to the free (monomer) value; ν_3 sharpens and shifts to 1545 cm^{-1}, a +25 cm^{-1} shift from the monomer value. These results indicate extensive hydrogen bonding in the blue solution. This hydrogen bonding will induce planarity of the chains, consistent with our analysis of the visible spectra suggesting increased planarity in the blue phase.

The role of hydrogen bonding is even more evident in Figure 5, where the N-H stretch region is shown. The N-H stretching frequency shifts from 3345 cm^{-1} (free N-H)[13] in the monomer to 3320 cm^{-1} in the polymer (X_c=0.66). Additionally, the integral of the N-H absorption changes substantially from 8850 lit/mol-cm^2 in the monomer, to 21500 lit/mol-cm^2 for the polymer in pure $CHCl_3$, to 34500 lit/mol-cm^2 in the polymer solution of X_c=0.66. These results are in excellent agreement with literature values: 8900 and 32000 lit/mol-cm^2 for free and strongly hydrogen-bonded N-H groups.[14] The magnitudes of the changes in frequency and intensity of the N-H stretch implicate significant hydrogen bonding in the yellow solutions and strong hydrogen bonding in blue solutions.

Note that according to the spectra shown in Figure 5 the $CHCl_3$ solution of poly3BCMU clearly contains a substantial number of free N-H groups. Analysis of the data suggests that about 1/4 of the N-H groups are not hydrogen bonded.(1) Rotation about the single bond in the PDA backbone would free two N-H groups and

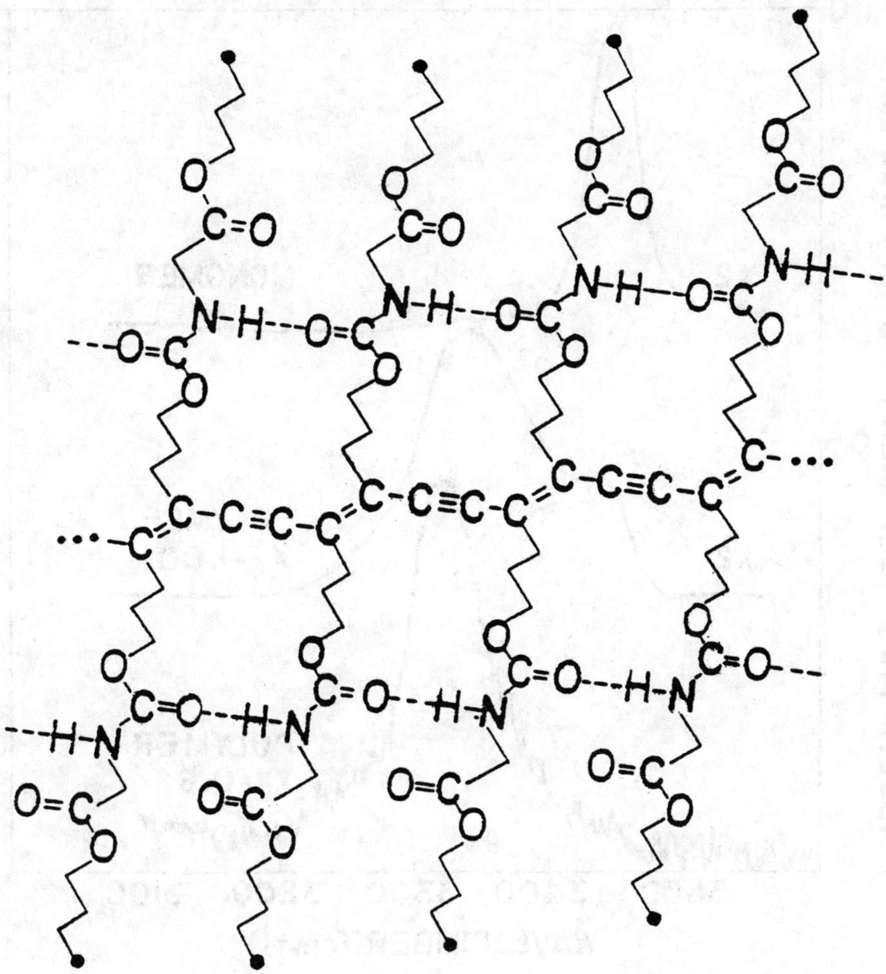

Figure 4. Planar hydrogen-bonded conformation of poly3BCMU as suggested by molecular models. Hydrogen bonds are indicated by dashed lines.

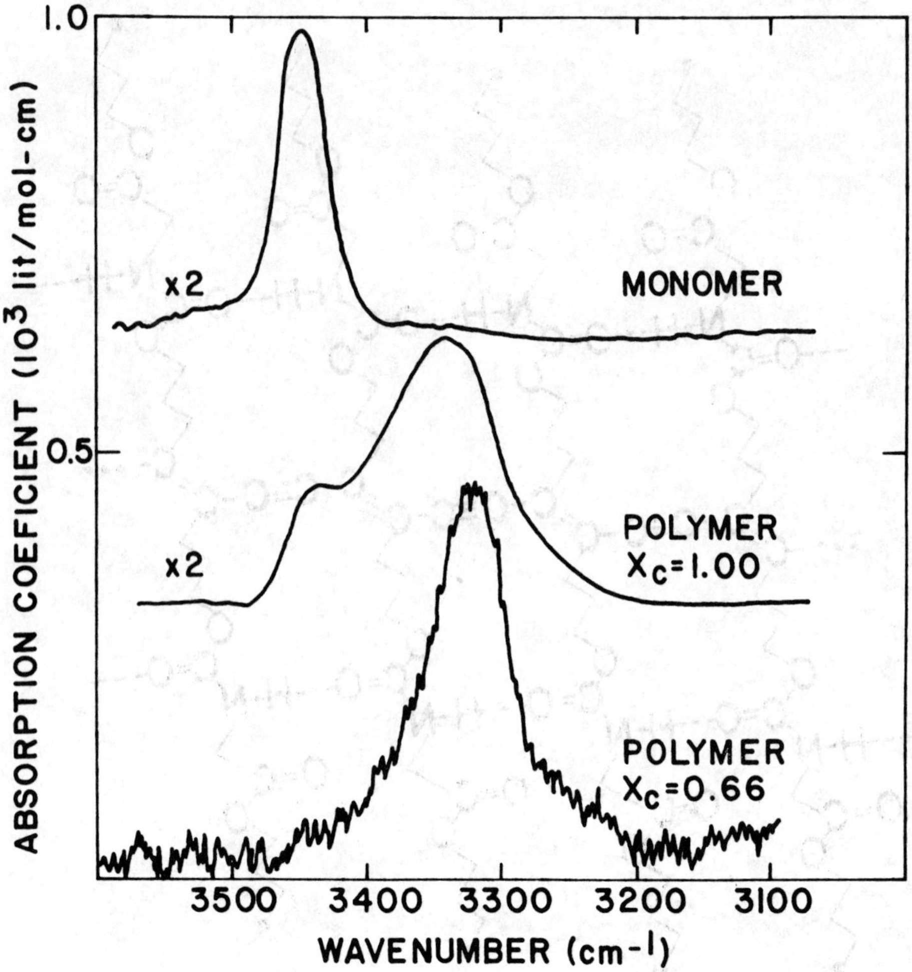

Figure 5. FTIR spectra in the region of the N-H stretch vibration. The monomer spectrum is that expected for "free" N-H. The polymer spectrum (X_c = 0.66) is that expected for strongly hydrogen-bonded N-H. The polymer spectrum in pure $CHCl_3$ is seen to be a mixture of free and hydrogen-bonded N-H with possibly some additional absorption in the intermediate region. (From reference 1)

interrupt conjugation at the site of this structural defect. Thus these results support the discrete defect model mentioned earlier. They are inconsistent with a systematic distortion of the backbone, including the "wormlike" chain suggested by Wenz et al (9), since a systematic distortion would produce a single value for the N-H frequency (or a broadened spectrum with a single peak).

Results for poly4BCMU are very similar, except in that case the color change is from yellow to red. The changes in infrared spectra are less dramatic suggesting more disorder in the red phase of poly4BCMU compared to the blue phase of poly3BCMU. Molecular models indicate that the hydrogen bonded configuration for poly4BCMU is more strained than that for poly3BCMU. Thus, the red color in the poly4BCMU case apparently results because of distortion of the polymer backbone in the ordered (hydrogen bonded) phase. This distortion is also thought to be involved in the thermochromic phase transitions observed in other urethane-substituted PDA crystals.(15)

III. KINETICS OF THE CONFORMATIONAL TRANSITION

The kinetics of the conformational transition are in some aspects rather sluggish. Though the transition from yellow to blue appears by eye to take place very rapidly, the spectra for the blue solutions continue to change at a detectable level for several hours after the addition of hexane. The blue-to-yellow transition appears to be slow even on a short time scale. These observations led us to investigate in some detail the time dependence of the spectroscopic changes--down to a time scale of a few milliseconds after rapid mixing. A Durrum-Gibson Stopped-Flow spectrophotometer was used to measure the time dependence of the absorption changes from 5 ms to 10 s after rapid mixing. The time scale for complete solvent mixing is 3-5 ms with this system. A Cary 219 UV/VIS spectrophotometer was used to record spectra for 10 seconds and longer after mixing was complete. To investigate the nonplanar (yellow $CHCl_3$ solution) to planar (blue $CHCl_3$/hexane solution) transformation, hexane was rapidly mixed into $CHCl_3$ polymer solution in an amount necessary to bring the final solvent composition to the desired point, typically $X_c = 0.62$. For the reverse transition (planar to nonplanar), $CHCl_3$ was added to a blue solution with typically $X_c = 0.62$. Absorption changes at three wavelengths were monitored: 465 nm for the nonplanar form, 630 nm for the planar form, and 506 nm. The third wavelength, 506 nm, corresponds approximately to the isobestic point for the transformation; absorption at this wavelength would be unchanged for a true isobestic point (no observable intermediates).

Results for the nonplanar-to-planar conformational transition are shown in Figure 6. The most remarkable aspect of these data is the fact that the changes in absorption vary roughly with the logarithum of time. This explains our earlier observation that although the transformation to the blue phase occurs vary rapidly on mixing, observable spectroscopic changes continue to take place on a much longer time scale. The rate of disappearance of the yellow phase (measured at 465 nm) is about the same as the rate of appearance of the blue phase (as measured at 630 nm). At the isobestic point (506 nm), a rapid increase in absorption occurs on mixing--too fast for us to measure. This additional absorption at 506 nm decreases rapidly on a time scale of about 100 ms and then slowly approaches its equilibrium value of about 9000 lit/mol-cm. Obviously if the conformational transition involves the production of long conjugated sequences from short ones, there must be species of intermediate conjugation length existing at some time scale. We associate the time dependence of the absorption at 506 nm with the observation of these intermediate species. The kinetics of this conformational transition are completely independent of polymer concentration over the range studied, 6×10^{-6} mol/lit to 6×10^{-4} mol/lit. This observation adds to the evidence for the single chain nature of the conformational transition and argues strongly against aggregation having a significant role at this stage. The reverse transition, planar-to-nonplanar, shows very similar kinetics to the forward transition except in the reverse order. The rate is slow at early times and speeds up toward the end of the conformational transition. The temperature dependence of the nonplanar-to-planar transition (studied in the long time region only) yields an activation energy of 2.5 kcal/mol over the temperature range 5 to 25 C. Room temperature kinetic measurements in two additional solvent systems, DMF/hexane and CH_2Cl_2/hexane, provide similar results.

The model we propose for the observed kinetics is based on a conjugation length dependent rate constant for the ordering of the individual chains in solution. Consider first the nonplanar to planar transformation. As the initial state we assume an infinitely long PDA chain with a randomly distributed concentration of structural defects which interrupt conjugation and break the hydrogen bonds between adjacent substituents along the chain. A 90 degree rotation about a single bond in the backbone would be a reasonable means for generating this kind of defect. The shortest segments in the conjugation length distribution would be the easiest to rearrange into a planar relationship with their neighboring segments, i.e. they would display the highest rate constant for the nonplanar-to-planar transformation. The longer segments in the distribution would be slower to rearrange.

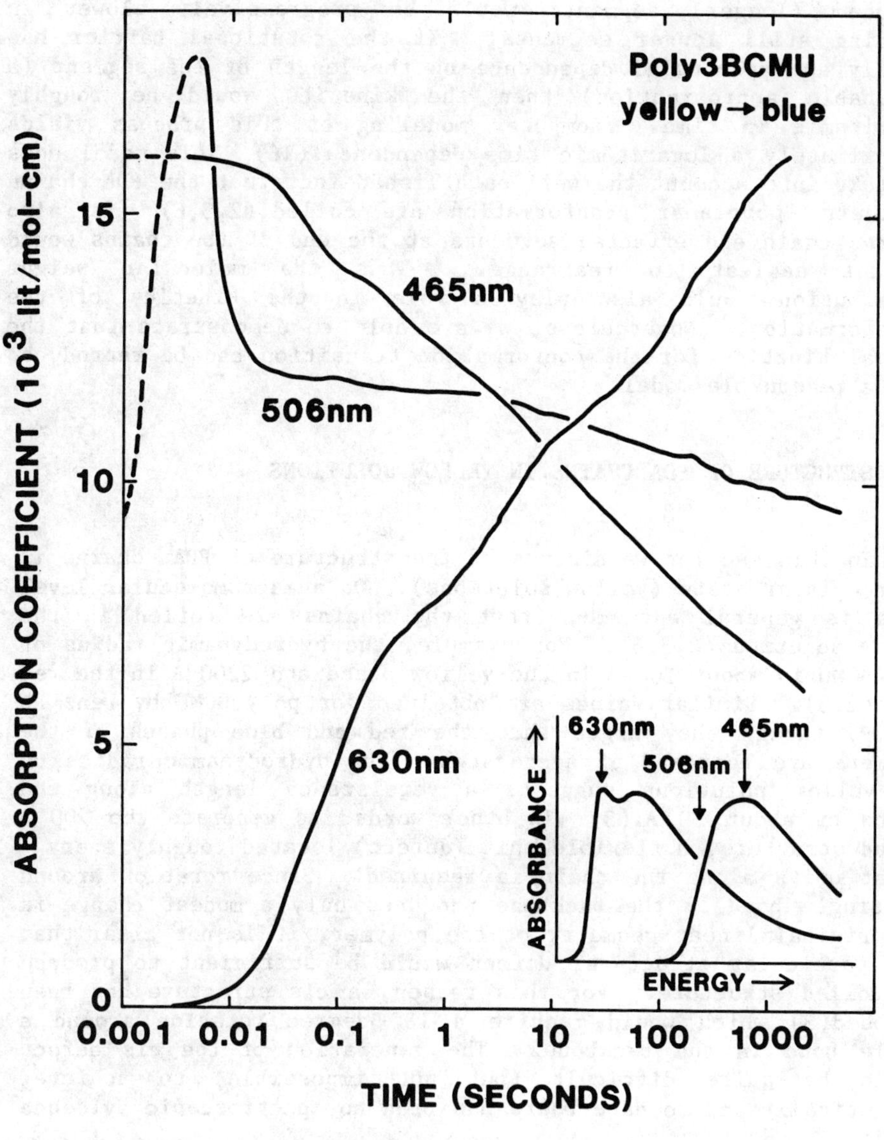

Figure 6. Kinetics of the yellow to blue transition. The concentration of the yellow component is monitored at 465 nm; the point for the color transition. The inset shows the absorption spectra for the two phases with arrows indicating the monitoring points.

Similarly as one proceeds through the transformation the newly produced (longer) segments would be progressively slower in yielding still longer segments. If the rotational barrier has roughly an exponential dependence on the length of the segment (a reasonable approximation) then the kinetics would be roughly logarithmic in time. Computer modeling of this process yields approximately a logarithmic time dependence.(16) This model does not take into account the well-established fact that the PDA chains in their "nonplanar" conformation are coiled.(2,3,6) It also ignores chain end effects; segments at the end of the chains would be the easiest to rearrange. Thus the molecular weight distribution could also play a role in the kinetics of the transformation. Nonetheless, we are able to demonstrate that the unusual kinetics for the conformation transition can be reproduced with a reasonable model.

IV. STRUCTURE OF PDA CHAINS IN YELLOW SOLUTIONS

In this section we discuss of the structure of PDA chains in the nonplanar state (yellow solutions). On a macromolecular level there is general agreement that the chains are coiled in the yellow solutions(2,3,6). For example, the hydrodynamic radius of poly4BCMU is about 200 A in the yellow phase and 2200 A in the red phase(2,3). Similar values are obtained for poly3BCMU by Wenz *et al*(6,9) though they argue that the red and blue phases of the polymers are composed of aggregates. The hydrodynamic radius of the yellow solutions suggests a persistence length along the chains of about 30 A.(3) In other words, to generate the 200 A coiled structure, a flexible unit (defect) located roughly every 6 repeat units along the chain is required. Since rotation around the single bond in the backbone produces only a modest change in the original linear geometry of the polymer, it is not clear that this particular structural defect would be sufficient to produce the coiled structure. For this reason, a cis structure has been proposed(3) which would require a 180 degree rotation around a double bond in the backbone. The generation of the cis defect should be quite difficult (but not impossible) to achieve, energetically and to date there has been no spectroscopic evidence for it.

We now turn to spectroscopic probes of the microstructure of the PDA chains. We have already noted from our analysis of the N-H stretch region of the infrared spectra (Figure 5) that about 25% of the N-H groups are not involved in hydrogen bonding in the yellow phase. This result suggests a defect every 4 repeat units--a defect which is likely to interrupt conjugation along the chain and produce an average conjugation length, ℓ, of about 4 repeat units. This analysis leads quite naturally to the ℓ distribution as a model for the microstructure of the PDA chains in the yellow solutions. If the polymer in the yellow solution consists of a distribution of ℓ values (i.e. chromophores), photoselection of the individual chromophores within the inhomogeneously broadened absorption band should be possible with resonance Raman scattering. In other words, when the incident laser frequency ω is in resonance with a particular chromophore (particular ℓ) in the distribution, the Raman spectrum obtained would be characteristic of that particular chromophore.

Raman frequencies (ν) associated the polymer backbone would be expected to vary with ℓ for basically the same reason that the optical absorption energy varies with ℓ, i.e. electron delocalization. In particular, we can expect that as electron delocalization increases with increasing ℓ, C=C and C≡C will become weaker and the Raman active vibrational frequencies associated primarily with these bonds, $\nu_{C=C}$ and $\nu_{C=C}$, will decrease. Thus the discrete defect model predicts a significant variation of Raman frequency with excitation frequency. This is precisely what is observed in Figure 7 where experimental results of $\nu_{C=C}$ versus excitation frequency ω are plotted.

We first consider the yellow solution and an explanation of the results based on Raman photoselection. At small ω, the incident laser samples the ensemble average of chromophores so that an average value of $\nu_{C=C}$ is obtained (~ 1520 cm^{-1}). As ω increases and approaches the polymer absorption edge, the laser samples the chromophores with the largest ℓ values present insignificant concentrations in the distribution. Hence, $\nu_{C=C}$ decreases in this region. As ω enters the absorption band, the laser samples chromophores with progressively shorter and shorter ℓ values, and $\nu_{C=C}$ increases continuously to the largest ω used in the experiment. Since the Raman line width is about 25 cm^{-1}, no discontinuities in $\nu_{C=C}$ versus ω can be expected. In fact, a more detailed analysis(17) shows that each Raman line has contributions from a number of chromophores within the distrubution with a peak frequency determined primarily by the resonance Raman effect.

For the blue solution of poly3BCMU, which is proposed to consist primarily of chains with effectively infinite ℓ,(1) we find $\nu_{C=C}$ of 1462 cm^{-1} at small ω; essentially the same value is found for poly3BCMU crystals for which $\ell=\infty$ may be assumed. At large ω, well within the absorption band, some shorter ℓ species are indicated by $\nu_{C=C}$ values in the 1490-1520 cm^{-1} range. At ω =

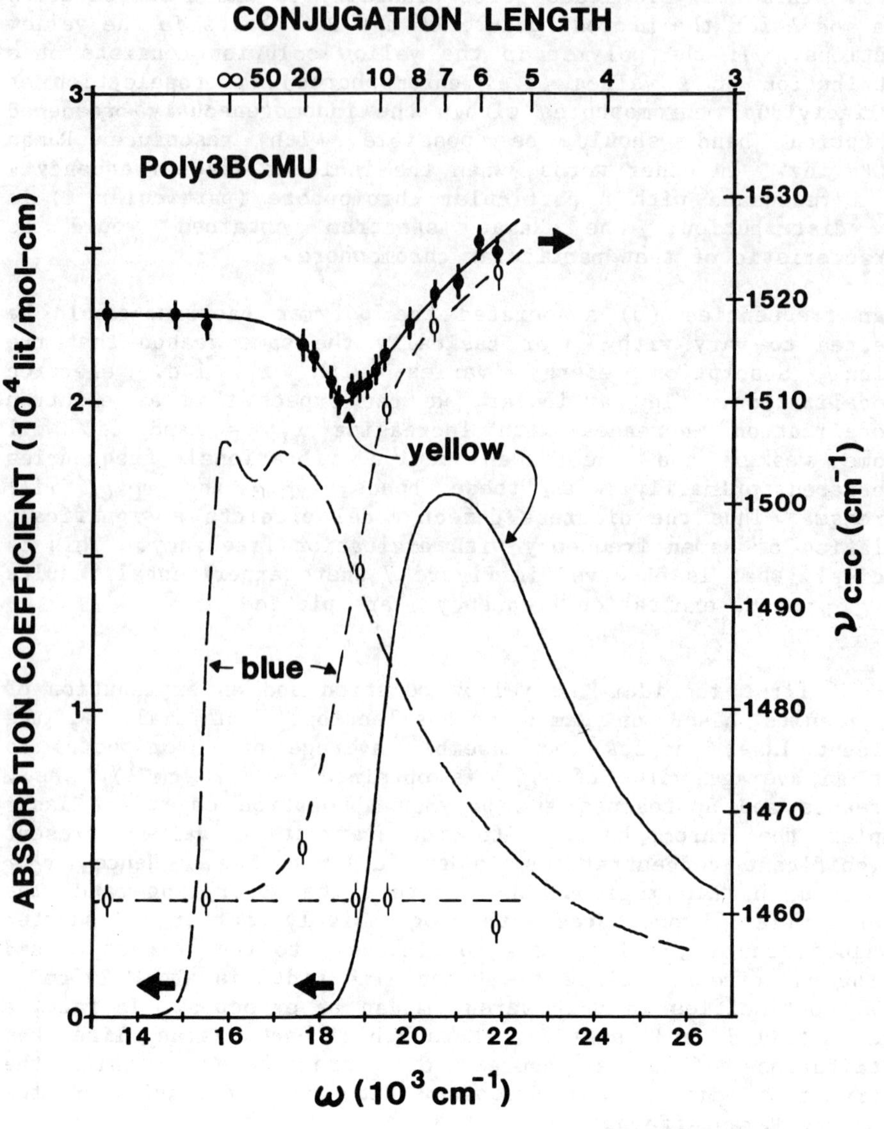

Figure 7. Variation of the peak Raman frequency with excitation energy for the mode associated primarily with the double bond stretch. Data are shown for both the yellow and blue solutions of poly3BCMU. (From reference 17)

18,800 cm^{-1}, two peaks are observed in the Raman spectrum, $\nu_{C=C}$ of 1461 and $\nu_{C=C}$ of 1491 cm^{-1}, with approximately equal intensity. Thus, these data indicate some degree of disorder in the blue solution, as is also suspected from a comparison of the absorption spectrum of the blue solution to that of the corresponding crystal (5). Results for $\nu_{C\equiv C}$ for both solutions show a behavior which parallels that of $\nu_{C=C}$

In summary, these Raman photoselection results strongly support a model for the polymer microstructure in the yellow solution based on a distribution of conjugation lengths produced by discrete defects. They appear in consistent inconsistent with a "wormlike" chain model.(9) A more quantitative analysis of these data has appeared elsewhere.(17)

A final comment should be made on the comparison of the infrared and visible spectra obtained on the yellow solutions. The infrared data provide an average ℓ of 4 repeat units while the visible absorption data of Figure 1 suggest an average ℓ of 7. These are not necessarily contradictory. The peak observed at 470 nm in the absorption spectrum would not actually correspond to the <u>average</u> ℓ value, since the spectroscopic width of the transition would also bear a major effect. In fact, we expect that the width would increase on decreasing ℓ, so that, the absorption peak occurs at ℓ values which are longer than the average value of the ℓ distribution.(17) With inclusion of this effect(17), the infrared and visible data can be brought into agreement with both interpreted in terms of an average ℓ value of 4 repeat units.

V. SUMMARY

We have shown that solutions of polydiacetylenes in good solvents have a disordered structure whose electronic properties can be described with a model based on a distribution of conjugation lengths. The conjugation length model has been shown to be adequate in interpreting infrared spectra, Raman spectra, and visible absorption spectra. The model has also been applied in a detailed study of the nonlinear optical properties of polyBCMU solutions.(4) The remarkable "visual" conformational transition has been shown to be a single chain phenomenum driven by intramolecular hydrogen bonding between substituent groups. Compatible analyses of the phase diagram for the transition and of the kinetics of the transition strongly support the single chain interpretation and rule out aggregation as the driving force for the transition. The kinetics of the conformational transition display a logarithmic time dependence which can be qualitatively explained with the nonplanar-planar, single chain model for the transition.

ACKNOWLEDGEMENTS

We acknowledge our collaborators in various phases of this work, especially M. L. Shand, M. Schott, R. Silbey, and R. H. Baughman. We are grateful to J. E. Frommer for her careful reading of the manuscript and her helpful suggestions. The kinetic measurements were carried out at the University of Michigan.

REFERENCES

1. G.N. Patel, R.R. Chance, and J.D. Witt, J. Chem. Phys. 70, 4387(1979).
2. G.N. Patel and E.K. Walsh, J. Polym. Sci. Polym. Lett. Ed. 17, 203(1979).
3. K.C. Lim and A.J. Heeger, preprint. See also, K.C. Lim, C.R. Fincher, and A.J. Heeger, Phys. Rev. Lett. 50, 1934(1983).
4. R.R. Chance, M.L. Shand, C. Hogg, and R. Silbey, Phys. Rev. B 22, 3540(1980.
5. R.R. Chance, G.N. Patel, and J.D. Witt, J. Chem. Phys. 71, 206(1979).
6. G. Wenz and G. Wegner, Makromol. Chem. Rapid. Commun. 3, 206(1982).
7. C. Plachetta, N.O. Rau, A. Hauck, and R.C. Schultz, Makromol. Chem. Rapid. Commun. 3, 249(1982).
8. C. Plachetta and R.C. Schultz, Makromol. Chem. Rapid. Commun. 3, 815(1982).
9. G. Wenz, M.A. Muller, M. Schmidt, and G. Wegner, Macromol. 17, 837(1984); M.A. Muller, M. Schmidt, and G. Wegner, Makromol. Chemie Rapid Comm. 5, 83(1984).
10. F. Wudl et al, these proceedings.
11. R.H. Baughman and R.R. Chance, J. Polym. Sci. Polym. Phys. Ed. 14, 2037(1976).
12. S. Kazunori, H. Ohnuma, and T. Kotaka, Polym. J. (Tokyo) 14, 895 (1982).
13. R.A. Nyquist, Spectrochim Acta 19, 509(1963).
14. L.R. Schroeder and S.L. Cooper, J. Appl. Phys. 47, 4310(1976).
15. R.R. Chance, Macromol. 13, 396(1980).
16. R.R. Chance and R.H. Baughman, unpublished results.
17. M.L. Shand, R.R. Chance, M. LePostollec, and M. Schott, Phys. Rev. B 25, 4431(1982).

Polydiacetylene Macromolecules in Solution: Rods, Coils and Gels

K.C. Lim, A. Kapitulnik, R. Zacher, S. Casalnuovo, F. Wudl, and A.J. Heeger
Institute for Polymers and Organic Solids
and
Department of Physics
University of California
Santa Barbara, CA 93106

Abstract

In this review of work at the Institute for Polymers and Organic Solids at UCSB, we summarize the results obtained from spectroscopic studies, quasi-elastic light scattering and field induced birefringence of dilute solutions. We then turn to more recent studies which focus on gels made up of networks of the rodlike molecules.

I. Introduction

Conjugated polymers in solution are of particular interest because of the coupling of the delocalized electronic structure to the polymer conformation. The discovery that with proper choice of side groups, some polydiacetylenes can be dissolved in common solvents opened the way for extensive experimental studies in this

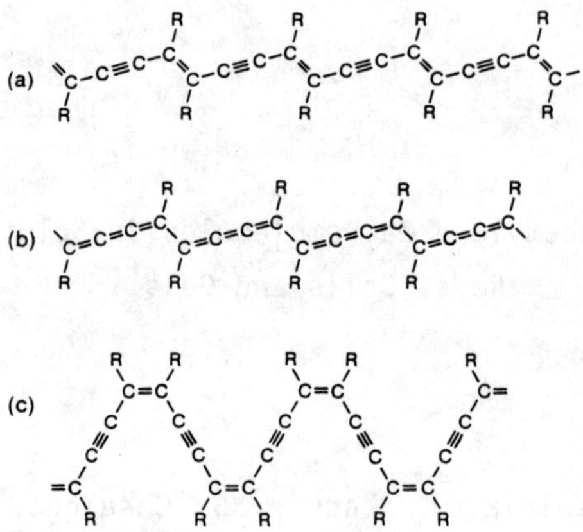

Figure 1. (a) Trans-polydiacetylene
(b) Polybutatriene
(c) Cis-polydiacetylene

area.[1] Polydiacetylene solutions were found to exhibit dramatic color change transitions when the solvent/nonsolvent ratio or the temperature is varied.[1 2] Patel et al[1] argued that these color-changes resulted from a planar -- nonplanar conformational transition of the polydiacetylene macromolecules in dilute solution. Direct evidence of the conformational transition came from the light scattering studies of Lim et al[3] which demonstrated a major change in hydrodynamic radius coincident with the spectroscopically observed transition. They interpreted their results in terms of a reversible rod-coil transition.

In a series of recent studies using quasi-elastic light scattering and field induced birefringence we have demonstrated that the polydiacetylene 4BCMU (where R = "4BCMU" = $-(CH_2)_4OCONHCH_2COOC_4H_9$; see Fig. 1) does inceed exist in two stable configurations: the "coil" and the extended "rod".

The light scattering and spectroscopic data,[4] taken as a function of the polymer concentration, have shown that the transition is a single chain (intramolecular) phenomenon. However, because of the large end-to-end length ($L \cong 1.2\mu m$) of the fully extended polymer, the dilute limit is not reached until concentrations below 10^{-5} g/cm^3. At higher concentrations evidence of cluster growth and aggregation are observed prior to gelation which occurs above a critical concentration, $c_o \cong 5 \times 10^{-4}$ g/cm^3. This cluster growth occurs as a result of the rodlike conformation of the individual molecules, but it is not the cause of the transition.

The field induced orientation[5] of poly-4BCMU in toluene in the red phase (<72°C) indicates an anisotropic polarizability tensor characteristic of a rodlike conformation. The field-induced signal from the polymer in its coil conformation (yellow phase) is extremely weak (smaller by a factor of at least 10^5) confirming the rod-coil transition. From studies of the transient response of the anisotropic light scattering from the red solution, we find a rotational relaxation time of 0.1 sec. This value is consistent with the rotational diffusion constant expected for rodlike polydiacetylene macromolecules.

II. Absorption Spectra of Polydiacetylene 4BCMU: Temperature Dependence

The absorption spectra for temperatures spanning the transition region are shown in Fig. 2. In the transitional region, the existence of an isosbestic point implies a coexistence of the two phases. In fact detailed analysis shows that the entire series of curves in Fig. 2 can be precisely described by the relation

$$\alpha(\omega) = f\alpha_R(\omega) + (1 - f)\,\alpha_y(\omega) \qquad (1)$$

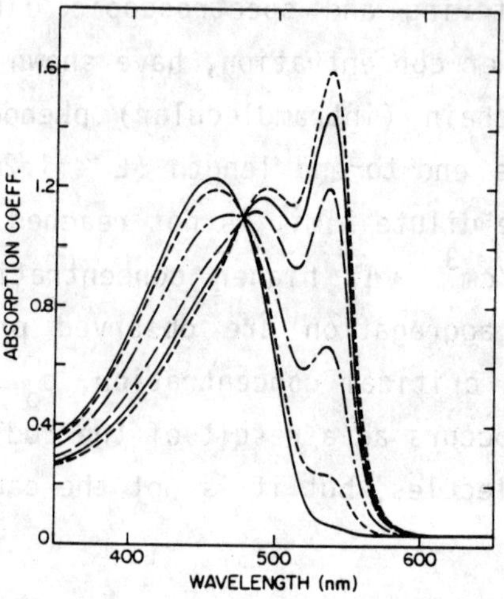

Figure 2. Absorption spectra for polydiacetylene 4BCMU in toluene at 72.0°C, 70.5°C, 67.5°C, 63.0°C, 59.2°C and 56.0°C. The highest temperature data are represented by the solid curve peaking at 450 nm; the lowest temperature data are shown as the dashed curve peaking at 530 nm.

where $\alpha_R(\omega)$ and $\alpha_y(\omega)$ are the limiting absorption curves for the red and yellow phases, respectively, and f(T) is the temperature dependent fraction which varies from 1 (below 40°C) to 0 (above 72°C). The thermally driven transition can also be observed for polydiacetylene 4BCMU in the chloroform/hexane solvent system, where for $X_c < 0.56$ the polymer is in the red phase at room temperature (see ref. 4).

Patel et al.[1 2 6] made a careful study of the phase diagram for polydiacetylene 4BCMU in hexane/chloroform. They were able to unambiguously distinguish the precipitation (cloud) point from the color transition. As expected the precipitation point varied considerably with polymer concentration whereas the conformational transition occurred at fixed X_c over a range of concentrations spanning nearly three orders of magnitude. We have extended the spectroscopic studies in chloroform/hexane (as a function of X_c) and in toluene (as a function of T) to concentrations two orders of magnitude even more dilute. These results demonstrate that the color transition (and the associated changes in conformation and electronic structure) is fundamentally a purely single chain phenomenon.

III. Quasi-Elastic Light Scattering From Polydiacetylene 4BCMU in Toluene: Temperature Dependence

In the case of quasi-elastic light scattering from a monodispersed polymer in solution, the time constant for the autocorrelation function appropriate to translational diffusion is given by[7]

$$\tau^{-1} = 2D_t q^2 \tag{2}$$

where $q = (4\pi n/\lambda)\sin\theta/2$. This relation is strictly valid for $qR_H < 1$. For larger q, internal modes can also contribute to the decay of the correlation function. Preliminary studies of the angular dependence (θ) verified that in our experiments $\gamma \propto q^{-2}$ so that the diffusion constant can be obtained directly from γ. The hydrodynamic radius is obtained from the Stokes equation,

$$D_t = k_B T/6\pi\eta R_H \tag{3}$$

where η is the solution viscosity, k_B is the Boltzmann constant, and T is the absolute temperature.

To prepare the solution samples for light scattering polydiacetylene 4BCMU was dissolved into toluene at elevated temperature and subsequently passed through a 0.22μm filter with the solution and filtering apparatus heated to approximately 80°C.

The solutions were then maintained at this temperature until needed for the light scattering measurements as a function of temperature. The measurements were carried out with the solution in a sealed cuvette. However, the high vapor pressure of toluene at elevated temperatures invariably resulted in some leakage and eventual loss of solvent by evaporation.

The light scattering data presented here were obtained from samples which gave high quality autocorrelation functions and consistent data from run to run and from day to day. In such cases, the scattering volume appeared clear and uniform. Moreover, no indication of large aggregates was observed after convective stirring following a change in the sample temperature. A good sample typically gave consistent results over a period of one to two weeks, at least until there was significant loss of solvent.

Even with careful filtering and sample handling we experienced considerable difficulty in obtaining good samples of solutions in the red phase. The problem appears to arrise from a tendency toward aggregation of the polymer molecules. Such samples yielded fluctuating diffusion constants from run to run, noise spikes in the intensity, etc. In such cases visual checking of the scattering volume invariably revealed clear evidence of inhomogeneity, and the resulting values for the hydrodynamic radius were typically a few times larger than those obtained from the good samples. Additional evidence of aggregation was the observation of a major increase in the noise after a change in temperature (which would be expected from convective stirring of the aggregated solution).

Our conclusion is that with care and persistence it is possible to prepare high-quality dilute molecular solutions despite a tendency toward aggregation of the rodlike molecules in the red phase.

Semi-log plots of the autocorrelation function taken at 24°C (red phase) and 73°C (yellow phase) are shown in Fig. 3. The change in slope (Dq^2) is clearly evident implying a corresponding change in R_H. The temperature dependence of R_H for poly-diacetylene 4BCMU in toluene is shown in Fig. 4.

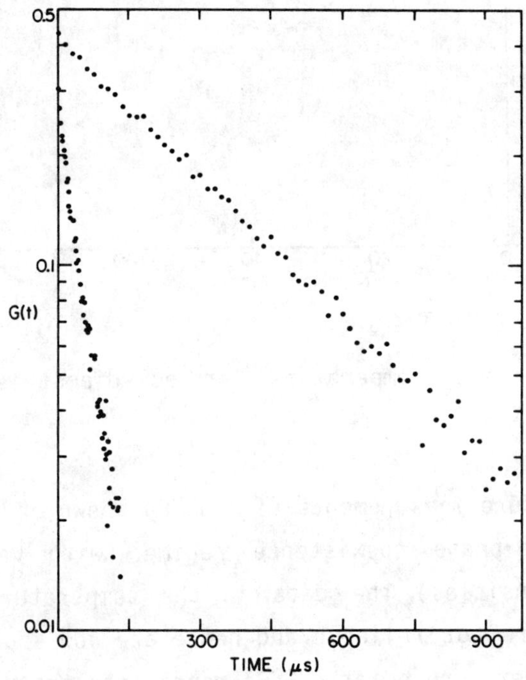

Figure 3. Semi-log plots of the autocorrelation functions (G(t)) vs. time for polydiacetylene 4BCMU in toluene. The longer time constant data correspond to the red solution at room temperature; the shorter time constant data were obtained from the yellow solution at 72°C.

The samples used to measure the diffusion constant of the rods in the red phase had a polymer concentration of about 10^{-6} gm/cm^3. However, because of the weak scattering from the yellow phase, a much higher concentration ($\sim 10^{-3}$ gm/cm^3) was used for the

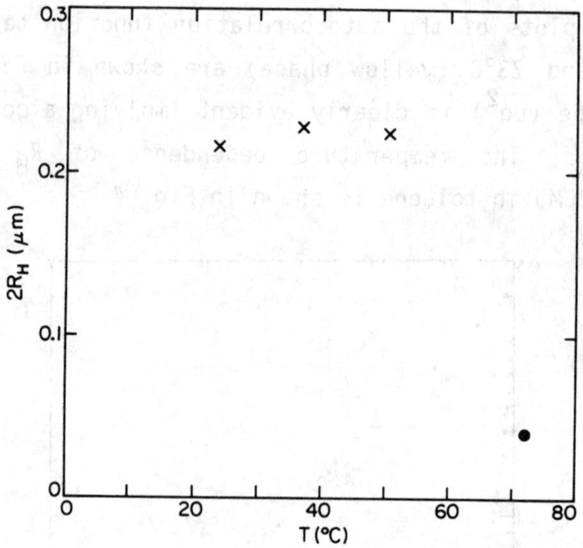

Figure 4. $2R_H$ vs. temperature for polydiacetylene 4BCMU in toluene.

highest temperature measurements (T > 65°C) shown on Figs 3 and 4. Due to the two-phase coexistence regime (with very different scattering intensities) the data in the temperature range from 50°C to 65°C were not reliable and hence are now shown on Fig. 4. The highest temperature point (73°C) represents the upper limit of the temperature controller unit of the light scattering apparatus.

The changes in R_H on going through the conformational transition from rod to coil as a function of T (in toluene) or X (chloroform/hexane) show the same general features. At low temperature (or in poor solvent) the polymer is fully extended with $2R_H \cong 0.22\mu m$; at high temperature (or in good solvent) the polymer collapses to a coil with $2R_H \cong 0.04\mu m$. Note that the value obtained for $2R_H$ is slightly different in toluene at T >

70°C ($R_H \cong 0.04\mu m$) from that obtained in chloroform ($2R_H \cong 0.06\mu m$) at room temperature. This is not surprising since the hydrodynamic radius of a coil is expected to vary somewhat from one solvent to another.

Since we know that gelation occurs[8] at concentrations above 6×10^{-4} gm/cm^3 it is important to establish the concentration below which the polydiacetylene 4BCMU rods can be considered truly dilute. The dependence of R_H on polymer concentration is shown in Fig. 5. For concentrations below 10^{-5} gm/cm^3, $2R_H$ is independent of concentration consistent with the interpretation of the data in terms of single polymer chains in dilute solution. At higher concentrations (> 10^{-5} gm/cm^3) the apparent hydrodynamic radius increases, indicative of the onset of cluster growth precursor to gelation. The dilute limit, therefore, is apparently achieved for concentrations below 10^{-5} gm/cm^3. The concentration independence of $2R_H \cong 0.22\mu m$ in the red phase below 10^{-5} gm/cm^3 confirms the conclusion that the resulting value is determined by the molecular conformation, and not by polymer aggregation.

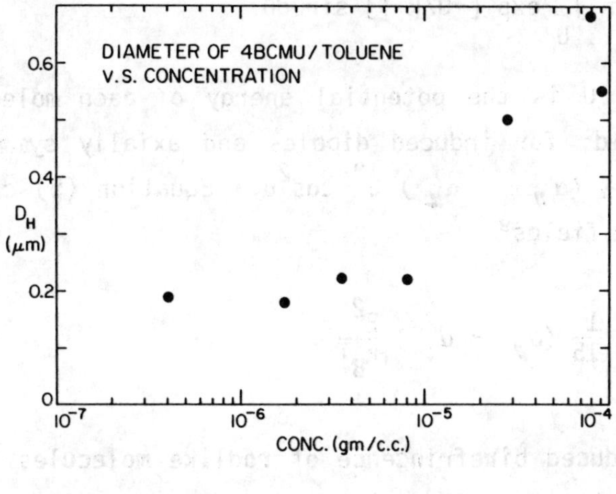

Figure 5. Hydrodynamic diameter (from quasi-elastic light scattering) of polydiacetylene 4BCMU as a function of polymer concentration (in toluene).

IV. Electric Field Induced Birefringence of Polydiacetylene 4BCMU

A. Theoretical Background (see ref. 5)

The field induced birefringence of polymer rods[9] is proportional to the average orientation order parameter $\bar{S} = \langle(3\cos^2\theta-1)/2\rangle$,

$$\Delta n = \frac{2\pi N}{n}(\alpha_{\parallel} - \alpha_{\perp})\bar{S} \equiv n_{\parallel} - n_{\perp} \tag{4}$$

where N is the number of molecules per unit volume, $\alpha_{\parallel}$ and $\alpha_{\perp}$ are the polarizabilities of the polymer (assumed axially symmetric) parallel and perpendicular to the chain axis, respectively, and n is the refractive index of the isotropic solution. The order parameter is calculated via

$$\bar{S}(E,T) = \int_0^{\pi} W(\theta)\, S(\theta)\, 2\pi\sin\theta d\theta \tag{5}$$

with the weight function

$$W(\theta) = \frac{\exp[-U/k_BT]}{2\pi\int_0^{\pi}\exp[-U/k\ T]\sin\theta d\theta}, \tag{6}$$

In eqn. (6), U is the potential energy of each molecule in the electric field; for induced dipoles and axially symmetric molecules, $U = -\frac{1}{2}(\alpha_{\parallel} - \alpha_{\perp})E^2\cos^2\theta$. Equation (5) can be evaluated for low fields[9]

$$\bar{S}(E,T) = \frac{1}{15}(\alpha_{\parallel} - \alpha_{\perp})\frac{E^2}{k_BT}.$$

The field induced birefringence of rodlike molecules in solution can therefore be written

$$\Delta n(\text{rods}) = \frac{2\pi N}{15n} (\alpha_{\parallel} - \alpha_{\perp})^2 \frac{E^2}{k_B T} . \quad (7)$$

For coiled polymers in solution, θ varies over the length of the macromolecule. If one models the coil as a polymer consisting of N randomly oriented straight segments connected by flexible joints, it has been shown[9] that the resulting field induced birefringence, Δn(coil), is smaller than the corresponding Δn(rod) by a factor of N.[10] Thus, if a given macromolecule exists in two forms, rod and coil,

$$\frac{\Delta n(\text{rod})}{\Delta n(\text{coil})} \cong N \quad (8)$$

The field induced birefringence is, therefore, a powerful probe for detecting and studying the rod-coil transition.

The time dependence of the free decay of the birefringence of a dilute solution is determine by the rotational diffusion constants of the molecules. Wegener et al.[11] have shown that for a monodispersed solution of axially symmetric rodlike molecules, the decay is a single exponential

$$\Delta n(t) = \Delta n_o \exp[-6D_r t] \equiv \Delta n_o \exp(-t/\tau_r) \quad (9)$$

where D_r is the rotational diffusion constant. Neglecting end effects[7,9]

$$D_r = \frac{3 k_B T}{\pi \eta L^3} \ \ell n \ L/d \quad (10)$$

where d is the effective diameter of the rod. The corresponding time constant for translational diffusion is given by Eqn. 2; for rods, D_t is as follows:

$$D_t = \frac{k_B T}{3\pi\eta_o L} \ell n \frac{L}{d} \tag{11}$$

Note that the ratio D_t/D_r is independent of the temperature, solvent viscosity, etc. In fact, for monodispersed rods, $D_t/D_r = (1/9)L^2$. More generally, when polydispersity is important

$$D_t/D_r = 1/9\ \bar{L^3}/\bar{L} \tag{12}$$

where $\bar{L^3}$ and $\bar{L}$ represent averages over the polydispersity distribution.

B. Experimental Results

The experimental configuration is typical for an electro-optic birefringence measurement.[9] An electro-optic cell with a set of parallel electrodes was placed between Glan-Thompson crossed polarizers. The axes of the polarizer and the orthogonal analyzer were set at $\pi/4$ with respect to the electrodes so that the electric field was at $\pi/4$ with respect to both the polarizer and the analyzer. A 5 mW He-Ne laser beam was incident through the first polarizer, through the electro-optic cell between the electrodes, then through the analyzer. The exit light intensity was monitored by a photodiode. The parallel electrodes were made of two thick brass plates spaced 3mm apart; the electrodes defined an optical path length of 3.7 cm. Care was taken to elimiate stress birefreingence in the windows of the electro-optic cell. The signal from the photodiode in zero applied field (resulting from light leakage, etc.) was below the noise level. Under these conditions, the detected intensity due to field induced orientation is proportional to $(\Delta n)^2$. To avoid electrophoresis, an ac electric field was used.[5] Concentrations used ranged from $\sim 5\times 10^{-6}$ gm/cm^3 to $\sim 5\times 10^{-5}$ gm/cm^3. Below 5×10^{-6} gm/cm^3, the birefringence signal was very weak; above 5×10^{-5} gm/cm^3, it is difficult to prepare large quantities of sample without aggregation (since the concentration is approaching the gelation point).

The sensitivity of the field-induced birefringence to the macromolecule conformation is expressed in eqn. (8). We, therefore, attempted to determine this ratio experimentally. Although large signals were easily detected (signal to noise of order 400:1 at 1.5 kV/cm) from the red phase at concentrations of order 10^{-5} gm/cm^3 (in toluene), any signal from the yellow phase at this concentration (in chloroform) was below the noise level. In order to detect any field induced birefringence in the yellow phase concentrations of order 10^{-3} gm/cm^3 were required and even then to obtain a signal of twice the noise level required an applied field more than a factor of 100 greater than that needed to obtain the same signal in the red phase. Normalizing to the same concentration, we conclude that Δn^2 is greater by a factor which is conservatively in excess of 10^5, so that $\Delta n(rod)/\Delta n(coil) > 300$. An independent check was carried out using the (more sensitive) standard configuration[9a] for measuring a signal linear in Δn. The observed ratio was 600. <u>Since the molecules involved are identical, this large ratio directly and unambiguously indicates a major conformational change</u>.

The large field induced birefringence in the red solutions could result either from orientation of individual rodlike molecules by the field or from field orientation of large aggregates in which the individual molecules are <u>fully</u> aligned. In the latter case, random growth of aggregates made up of a large number of extended and aligned polydiacetylene 4BCMU molecules would result in a particle with dimensions much greater than those of the indivdiual polymer chain. To distinguish between these two cases, we have carried out measurements of the transient decay of the birefringence after switching off the applied field. For rodlike particles, D_r is given by eqn. (10); note that D_r varies as L^{-3} and is therefore extremely sensitive to the particle dimension.

The transient response is shown in Figure 6. In switching on the ac field, the light intensity detected was observed to build up to an average intensity I_o (these experiments were carried out at $\omega\tau_r > 1$ so that the ac component on I_o was small). Similarly, the decay of the birefringence is evident for times after switching off the applied field.

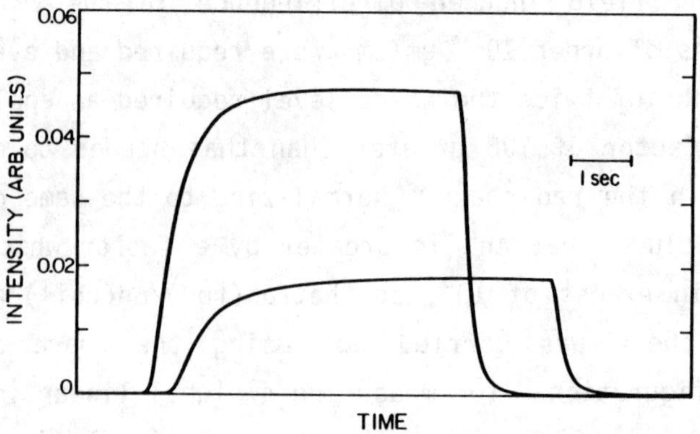

Figure 6. Transient response of the birefringence of polydiacetylene 4BCMU for a chopped ac electric field with applied rms field strength of 900 V/cm (lower curve) and 1400 V/cm (upper curve), showing the build-up and decay of orientational order.

The field-free decay of $(\Delta n)^2$ is shown in more detail on a semilog plot in Figure 7; the behavior is exponential with a time constant of $\tau_r/2 \cong 0.11$ second. This value was checked independently by using the standard configuration[9a] for measuring a signal linear in Δn; we found $\tau_r \cong 0.23$ s, in excellent agreement. The time constant for the decay is independent of the magnitude of the initial applied field (ranging from 700 V/cm to 1500 V/cm). Over the range of the concentrations used (5×10^{-6} gm/cm^{-3} < c < 5×10^{-5} gm/cm^3) we found τ_r to be independent of concentration.

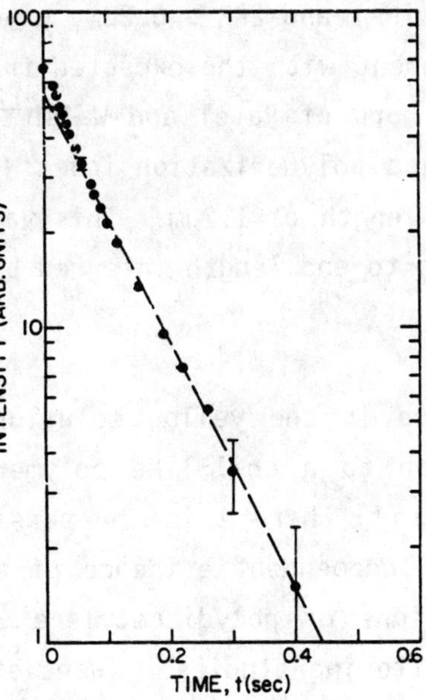

Figure 7. Semi-log plot of the intensity vs. time after switching off the applied field. The decay is due to the loss of orientation due to rotational diffusion. The dashed lines corresponds to a time constant of 0.1 sec; the initial slope corresponds to a time constant of 0.05s.

Experiments were carried out to investigate the effect of aggregation on the orientational relaxation time. In the dilute regime freshly prepared samples gave $\tau_r/2 \cong 0.1$s; deviations of less than ∿50% were observed from samples that had been allowed to stand for long periods (∿ a few days). Even in cases where visual examination showed clear evidence of significant aggregation, we found $\tau_r/2 \cong 0.25$ second.

V. The Rod-to-Coil Transition in Dilute Solution: Data Analysis

If the red phase is actually rod-like, the length of the rod L is related to $2R_H$ (for an ellipsoid of revolution) by $2R_H \cong$

$L/\ell n(L/d)$ (see eqns. 3 and 10). Taking $d = 40\AA$ (twice the length of the 4BCMU side chains) and $2R_H \cong 0.2\mu m$, gives $L \cong 1.2\mu m$. This is in excellent agreement with the expected fully extended length as inferred from the work of Patel and Walsh[12] who characterized the polymer and found a polymerization index $N \cong 2400$ corresponding to an end-to-end length of 1.2μm. This value is also in good agreement with the end-to-end length inferred by Wenz et al.[13] for polydiacetylene 3BCMU.

The smaller value in the yellow solutions implies a major change in conformation to a coil-like polymer. Assuming $2R_H = (L\ell)^{\frac{1}{2}}$ (for a random coil) where ℓ is the persistence length, one finds $\ell \sim 25 - 30\AA$. Independent evidence of a coil conforamtion for the yellow solutions of polydiacetylene 3BCMU has been obtained from light scattering studies by Wenz et al.,[13] who interpreted their data in terms of a continuously twisted worm-like coil. The question of whether the coil is worm-like or relatively sharply kinked must be left open since the data currently available do not distinguish between the two. We note, however, that a conjugated chain containing soliton-like twistkinks might be expected to have a lower energy than a continuously deformed conjugated chain.

The field-induced birefringence results demonstrate that the transient birefringence is relatively insensitive to aggregation and that the decay time in the dilute limit is independent of the concentration. Since the orientational relaxation time is a strong function of particle size ($\tau_r \sim L^3$, see eqn. 10) these results indicate that aggregates do not contribute significantly to the field induced birefringence. Evidently, the aggregates which form prior to gelation consist of polydiacetylene 4BCMU molecules which are essentially randomly oriented. Such aggregates would contribute to the light scattering intensity at low angles, but would not contribute significantly to the magnitude of the field induced birefringence nor would they be observed in the decay.

Since the field-induced birefringence is insenstiive to aggregation, we can unambiguously interpret the ratio of Δn (red phase and yellow phase) in terms of the rod-coil transition. The small value obtained from the coils implies (see eqn. 8) a large number of segments and a small persistence length. Assuming that the straight segments are of order 3 to 4 monomers (as inferred from analysis of the spectroscopic shift and the observation of one hydrogen bond in four being broken in the yellow phase[1,2,6] and assuming a typical molecular weight of 2400 monomers[2] then N(coil) $\sim$ 600-800. The large ratio obtained for Δn(red)/ Δn(yellow) implies, therefore, that in the red solutions the polymer is nearly fully extended with at most a few kinks or bends; N(rod) $\sim$ 1. This is in agreement with our earlier conclusion based on comparison of the absorption spectra of the red solutions with the spectra obtained from nearly completely aligned solid films of the same polymer.[14]

The birefringence decay and the autocorrelation function decay (from quasi-elastic light scattering) can be used to determine $\bar{L}$. For a Gaussian distribution, $\exp[-(L-\bar{L})^2/2\sigma^2]$, eqn (12) can be wrriten as

$$D_t/D_r = 1/9\ \bar{L}^2\ (1 + 3\sigma^2/\bar{L}^2) \qquad (13)$$

so that

$$\tau_r/\tau_t = 1/27\ \bar{L}^2\ q^2\ (1 + 3\sigma^2/\bar{L}^2) \qquad (14)$$

Finally, including the polydispersity factor:

$$\frac{1}{\tau_r} \cong \frac{18k_BT \quad \ln \bar{L}/d}{\pi\eta_o \; \bar{L}^3 \; (1+3\sigma^2/\bar{L}^2)} \tag{15}$$

Previous characterization studies[12] gave information on both $\bar{L}$ and the degree of polydispersity for the soluble polydiacetylenes with BCMU side groups. For poly-4BCMU, $\bar{L} \sim 2400$ monomers ($\sim$ 1.2μm) with a polydispersity index ($\bar{M}_w/\bar{M}_n$) of 5. The latter value yields $\sigma^2/\bar{L}^2 \cong 4$ (assuming a Gaussian distribution, $\bar{M}_w/\bar{M}_n = 1 + \sigma^2/\bar{L}^2$).

Using eqn. (15) with a polydispersity index[8] $\bar{M}_w/\bar{M}_n = 5$ we find $\bar{L} \cong 1.5\mu m$, in good agreement with the earlier estimates.[12] Using eqn 14 for the ratio τ_r/τ_t together with the experimental values ($\tau_r \cong 0.2s$ and $\tau_t \cong 500\mu m$) yields $\bar{L} = 2\mu m$. Thus the transient birefringence decay, independently and when combined with light scattering data, leads to values of $\bar{L}$ consistent with the length of individual rodlike poly 4BCMU macromolecules.

The experimental results obtained for τ_r are not consistent with large aggregates made up of 500 to 700 extended and aligned polydiacetylene molecules.[13] If such aggregation were the origin of the conformation change, the length of the typical aggregate would be larger than that of the extended molecules (yielding a much larger τ_r) and would grow with increasing concentration. On the contrary, we find τ_r to be in good agreement with the magnitude expected for single molecule rotational diffusion, to be concentration independent in the dilute limit, and to be explicitly insensitive to aggregation at higher concentrations. We conclude, therefore, that the conformational change observed in polydiacetylene 4BCMU is a single chain phenomenon and not caused by aggregation. The field induced birefringence data independently confirm that the polymer is tightly coiled in the yellow solution and nearly fully extended in the red solution; in other words the transition is indeed rod to coil.

VI. Gels of Polydiacetylene 4BCMU in the Rod-like Conformation

The existence of extended rodlike molecules implies a tendency for intermolecular interaction which leads to aggregation and gelation at relatively low polymer concentrations. Consistent with these ideas, the gel-sol transition has been observed and studied[8] (see following section). It was found that for gelation to occur, the polymer must be in its fully extended rodlike conformation; upon transformation to the flexible coil, the molecules diffuse readily and are unable to sustain a gel. The very existence of a critical concentration for gelation[8] of $C_o \cong 5\times10^{-4}$ gm/cm^3 (in the red solutions) rules out the many chain aggregates proposed by Müller et al.[13] In fact, with nearly spherical particles consisting of ∿700 polymer chains,[13] the gelation would occur only at concentrations above ∿0.3; three orders of magnitude greater than the observed value.[8] Thus, the cluster growth and aggregation occurs as a result of the rodlike conformation of the indivdiual molecules, but it is not the cause of the conformational change.

A. Spontaneous Nematic Alignment of Polydiacetylene 4BCMU Gels

Competing tendencies in polymer solutions restrict the types of polymers for which lyotropic liquid crystalline phases can form. On the one hand, a rigid rodlike polymer will produce an ordered phase at a lower concentration than will a flexible polymer.[15]. On the other hand, a rigid polymer is generally less soluble than a flexible one,[16] making it difficult to achieve even the reduced concentrations necessary to produce the ordered phase.

Since polydiacetylene 4BCMU has a rodlike conformation with a large axial ratio (length/diameter) in certain solvents, it is a likely candidate for a lyotropic polymer. However, it has low solubility in those solvents and no liquid crystalline phase has been observed. As described above, a rod-to-coil conformational transition can be induced in the PDA 4BCMU by changing the solvent quality[4] (e.g., by increasing the temperature) with a marked increase in polymer solubility. Cooling a sufficiently concentrated solution (> 0.06% by weight) from the high temperature, high solubility, coil phase back to the low temperature, low solubility, rod phase, produced a gel;[2,8] a random network of the rodlike polymer molecules embedded in the solvent. One might suspect, for such a polymer, that it would be possible to prepare a solution in the (isotropic) coil phase which, on cooling to the rod phase, would form an orientationally ordered gel.

We have recently reported[17] observations of macroscopic optical birefringence in PDA 4BCMU gels that indicate spontaneous, long-range, nematic alignment in these gels. Orientational order has been observed in polymer gels previously and is of considerable technological importance.[16] However, orientation is produced in those gels either by stretching the gel network or by shearing the polymer solutions prior to gelation. In contrast, PDA 4BCMU gels exhibit <u>spontaneous</u> birefringence, i.e., without subjecting them to external fields. We know of no previous reports of the spontaneous formation of a nematic gel phase.

The gels exhibit phenomena characteristic of birefringent media. Figure 8 demonstrates birefringent double refraction in the gel. A single, well-collimated, linearly polarized He-Ne laser beam (5mW, 6328Å) passes through a PDA 4BCMU gel ($c = 1.7 \times 10^{-2}$ g/ml) at a small angle to the gel surface and strikes a screen several meters away. The front and back surfaces of the

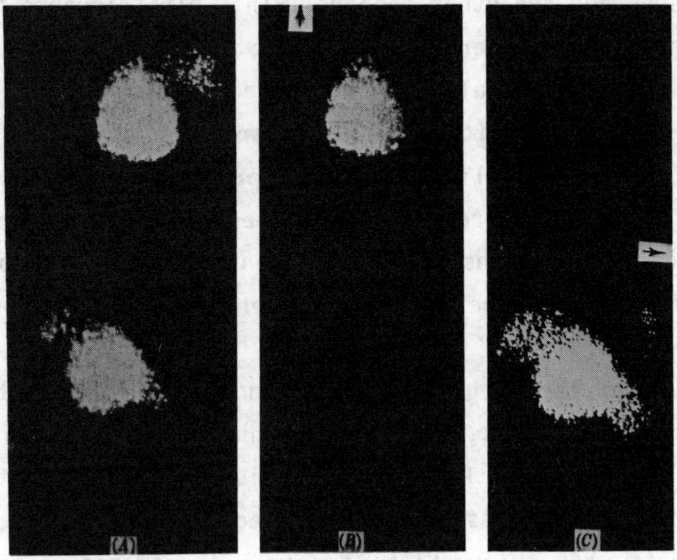

Figure 8. A linearly polarized He-Ne laser beam is decomposed into two beams on passing through the gel. The photographs show the two beams as they strike a screen:

a) Two beams resulting from double refraction by the gel;
b) A polarizer after the gel is rotated to extinguish one beam (polarization indicated by arrow);
c) An additional rotation of polarizer by 90° extinguishes the other beam.

gel are not parallel in order to enhance the separation of the two refracted beams. In Fig. 8a, we see the single beam has been decomposed into two separate beams. A polarizer is placed in these beams and rotated until one of the beams vanishes (Fig. 8b). When the polarizer is rotated 90°, the second beam vanishes, and the first reappears (Fig. 8c). It is also possible to rotate the incident polarization so that only one spot appears on the screen.

On rotating the incident polarization by an additional 90°, that spot vanishes and the other appears. This birefringence indicates net orientational order in the gel; i.e., the polymers are aligned, on average, along some preferred direction. The net alignment, coupled with the anisotropic molecular polarizability[6b,18] of the rodlike molecules, produces the birefringence. Moreover, this double refraction implies net orientational order over macroscopic lengths. If the orientation occurred only over microscopic domains, the net birefringence would average to zero.

Because of the large refractive index difference between the polymer and the solvent, the size of the polymer, and the random nature of the gel network, PDA 4BCMU gels scatter light strongtly. This property makes possible a novel technique for monitoring the birefringence when observations are made transverse to the direction of light propagation. In Fig. 9, we present photographs of a He-Ne laser beam (6328Å) as it passes through a PDA 4BCMU gel with concentration 5×10^{-3} g/ml. The viewing direction is normal to the direction of beam propagation and the incident beam is normal to the front surface of the gel. In Fig. 9a the laser is linearly polarized; in Fig. 9b, unpolarized. That the striae seen in Fig. 9a are not found in Fig. 9b rules out periodic density variations in the gel as their cause. Rather, the striae result from the interference of the light scattered from the ordinary wave with that scattered from the extraordinary wave. When linearly polarized, monochromatic light is incident normally on a birefringent medium, it can be considered as two superimposed waves, the ordinary (o) and the extraordinary (e) waves, propagating with orthogonal polarizations at two different velocities. At any point in the medium, the two waves differ in phase by $\delta = 2\pi\Delta n x/\lambda$, where $\Delta n = n_o - n_e$ is the difference in the refractive indices, λ is the wavelength of the light in vacuo, and x is the distance travelled by the light in the medium. The waves scattered by fluctuations in the dielectric constant retain this same phase

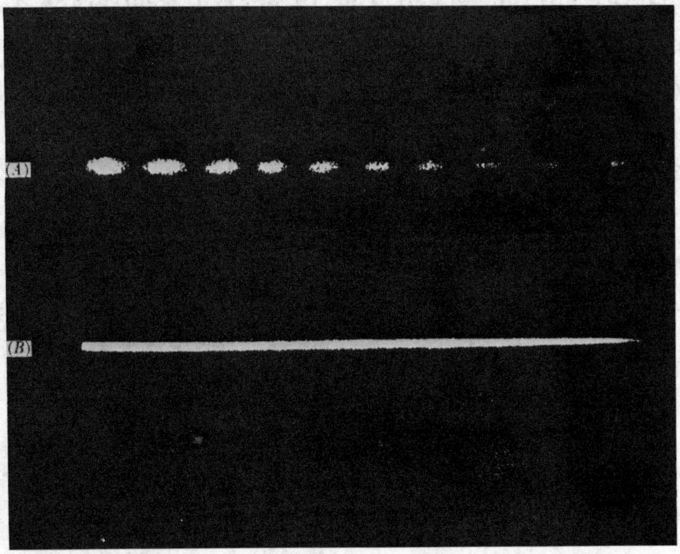

Figure 9. Photograph of He-Ne laser beam as it passes through the PDA 4BCMU gel; viewing direction is normal to propagation direction.

a) Laser is linearly polarized

b) Unpolarized. The dark fringes in a) are separated b 1.5 mm.

shift. The total scattered electric fields, $\vec{E}_s$, at a given point along the beam is then $\vec{E}_s = \vec{E}_{so} + \vec{E}_{se} \exp(i\delta)$, where $\vec{E}_{so}$ and $\vec{E}_{se}$ are the fields scattered from the ordinary and extraordinary waves, respectively. The scattered intensity, I_s, is proportional to E_s^2 or $I_s \alpha E_{so}^2 + E_{se}^2 + 2E_{so}E_{se} \cos \delta$. Therefore, I_s is modulated as a result of the birefringence of the gel. This explanation can be tested by rotating the polarization of the incident beam. We find two orientations, separated by 90°, for which the dark striae vanish. These orientations are the same as those for which the double refraction in Fig. 8 disappears and correspond to the incident polarization being parallel to one of the optical axes.

The spatial period, d, of the intensity modulation is related to Δn by $d = \lambda/\Delta n$, while Δn is related to the degree of orientational order in the gel and can be used to estimate the order parameter, $\bar{S} = (3\cos^2 \theta - 1)/2$. In Fig. 9a, we measure $d \cong 1.5$ mm, yielding $\Delta n \cong 4 \times 10^{-4}$. Measurement of the reflectivity of PDA 4BCMU crystals[6] gives $\Delta n \cong 3.8$ at $\lambda = 6328\mathring{A}$. If we let the crystal represent complete alignment in a system of unit density, then our value of Δn implies $\bar{S} \cong 0.02$. The striae are uniformly spaced along the 15 mm path of the beam, implying a constant order parameter over the sample along that direction.

In the absence of a preferred direction in the gel, one would expect nematic alignment only in microscopic domains oriented randomly with respect to one another. Since the principal axes are found to be parallel and perpendicular to the container walls, we believe the walls imposed this preferred direction. Moreover, the striae in Fig. 9 are always more closely spaced when the laser beam passes close to one of the container side walls. As the beam is translated away from the wall, $\bar{S}$ decreases (by a factor of 2-3 on going $\sim$ 0.5 cm from the wall). Once the beam passes the middle of the container and approaches the opposite side wall, $\bar{S}$ again increases.

Since x-ray scattering does not indicate the presence of microcrystals in the PDA gels, we suggest a picture based on a gel network of interconnected rods[8] to account for the oriented gel. Steric interactions among the rodlike polymers would tend to produce an aligned solution phase. An additional, attractive, short-ranged interaction (possibly hydrogen bonding, in this case),[1,2,6] insensitive to the orientation of neighboring rods, would produce the essentially random gel network. Whether this sytem exhibits orientational order then depends on the relative kinetics of gelation and orientation.

B. Slow Elastic Modes in Polydiacetylene Gels: Direct Coupling to Gel Network by an External Field[19]

Since a polymer gel is a network embedded in a solvent, the gel medium exhibits two types of propagating elastic modes. The first (longitudinal) originates from the compressional mode of the liquid and has a fast sound velocity, ~ 10^5 cm s^{-1}. The second is a transverse mode which originates from the elastic stiffness of the interconnected polymer network coupled to the solvent. The velocity of this slow mode, analogous to second sound, goes to zero as the newtork softens and comes apart at the gel-sol transition.

This slow mode can be excited and detected in polydiactylene gels through the interaction of a low-frequency ac electric field with the anisotropic polarizability of rod-like polydiacetylene macromolecules. The specific coupling of the ac field excitation and the birefringence detection to the network allows direct measurements of the fraction of the sample for which there is an elastic restoring force (the "infinite cluster").

Gels were prepared as described in Ref. 8. A He-Ne laser beam (5 μW) is passed through the gel, which is placed in a cell between crossed polarizers. The transmitted light is detected with a photodiode. By use of two parallel electrodes (1 cm x 1 cm) separated by d = 0.3 cm, a sinusoidal electric field was applied across the sample at 45° to the incident polarization. Turning on the field deforms the gel (via the network interaction with the field) leading to resonant birefringence when the frequency of the ac field coincides with one of the resonance frequencies, ν_n, of the elastic gel between the electrodes.

Two scanning techniques were used. First, a wave analyzer (HP-3581) was used to drive the system as well as to detect the signal from the photodiode. Thus, we measured th response of the system at the frequency of the oscillating electric field. In Fig. 10 we show the result for a ≅ 0.5-wt.% gel of PDA (4BCMU) in toluene. One sees five equally separated resonances. Using the resonance condition for zero displacement at the boundaries $n(\lambda_n/2) = d$ where λ_n is the wavelength of the nth mode, we obtain the slow-wave sound velocity (v_s) from the simple normal-mode equation ν_n $(v_s/\lambda_n) = n\ (v_s/2d)$. From Fig. 10, $v_s \cong 33$ cm s^{-1}. We interpret this as a direct measurement of transverse shear wave, excited by the ac electric field via a small displacement of the network due to small local rotations of the rodlike macromolecules.

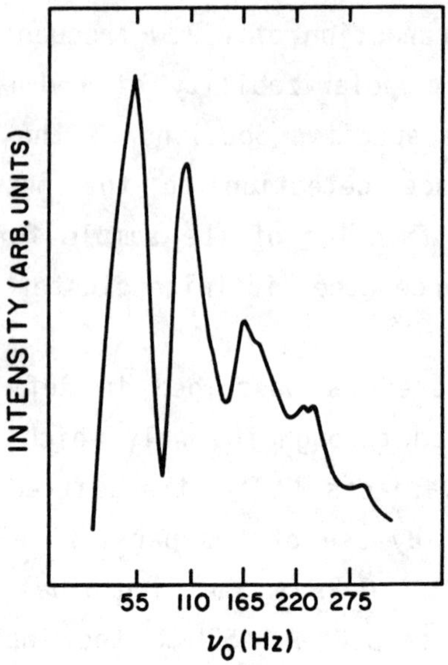

Figure 10. Resonant birefringence detection of slow elastic modes in a gel of ∿0.5 wt% polydiacetylene in toluene.

The second technique used the wave analyzer as a spectrum analyzer for the detected birefringence signal as the gel is griven at a fixed low frequency, ν. A typical result of such a scan is shown in the inset of Fig. 11 (ν = 100 Hz). The two curves in the inset are for the two different stages in the gelation process. One observes response peaks at ν and 2ν, the relative amplitude at ν being larger in the later stage of gelation. We argue below that the peak at ν represents the response of the infinite polymer network, while the peak at 2ν results from a residue of molecules (or finite clusters) moving freely in this complicated porous medium in equilibrium with the infinite network.

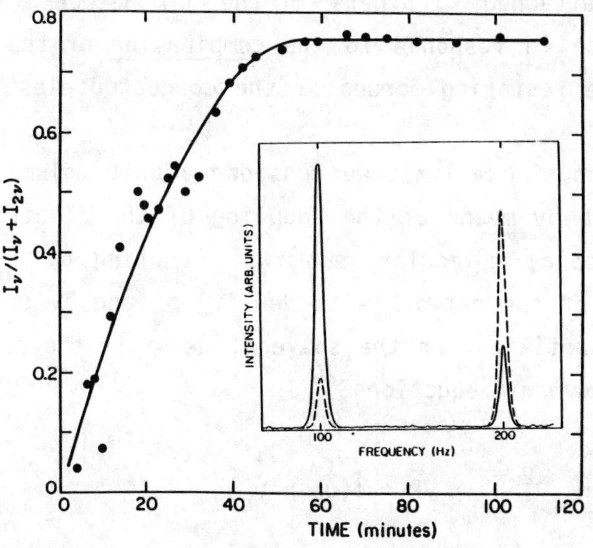

Figure 11. Kinetics of gelation; the ratio I_ν $(I_\nu + I_{2\nu})$ is plotted vs time (the solid curve is drawn to guide the eye). The inset shows the spectrum analysis of electric-field-induced birefringence; ac driving field at 100 Hz. The dashed curve was taken 8 min after quenching from the high temperature solution; the solid curve was taken after 1 h.

The equation of motion for a free rigid molecule having an <u>induced</u> <u>dipole</u> <u>moment</u> $\alpha\vec{E}(t)$ in a time-dependent electric field $E(t) = E_0 \cos\omega t$ (for simplicity we set $\alpha_{\parallel} = \alpha$ and $\alpha_{\perp} = 0$) is written

$$I\ddot{\theta} + \Gamma\dot{\theta} + \alpha E_0^2 \cos^2\omega t \sin\theta \cos\theta = 0 \qquad (16)$$

where I is the moment of inertia of the molecule, $\tau = I/\Gamma$ is the characteristic rotational diffusion decay time, and $\omega = 2\pi\nu$. In the regime of linear response to the field, a solution of such molecules will show a field-induced optical anisotropy with a dc component (for $\omega\tau > 1$) and a component oscillating at 2ν (see Sec. IV). If the concentration is high enough that a gel is formed, the network has a finite bulk and shear moduli. Since each molecule is attached to others in the gel network, the network will oscillate in response to the <u>combination</u> of the ac driving field and the restoring forces of the connected elastic network.

In the continuum limit we consider a unit volume of the gel and deform it by means of the coupling of the electric field to the (microscopic) molecular network. Denoting by $\vec{u}_n$ the small displacement of the network with density ρ_n and by $\vec{u}_s$ and ρ_s the equivalent quantities for the solvent, we write the coupled elastic and hydrodynamic equations,[20]

$$\rho_s \ddot{\vec{u}}_s = \eta_o \nabla^2 \dot{\vec{u}} + \nabla P + \phi(\dot{\vec{u}}_n - \dot{\vec{u}}_s), \qquad (17)$$

$$\rho_n \ddot{\vec{u}}_n = \mu\nabla^2\vec{u}_n + (K + \tfrac{1}{3}\mu)\,\nabla\,(\nabla \cdot \vec{u}_n)$$

$$+ \phi(\dot{\vec{u}}_s - \dot{\vec{u}}_n) + \vec{f}\,(\vec{E}\,(t)), \qquad (18)$$

where η_o is the viscosity of pure solvent, P is the pressure, ϕ is the friction between the network and the solvent, and K and μ are the bulk and shear moduli of the network, respectively. The external force term, $\vec{f}$, is proportional to $E(t)^2$ (induced dipole moment); we argue below that it is also proportional to $\vec{u}_n$.

Microscopically, each molecule is represented by a rod (length L) attached to other rods. Thus, the microscopic stiffness of the network comes from a combination of bending the rods plus bending the joints (we assume these to be of the same magnitude). The corresponding spring constant[21] is $\chi_o = (c/L^3)Y\tilde{I}$, where Y is the Young's modulus of the molecule, $\tilde{I}$ is the static moment of inertia around the axis of the cross section, and c is a number dependent on the boundary conditions (typically c ~ 10). Thus the elastic energy is $U_o = \frac{1}{2}\chi_o\ (L\delta\theta)^2$, where $\delta\theta$ is a small change in the angle between the molecular axis and the electric field.

The change in network energy resulting from the interaction with the electric field is $U_1 = \frac{1}{2}\chi_1 \times 2\cos^2(\omega t)(L\ \delta\theta)^2$, where $\chi_1 \cong \alpha E_o^2/L^2$ is an effective spring constant. To show this, we calculate the contribution to U_1 from the rth molecule and subsequently average over all molecules. This contribution is proportional to $[\cos^2(\theta_r - \delta\theta_r) - \cos^2\theta_r]$ which, for small $\delta\theta_r$, is approximately $[\cos(2x\theta_r)(\delta\theta)]$, where θ_r is a random variable between 0 and π and $\delta\theta_r = \delta\theta g(\theta_r)$. For deformation wavelengths much larger than the polymer size, the random gel requires that $g(\theta_r) = g(\pi - \theta_r)$. Thus the averaging procedure leaves only the even term, and $U_1 \propto (\delta\theta)^2$. This implies (on the macroscopic level) a u_n^2 dependence of the electric field energy, or in Eq. (18) $\vec{f} \propto \vec{u}_n$. In particular, for a transverse mode with u defined as the transverse displacement, $f_t = K_1\cos^2\omega t$, where K_1 is a constant proportional to E_o^2. Taking the spatial Fourier transform for the transverse component of Eq. (18) for each mode q, we obtain

$$\rho\ddot{u}(q) + \eta\dot{u}(q) + [K_o(q) - \tfrac{1}{2}K_1] \times [1 - s\cos(2\omega t)]\ u(q) = 0, \qquad (19)$$

where ρ is the gel density and η is an effective viscosity which in principal depends upon η_o, ϕ, and q. $K_o(q) = q^2\mu$, and $s = K_1/(2K_o - K_1)$. Provided $s << 1$, this equation yields a resonant response at the frequency $\omega_o(q) = (K_o/\rho)^{\frac{1}{2}} = q(\mu/\rho)^{\frac{1}{2}}$ <u>when driven parametrically</u>, i.e., when $\omega = \omega_o(q)$. Alternative explanations (e.g., based on permanent dipoles) would require a different frequency response from that observed experimentally.

Our analysis[19] leads to the conclusion that in the gel the dominant response will be $u(t) \sim \cos(\omega_o t)$; harmonics will be observd but with much smaller amplitude (of order s). Moreover, since u(t) scales with $L\ \delta\theta$, the distortion-induced birefringence should oscillate at ω_o. Thus the resonances observed in the birefringence (Fig. 1) are indeed due to excitation of the transverse elastic modes of the gel. Using $v_s^2 = \mu/\rho$ with $\rho \sim 0.9$ g/cm^3, we find $\mu \cong 900$ dyn/cm^2 for the 0.5% gel of Fig. 10, in agreement with previous measurements[8] at this concentration. The equal spacing of the elastic resonances implies a linear dispersion relation.

Under the assumption that $s << 1$ (for the infinite cluster) throughout the gelation process, the second harmonic that is observed at early times is due to free molecules and small clusters which contribute to the birefringence at 2ν. Thus the relative intensities in the detected birefringence at ν and 2ν provide a direct means of monitoring the gelation process. Starting from a red solution rapidly quenched from the yellow (coil) phase, we scanned the harmonic content (I_ν and $I_{2\nu}$) as a function of time after quench during the gelation process (Fig. 11). Throughout this period intensity at higher-order frequencies is negligible, implying $s << 1$. In Fig. 11, at long times, the ratio saturates at a value less than unity indicating the existence of finite clusters and dangling molecules which contribute to $I_{2\nu}$ even in the gel at quasiequilibrium (the fluctations at early times may result fromtime variation of the elastic resonance frequencies as the gel stiffens).

In summary, using the electric field coupling to the anisotropic rodlike polydiacetylene molecules we were able to develop a new method for exciting and detecting elastic modes in polydiacetylene gels by direct coupling to the polymer network. The success of this technique depends upon the large anisotropic polarizability of the rodlike molecules, a feature which will be a property of all gels made from conjugated polymers.

VII. Conclusion

The study of solutions and gels of conjugated polymers offers special opportunities. The direct coupling of the electronic structure to the polymer conformation leads to dramatic changes in the optical absorption and in the field induced birefringence when the polymer conformation is changed from rod to coil (either by changing solvent quality or temperature). The anisotropic rodlike conformation leads to novel gels made up of a network of anisotropic molecules. This molecular anisotropy can be exploited by using electric fields to couple to and study the elastic properties of the network and they can lead to a novel nematic gel; a broken symmetry state with macroscopic order. Since the synthesis of a large number of soluble polydiacetylenes has been demonstrated, this system promises to continue to provide important insights into the nature of rodlike conjugated polymers in solution.

Acknowledgement

This research was supported by the National Science Foundation through Grant No. DMR83-12725. The polydiacetylene 4BCMU was provided by T. Prusik of Allied Corporation.

References

1. G.N. Patel, R.R. Chance and J.D. Witt. J. Chem. Phys. 70, 4387 (1979).

2. G.N. Patel, J.D. Witt and Y.P. Khanna. J. Polymer Sci., Polym. Phys. Ed. 18, 1383 (1980).

3a. K.C. Lim, C.R. Fincher and A.J. Heeger. Phys. Rev. Lett. 50, 1934 (1983).

b. K.C. Lim, C.R. Fincher, S.A. Casalnuovo and A.J. Heeger. Molec. Cryst. Liq. Cryst. 105, 329 (1984).

4. K.C. Lim and A.J. Heeger. J. Chem. Phys. (in press).

5. K.C. Lim, A. Kapitulnik, R. Zacher and A.J. Heeger. J. Chem. Phys. (in press).

6a. G.N. Patel. J. Polym. ci., Polym. Lett. Ed. 16, 607 (1978); 17, 203 (1978).

b. R.R. Chance, G.N. Patel and J.D. Witt. J. Chem. Phys. 71, 206 (1979).

c. R.R. Chance, M.L. Shand, C. Hogg and R. Silbey. Phys. Rev. B 22, 3540 (1980).

7. B.J. Berne and R. Pecora. Dynamic Light Scattering (John Wiley & Sons, 1976). In general, the decay of G(t)is determined by a combination of translational and rotational diffusion. Because of the length of the rodlike PDAin solution, rotational diffusion is very slow (see accompanying

paper; ref. 10) and does not affect the light scattering data.

8. M. Sinclair, K.C. Lim, A.J. Heeger. Phys. Rev. Lett. 51, 1768 (1983).

9a. Measurement of Suspended Particles by Light Scattering, edited by B.E. Dahneke (Wiley Interscience, John Wiley & Sons, New York (1983). See Chapt. 15 by D. Eden and J.G. Elias.

b. Molecular Electronics, Part I (Theory and Experiment), edited by C.T. O'Konski (Marcel Dekker, Inc., New York and Brussel 1976). See Chapt. 3 by C.T. O'Konski and S. Krause.

10. Molecular Electronics, Part I (Theory and Experiment), edited by C.T. O'Konski (Marcel Dekker, Inc., New York and Brussel 1976). See Chapt. 5 on Flexible Polymers by R.L. Jenigan and D.S. Thompson.

11. W.A. Wegener, R.O. Dowben and V.J. Koester. J. Chem. Phys. 70, 622 (1970).

12. G.N. Patel and E.K. Walsh. J. Polym. Sci. Polym. Lett. Ed. 17, 203 (1979).

13a. G. Wenz and G. Wegner. Makromol. Chem. Rapid Commun. 3, 231 (1982).

b. G. Wenz, M.A. Muller, M. Schmidt, G. Wegner. Macromolecules (in press).

c. M.A. Muller, M. Schmidt and G. Weger. Makromol. Chem. Rapid Commun. 5, 83 (1984).

14. S.A. Casalnuovo, K.C. Lim and A.J. Heeger. Makromol. Chem., Rapid Commun. 5, 77 (1984).

15. P.J. Flory. Proc. R. Soc. London, Ser. A 234, 73 (1956).

16. D.C. Prevorsek, "Recent Advances in High-Strength Fibers and Molecular Composites," in Polymer Liquid Crystals, A. Ciferri, W.R. Krigbaum, R.B. Meyer, eds., Academic Press (1982).

17. S.A. Casalnuovo and A.J. Heeger. Phys. Rev. Lett. 53, (Dec. 3, 1984).

18. S.A. Casalnuovo, K.C. Lim and A.J. Heeger. Makromol. Chem. Rapid Commun. 5, 77 (1984).

19. A. Kapitulnik, S. Casalnuovo, K.C. Lim and A.J. Heeger. Phys. Rev. Lett., 53, 469 (1984).

20. P.G. deGennes. J. Chem. Phys. 9, 594 (1976).

21. L.D. Landau and E.M. Lifschitz. Theory of Elasticity (Pergamon, Oxford, 1970), Chaps. 2 and 18.

1,4- DIPHENYLBUTENYNE: A MODEL COMPOUND FOR POLYDIACETYLENES (PDA)

N. BALCIOĞLU, C. ÜNALEROĞLU, and S. AYDIN
Hacettepe University, Chemistry Department.
Beytepe-Campus, ANKARA

Abstract: From the optical and thermal properties of 1,4-diphenylbutenynes a model is deduced, which is enable to explain the behaviour of PDA's in solution.

I. INTRODUCTION

The optical properties of PDA's attract interest in the last decade[1,2,3]. A dramatic solvent and temperature effect in the absorption spectra of these polymers were observed[4]. When, the blue coloured solid polymer is dissolved in a good solvent a yellow solution is formed at about 25°C, whereas at lower temperatures or in a poor quality solvent a blue solution is obtained. A temperature effect is also observed in the crystalline state: the initial red or blue colour of the polymer is changed into yellow when heat is applied[6].

So far no generally accepted conclusion has been disclosed to explain the relation between the exact structure of the polymer and the colour transition. We considered to choose a suitable model compound to examine the optical and thermal properties of the polymer. E-1,4-diphenylbutenyne (1) seems to be appropiate for this purpose. It has similar structure as the repetetive unit of the polyacetylene backbone, 2.

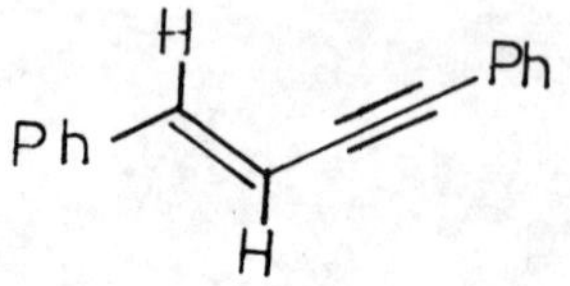

1

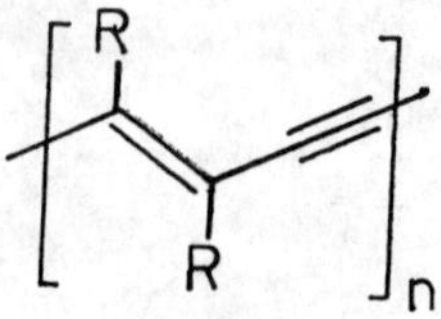

2

II. cis-trans-ISOMERISATION OF 1,4-DIPHENYLBUTENYNES

E-1,4-Diphenylbutenyne 1 undergoes a reversible cis-trans isomerisation, thermally[6] at about 230° in boiling decanol, or upon irradiation with UV-or sunlight[6,7] at room temperature, forming a yellow solution.

The observed facile cis-trans isomerisation of the double bonds is not unique for the enyne systems. Polyconjugated olefins behave similarly. 1,4-Diphenyl-1,3-butadienes[8] undergo thermal isomerisation above 200°, and the carotenoids[9] isomerise even in boiling hexane. The isomerisation of these compounds can also be induced, when they are exposed to sunlight[8,9].

It is generally accepted that the optical behaviour of the polyenynes is similar to that of the polyenes[1,3]. However, trans-to-cis isomerisation of the double bond causes a red shift in the absorption band of the 1,4-diphenylbutenyne, in contrast to the polyene systems[9]. In the spectrum of the irradiated solution of 1, which contains ca 50% of the Z-isomer, two new bands at 430 nm ($E_1^{\%}$: 1.25) and 260 nm ($E_1^{\%}$: 415) appear[6].

On the other hand, Z-Stilbene absorbs at a shorter wavelength than the E-isomer, although the opposite is predicted by the theoretical calculations[10]. The discrepancy between the experimental and theoretical results concerning the absorption bands of the E- and Z-sitlbenes is attributed to the deviation from the coplanarity of the phenyl groups due to the steric interaction of the o-hydrogenes in the cis-isomer. A similar steric interaction has been predicted[8] for the planar E,Z-1,4-diphenyl-1,3-butadiene.

Our CNDO/S calculations[11] on E-and Z-3-hexen-1,5-diynes show that the Z-isomer should absorb at a longer wavelength than the E-isomer, which means that cis-trans isomerisation should have a similar effect in the absorption properties of the enyne-and the polyene systems, as well. The prediction and the observation are consistent in the case of 1,4-diphenylbutenynes. Both isomer may have a stable fully conjugated planar conformation, since no repulsion is expected between the phenyl group and the triple bond in these compounds.

III. PDA's, AS AN EXTENDED ENYNE-SYSTEM

Three different models i.e. worm-like structure by Wegner et al[3]., coil-like all-cis structure by Heeger et al[12], and random flight chain model by Chance et al[4]. are proposed to describe the structure of the polymer in yellow solution. None of these models can give a statisfactory explaination for the "peculiar" properties of the dissolved PDA's and they are critisized in the literature[3,14,15].

We believe that our foregoing discussion about the thermal and optical properties of the 1,4-diphenylbutenynes and 1,4-diphenyl-1,3-butadienes can be extended to polydiacetylene systems. These polymers, with their extended π - system should undergo a facile thermal or photochemical cis-trans isomerisation even at room temperature and daylight. cis-trans isomerisation requires a 180° rotation around a double bond. Certainly this is not possible in the rigid crystalline state or in the aggregate form of the polymer at room temperature.

It is known from X-ray diffraction analysis[16] that the polymer backbone has a fully conjugated planar all-trans structure in the crystalline state. Introduction of a cis-double bond in the polymer backbone is expected when the solid polymer is heated or taken into solution, which has never been observed so far.

A 180° rotation of a large dimensional polymer backbone around the double bond should be difficult even in the dissolved form of the polymer. But there is still an another way to achieve a *cis-trans* isomerisation in the PDA chain: Rotation of a polymer segment of certain repetetive units around the double bonds at each end of the segment is sufficient for the simultaneous configurational isomerisation of these two double bonds without complete rotation of the backbone. The fully conjugated planar structures of the all-*trans* PDA backbone and its isomerised form are shown in Figure 1a an 1b, respectively.

We have calculated[11] the energy difference between E-and Z-3-hexen-1,5-diyne byCNDO/2 method in order to estimate the length of the segment located between two consecutive *cis*-double bonds in the polymer chain (cf.Fig.1b). The computed value of 1,14 kcal/mol corresponds, to a cis/trans ratio of 1/7 in the polymer. This means that in a free rotating PDA chain seven repetetive units with *trans*-double bonds should be located between two consecutive *cis*- double bonds, at room temperature. That is, however, exactly equal to the value which is estimated from UV-VIS. Spectra[3,4] as the effective conjugation length in the yellow solution. This coincidence implies the possibility that the decrease in conjugation can be attributed to the formation of the *cis*-bonds in the polymer chain.

In account of the foregoing discussion, however, *trans*-to-*cis* isomerisation should result a red-shift in the absorbtion spectra of the planar enyne systems. Hence, the structure shown in Fig. 1b is not consistent with the observed yellow colour of the polymer solution. This apparent discrepancy can be abolished if we focus our attention at the possible locale structures of the polymer around the *cis*-double bonds. Examination of the molecular models shows, that in all three possible conformations a dramatic repulsion is expected between the substituents on the adjacent *trans*-double bonds in the planar structure (in Fig.1b only one of the three possible conformations is depicted). The repulsion can be minimized by the rotation of the two segments attached to a

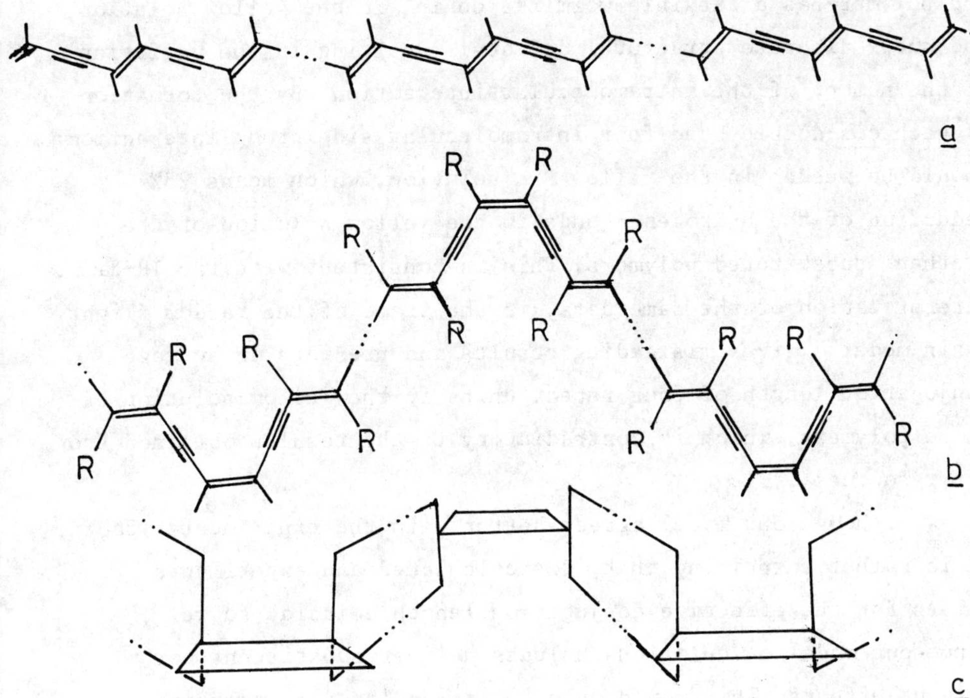

a

b

c

Fig.1. The predicted structures of PDA in solution
a) In the blue solution (also in crystalline state[16])
b) Unstable intermediate, c) In the yellow solution.

cis-double bond in a controtatory or disrotatory manner. In a conformation obtained by 90° rotation of the segment the repulsive effects are reduced, meanwhile conjugation between the segments are interrupted by the perpendicular π-orbital of the cis-double bond. Although the segments can now be in conjugation with the initial non-planar π-orbitals of the triple bond, resulting an effective conjugation length of fifteen π-bonds in each segment, at room temperature. This number is linked to the equilibrium constant of the cis-trans isomerisation. Therefore it varies with temperature.

Our suggested model combines further experimental facts that the polymer has a flexible wormlike chain[3] in the yellow solution, and shows discrete structural defects[4] resulting in the variation of the number of the intramolecular interaction. By the formation of each cis-double bond four intramolecular side group interactions should be weeken in the yellow PDA solution, which means 25% reduction of the hydrogene-bonds in the yellow solution of the urethane substituted polymers. This is consistent with the IR-data[4]. Interpretation of the same data, in the frame of the random flight chain model[15] gives misleading results and predicts an average conjugation length of four repeat units in the yellow solution of these polymers, which is contradictory to the results obtained from other sources[3].

Although our model agrees better with the experimental facts, it is rather surprising that, the calculated and experimental values for the effective conjugation length coincide so well, since our model calculation includes neither substituent nor solvent effects. This could only be rationalised by assuming mutual cancellation of these effects.

Hence, the cis-trans ratio of the double bonds should not be effected by the side groups of the polymer in a good solvent. By addition of a nonsolvent to the yellow polymer solution, the magnitude of polymer-solvent interaction will be decreased, and with the increased intramolecular interaction the polymer chain should aligne itself in the most table planar all-trans structure, yielding a blue solution. Assuming that the necessary activation energy will be supplied by room temperature or by daylight, as discussed above. Hence, not only the temperature, but also the solvent quality can affect the cis-trans ratio of the bonds (consequently the effective conjugation length) in the polymer chain.

The rod-like structure, given in Fig.1a, is also suitable for a chain-chain interaction leading to formation of aggregates or precipitates. Hence, the observed aggregate formation during the colour transition is not contradictory to the accepted single chain phenomenon.

Although the success of the proposed model, there is no spectral evidence for the presence of cis double bonds in the yellow solution. Moreover, all attempts[3] to produce the cis-structure by photo-induced or catalytic cis-trans isomerisation led to a random chain scission rather than the intended structure. Probably increased amount of the cis-double bonds decrease the stability of the polymer chain due to the repulsive interaction between the side groups and cause the observed chain scission.

According to our model only 6% of the backhone carbons (less than 1% of the total carbon in 3-BCMU) are located on the cis-bonds. Their detection therefore should be very difficult by the spectroscopic methods.

IV. CONCLUSIONS

The model deduced from the optical and thermal properties of the 1,4-diphenylbutenynes, is shown in Fig.1, explains solution properties of the PDA's satisfactorily, and unifies the controversial opinions in this field. Further experiments and theoretical work are needed on model compounds and on specific PDA polymers, for additional proof.

Acknowledgement: We would like to thank Prof.Dr. O. Tarhan and Assoc. Prof.Dr. B. Peynircioğlu for their valuable comments and help to prepare the manuscript in english.

REFERENCES and NOTES

1. R.H.Boughman, and R.R.Chance, J.Polym.Sci.Polym. Phys.Ed. 14, 2037 (1976).
2. R.R.Chance, Macromol, 13, 396 (1980).
3. G.Wenz, M.A.Müller,M.Schmidt, and G.Wegner,Macromol 17, 837 (1984).
4. G.N.Patal, R.R.Chance, and J.D.Witt, J.Chem.Phys. 70, 4387 (1979).
5. R.R.Chance, G.N.Patel, and J.D.With, J.Chem.Phys. 71,206 (1979).
6. N.Balcıoğlu and S.Aydın, unpublished results.
7. F.Straus, Lieblas Ann. 342, 190 (1905).
8. J.H.Pinckard, B.Wille and L.Zechmeister, J.Amer. Chem.Soc. 70, 1938 (1948).
9. L.Zechmeister, A.L.Le Rosen, W.A.Schroeder, A.Polgar and L.Pauling, J.Amer. Chem.Soc. 65, 1940 (1941).
10. H.Suzuki "Electronic Absorption Spectra and Geometry of Organic Compounds" A.P. London, 1967. p. 315.
11. N.Balcıoğlu, and C.Ünaleroğlu, to be published. The computed λ_{mak} values are 296 and 321 nm for 1 and for its Z-isomer respectively.
12. K.C.Lim, C.R.Fincher and A.J.Heeger, Phys. Rev.Lett. 50, 1934 (1983).
13. G.N.Patel and E.K.Walsh, J.Polym.Sci.Polym.Lett Ed. 17, 203 (1979).
14. M.A.Müller, M.Schmidt, and G.Wegner, Makromol. Chem.Rapid Commun. 5, 83 (1984).
15. R.R.Chance, M.W.Washabaugh, and D.J.Hupe, these proceedings.
16. E.Hadicke, K.Penzien, and W.Schwell. Angew. Chem. 83, 1024 (1971).

TOWARD POLYDIACETYLENES AS UTILE OPTICAL AND ELECTRONIC ELEMENTS

D.J. Sandman, G.M. Carter, Y.J. Chen, B.S Elman, M.K. Thakur, and S.K. Tripathy

GTE Laboratories Incorporated
40 Sylvan Road
Waltham, MA 02254 U.S.A.

Motivations for the study of nonlinear optical phenomena in polydiacetylenes (PDAs) are discussed. Recent progress in the fabrication and characterization of PDA waveguides is summarized along with approaches to the interpretation of the observed nonlinear optical phenomena. Attempts to create a carrier population in PDAs via charge transfer chemistry are reviewed. The implantation of PDAs with ^{75}As ions has been studied by several spectroscopic techniques and converts these insulators into highly conducting materials.

INTRODUCTION

While there is substantial theoretical interest in the electronic structure of forms of conjugated carbon beyond graphite (1), at the present time the polydiacetylenes (PDAs) provide the most accessible experimental systems with one-dimensional conjugated polymer chains available in fully ordered single crystal form. As such the PDAs occupy a key position among organic materials which exhibit phenomena indicating delocalized interactions in that quantum chemical calculations (2) estimate bandwidths substantially greater than those of conducting anisotropic ion-radical solids such as TTF-TCNQ or $(TMTSF)_2X$ which are regarded as tight-binding structures. As fully ordered crystals, the PDAs contrast with partially ordered conjugated polymers such as the currently available forms of polyacetylene, $(CH)_x$, or poly-p-phenylene, $(C_6H_4)_x$. While these latter materials exhibit many interesting electrical, magnetic, and optical properties, the lack of fully ordered materials has retarded the fundamental study of these materials (3,4).

The description of PDAs as wide bandwidth systems provides an efficacious perspective for our present itinerary through their optical and electronic properties, particularly by at least superficial analogy to well known inorganic solids. Not surprisingly, seminal reports of large carrier mobilities (5) and nonlinear optical coefficients (6) have appeared and have generated substantial interest in their verification and elaboration. Our collective efforts in these directions will be the major focus of this report.

POLYDIACETYLENES IN NONLINEAR OPTICAL ELEMENTS: PROGRESS IN MATERIALS AND MEASUREMENTS

Current discussions of potential applications of nonlinear optical signal processing in computers (7), both general and special purpose, and in communications for functions such as switching, amplifying, and multiplexing are primarily focussed on devices for the manipulation of laser beams in thin transparent films. With particular reference to bistable optical devices, the fundamental physical mechanisms which provide limits on speed, power dissipation, and size of optical switching elements have been discussed (8).

In bulk media, there is a nonlinearity in the constitutive relationship between the induced polarization (P_i) proportional to the amplitude of the light field which may be expanded in terms of a power series of the electric field components:

$$P_i = \chi_{ij}^{(1)} E_i + \chi_{ijk}^{(2)} E_j E_k + (1)\ \chi_{ijkl}^{(3)} E_j E_k E_l + \ldots \qquad (1)$$

$\chi^{(1)}$ represents the linear optical properties, and $\chi^{(2)}$ and $\chi^{(3)}$ are, respectively, the second and third order nonlinear susceptibilities (9). At the present time, there is considerable experimental and theoretical interest in organic materials in general and PDAs in particular which have large values of $\chi^{(2)}$ and $\chi^{(3)}$ (10).

Second harmonic generation, optical rectification, and the linear electro-optic effect involve $\chi^{(2)}$, which occurs only in noncentrosymmetric media. Large values of $\chi^{(2)}$ in appropriately substituted PDAs have been inferred from measurements on polycrystalline samples, (11) and this will not be discussed further herein. For a theoretical treatment of the origins of $\chi^{(2)}$ in organic solids, the interested reader may consult (11). The remainder of this section will be devoted to studies of $\chi^{(3)}$ in both free standing crystals and thin film waveguide structures of PDAs.

The optical devices of interest are based on a nonlinear refractive index (n_2), i.e., a refractive index $n(\omega)$ that increases or decreases with light intensity I:

$$n(\omega) = n_0(\omega) + n_2 I \tag{2}$$

The origin of n_2 can be: a) a saturable absorption (i.e. absorption that decreases with increasing light intensity) or b) n_2 associated with a virtual transition in the transparent region. Since no reports of saturable absorption in PDAs have appeared to date, the focus of PDA research has been on the assessment of the magnitude and temporal response of n_2 which is related to $\chi^{(3)}$ by equation 3 (12):

$$n_2 = \frac{16\pi^2 \chi^{(3)}}{C\varepsilon_1} \tag{3}$$

The nonlinear index n_2 contributes to such well-known effects as self-focusing, self-trapping, phase conjugation, optical bistability, etc. which are fundamental to all-optical signal processing applications. In the seminal report of optical nonlinearities in PDAs (6), single crystals of poly-PTS were found in frequency tripling experiments with light polarized parallel to the chain direction to have values of $\chi^{(3)}(3\omega)$ in the infrared comparable to germanium and gallium arsenide, i.e. ca. 10^{-10} esu, while poly-TCDU had a value of $\chi^{(3)}(3\omega)$ about an order of magnitude lower than that of the PTS polymer. For poly-PTS, a value for n_2 of 1.8×10^{-6} MW^{-1} cm^2 was deduced. If this nonlinearity originates in the intrinsic electronic structure, it may have a response time of the order of 10^{-14} seconds (8). Additionally, it was inferred (6) that $\chi^{(3)}(3\omega)$ showed a resonant effect as 3ω approached the poly-PTS absorption edge. A similar large value of $\chi^{(3)}(\omega) \propto n_2$ was deduced for poly-PTS crystals at 1.9 μm (13); n_2 was negative in sign in this experiment (14).

In further work with single crystals of poly-PTS and -TCDU, the two-photon absorption spectrum was measured using 1.06 μm laser light (15). It resembled the low energy one photon absorption tail of a PDA, and was interpreted as arising from defect creation by the laser light. Picosecond regime lifetimes were inferred for the defects.

The studies of the free standing crystals of poly-PTS (and other PDAs) have been critical in forming the scientific basis of $\chi^{(3)}$ in PDAs and are of continuing interest. However, poly-PTS crystals grown to date do not have optically flat surfaces, and their study has been facilitated by immersion in diiomethane, which has a refractive index comparable to the polymer in the near infrared region (13). Thin polycrystalline films of poly-PTS (16) do not improve the situation in that they have significant transmission losses in the near infrared (17).

Our initial studies (18) of $\chi^{(3)}$ in PDA waveguides were performed using the PDA from $CH_3(CH_2)_{15}(C{\equiv}C)_2(CH_2)_8{-}CO_2H$ in Langmuir-Blodgett (L-B) multilayers coated on silver gratings. This structure allows coupling of an incident laser photon to a planar guided wave mode in the PDA waveguide or a surface plasmon mode at the silver-PDA interface (Figure 1). Physical aspects of the nonlinear coupling between radiation and confined modes (either guided waves or surface plasmons) have been discussed (19).

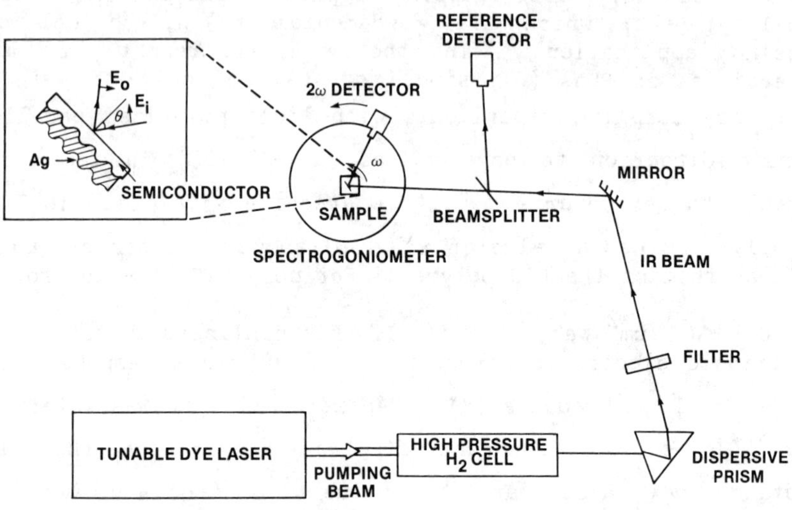

Figure 1. Experimental setup for the intensity-dependent nonlinear coupling measurement. The sample, as shown in the inset, is mounted on an x-ray spectrogoniometer and the polarized light is incident on the PDA film surface at angle θ (from (18) with permission).

With reference to Figure 1, one can relate the change in coupling angle $\Delta\langle\theta\rangle$, to n_2 by:

$$n_2 \Delta\langle I\rangle = \Delta\theta \cos\theta \qquad (4)$$

where $\langle\ \rangle$ indicates an appropriate average of the intensity over the mode volume. Both the sign(relative to $\chi^{(1)}$) and magnitude of $\chi^{(3)}$ are obtained using this technique. Moreover, this technique, in a surface plasmon mode, yielded values of $\chi^{(3)}$ for silicon and gallium arsenide at 1 μm in accord with previously reported experiments (12).

For a 5000 angstrom thick L-B PDA multilayer of the acid given above, Figure 2 shows the change in θ with incident beam intensity at 7550 angstroms. Analysis of data such as that shown in Figure 2 leads to an estimate of
$n_2 \simeq 10^{-6}\ MW^{-1} cm^2$, for $0.75\ \mu m < \lambda < 1.05\ \mu m$, (20) comparable to the value of n_2 reported (6) for poly-PTS crystals. Further, a large increase in $\chi^{(3)}$ is observed for the PDA absorption region (20).

Examination of Figure 2 reveals the shift of θ to larger angles with increased light intensity. This indicates that $\chi^{(3)}$ is negative with respect to $\chi^{(1)}$, a situation also noted for poly-PTS crystals (13,14). This is noteworthy since a one-electron tight-binding treatment of PDAs as one-dimensional semiconductors in the delocalized limit indicated $\chi^{(1)}$ (21) and $\chi^{(3)}$(22) to have the same (positive) sign.

Use of a discrete three level model (23) facilitates further discussion of the experimental results (13,14,20) for n_2. This model yields two contributions to n_2; one is positive with a two-photon resonant denominator, and the other is negative with only single-photon resonant denominators. At present, there is only one estimate of two photon absorption (TPA) energy in a solid PDA, namely the observations in poly-PTS at 2.34 eV (24), with TPA energies in solution (25) at higher energy. In the present model (23), a two-photon contribution changes sign when the wavelength is above the two photon resonance. There is no manifestation of such an effect in existing data (13,14,20), and we conclude (26) that two photon processes are not major contributors to existing measurements of n_2 in solid PDAs. Given the general importance of two photon processes in nonlinear optical phenomena, a detailed study of such processes in solid PDAs is clearly of interest.

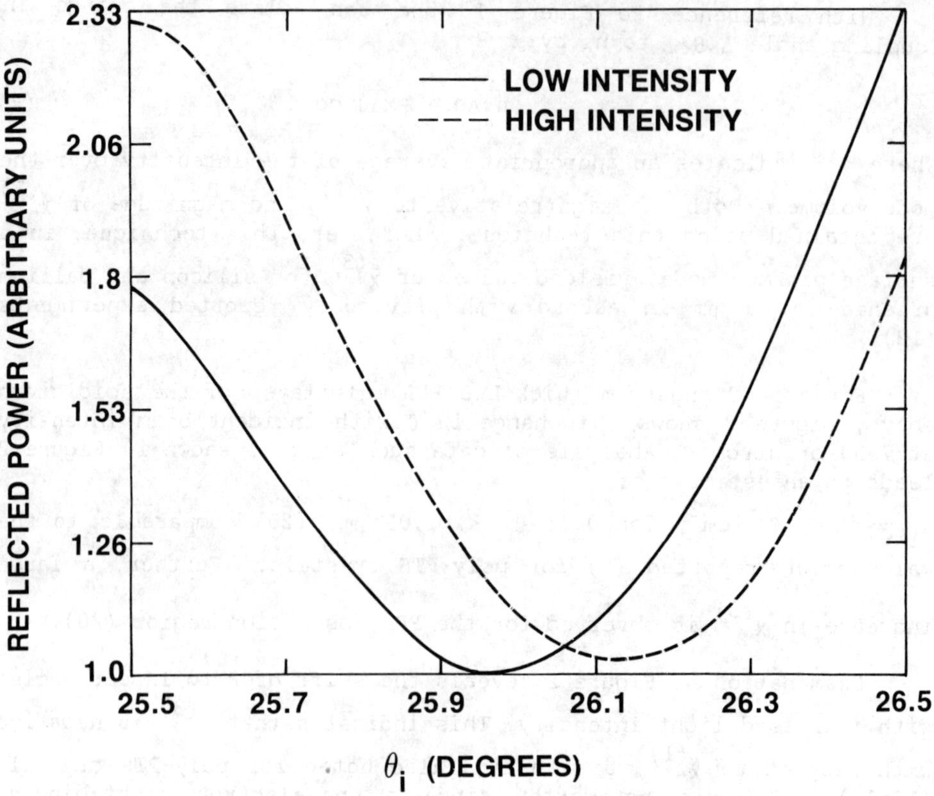

Figure 2. Reflected power from the L-B PDA acid at 7550 angstroms as a function of incident angle for two input laser energies, $E_i = 3 \times 10^{-6}$ J/pulse (——) and $E_i = 1.3 \times 10^{-5}$ J/pulse (----) (from (20) with permission).

A value for $|\chi^{(3)}|$ of ca. 10^{-12} esu was inferred for L-B multilayers of the cadmium salt of the PDA discussed above from third harmonic generation using 1.06 μm laser light (27).

The results of the studies of free standing crystals of poly-PTS (6,13) and the L-B PDA acid waveguides (18,20) indicate that n_2 in PDAs can be very large. It is now of interest to study the temporal response of the nonlinearity in picosecond and subpicosecond regimes (17). Theoretical arguments concerning intrinsic conjugation defects in a one-dimensional system (i.e. solitons or polarons) suggest that their creation will be in the picosecond regime (28). Figure 3 displays a bond representation

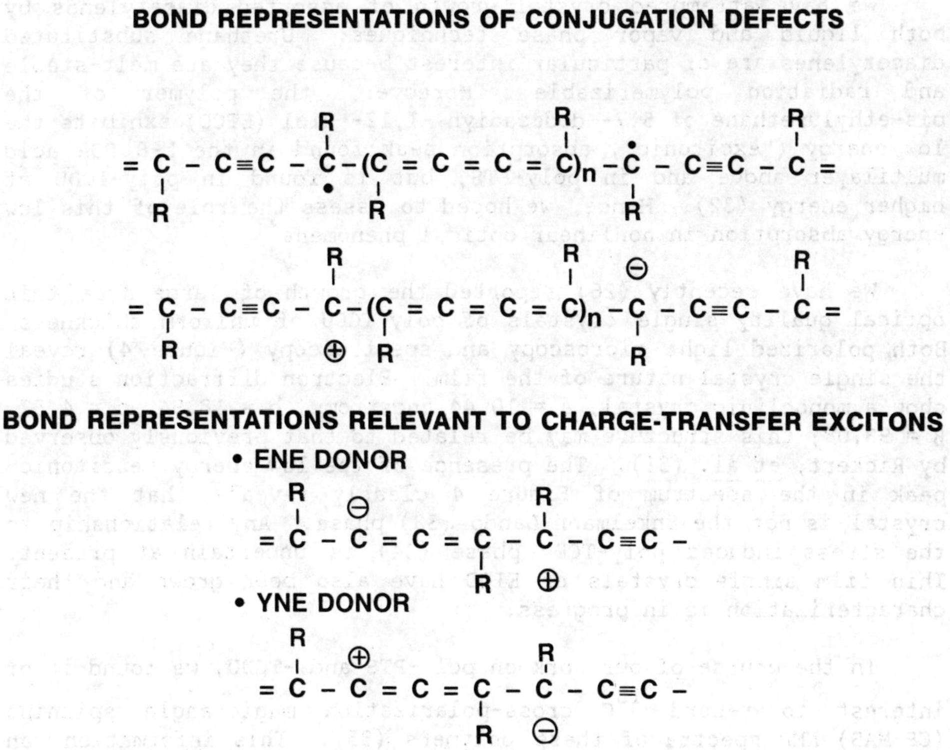

Figure 3. Bond representations of nonground state forms of PDAs, as described in the text.

(29) of such a defect, along with that of a charge-transferred defect. The latter should be distinguished from those which may be written for charge transfer excitons (30), and these representations are also presented in Figure 3.

While L-B PDA multilayers have surfaces of satisfactory optical quality, their labor-intensive fabrication and questions concerning registry between layers and lack of orientation between domains motivate investigation of other approaches to large area single crystal PDA thin films (17). In this connection, it is relevant to note that microscopic size films of the polymers from 2, 4-hexadiyn-1,6- diol and TCDU have been grown on alkali halide substrates and their lattice constants were determined (31). PTS single crystals of unspecified dimensions were grown and polymerized between glass slides and characterized by polarized transmission spectroscopy (16).

We have attempted crystal growth of assorted diacetylenes by both liquid and vapor phase techniques. Urethane substituted diacetylenes are of particular interest because they are melt-stable and radiation polymerizable. Moreover, the polymer of the bis-ethylurethane of 5,7- dodecadiyn -1,12- diol (ETCD) exhibits the low energy ("excitonic") absorption peak found in the L-B PDA acid multilayer above and in poly-PTS, but is found in poly-TCDU at higher energy (32). Hence, we hoped to assess the role of this low energy absorption in nonlinear optical phenomena.

We have recently (26) reported the growth of large area thin optical quality single crystals of poly-TCDU of uniform thickness. Both polarized light microscopy and spectroscopy (Figure 4) reveal the single crystal nature of the film. Electron diffraction studies show a monoclinic crystal, $\underline{a}$ = 10.44 angstroms, $\underline{b}$ = 18.94, $\underline{c}$ = 4.87, β = 98.0°; this structure may be related to that previously observed by Rickert, et al. (31). The presence of the low energy "excitonic" peak in the spectrum of Figure 4 clearly reveals that the new crystal is not the Enkelmann-Lando (33) phase. Any relationship to the stress induced poly-TCDU phase (34) is uncertain at present. Thin film single crystals of ETCD have also been grown and their characterization is in progress.

In the course of our work on poly-PTS and -TCDU, we found it of interest to record ^{13}C cross-polarization magic-angle spinning (CP-MAS) NMR spectra of these polymers (35). This information can be used to deduce the nature of the bonding of the conjugated backbone, a topic of interest especially for poly-TCDU. While only low precision structural data is available for poly-TCDU, (33) this data has been used to infer that poly-TCDU is a butatriene phase (31). Poly-PTS and -TCDU exhibit acetylene resonances at 106.4 and 104 ppm, respectively, downfield of TMS, comparable to acetylene resonances of other PDAs (36). Both polymers lack resonances in the range 136-171 ppm assignable to a digonal butatriene carbon (37). The solid state ^{13}C CP-MAS NMR spectrum of the structurally established (38) phase of 1,1,4,4-tetraphenylbutatriene exhibits resonances at 124.1 and 152.9 ppm, values comparable to solution data (37), for the α and β cumulene carbons, respectively. Thus, we conclude that the solid state ^{13}C NMR data summarized herein indicate the usual acetylenic structure for both poly-PTS and -TCDU, and that these spectra provide no evidence for a significant contribution from a butatriene structure.

TOWARD POLYDIACETYLENES IN ACTIVE ELECTRONIC ELEMENTS

Despite the metal-like reflectance of PDA crystals, dc conductivity measurements indicated that the materials were insulators (29). Initial studies of carrier motion in PDA crystals,

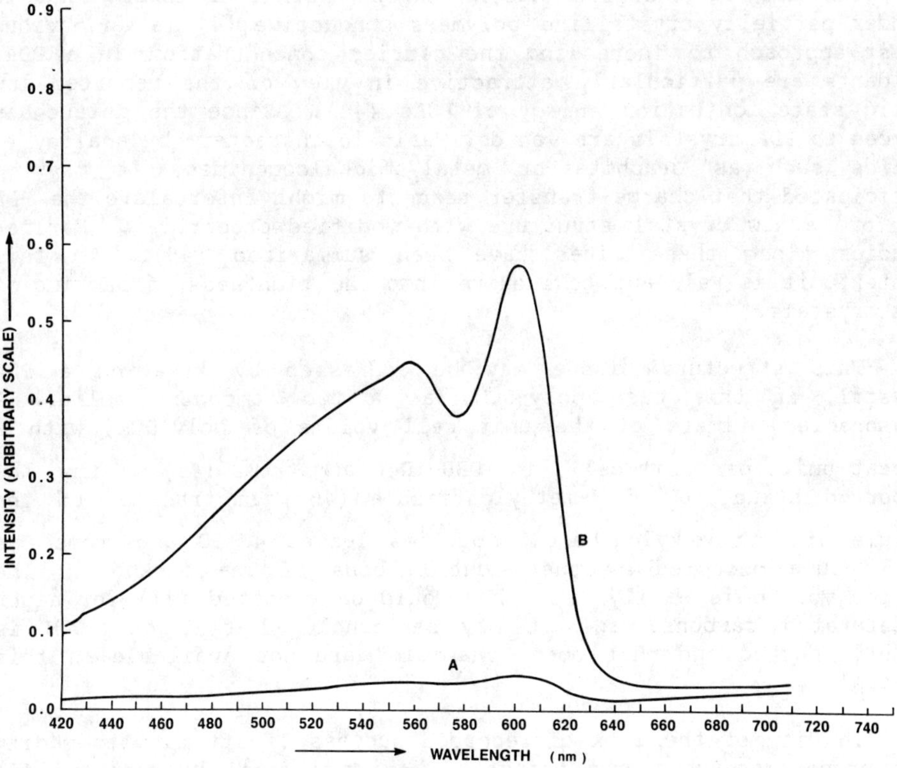

Figure 4. Electronic spectrum of poly-TCDU for polarization perpendicular (A) and parallel (B) to the chain axis (from (27) with permission).

which have been summarized, (39) came from photoconductivity measurements which indicated a mobility of ca. 5 $cm^2V^{-1}s^{-1}$ along the chain direction, modestly higher than a molecular solid such as anthracene. Subsequently, Donovan and Wilson(5) reported an electron mobility greater than $10^5cm^2V^{-1}s^{-1}$ in poly-PTS, and high mobilities were also deduced from electron injection (40) and electroreflectance (41) studies of DCH polymer. While reservations concerning the interpretation of these experiments have been recorded (42,43), nevertheless it is of interest to attempt to create large carrier concentrations in PDAs.

The use of charge-transfer reagents such as those used to render partially crystalline polymers conductive (4) is an obvious first approach to increasing the carrier concentration in a PDA. Oxidants are particularly attractive in view of the reported low solid state ionization energy of PDAs (44). Since the interchain forces in PDA crystals are van der Waals in character, by analogy to solids such as graphite or metal dichalcogenides, it may be anticipated that charge-transfer reagents might intercalate the PDA to form a new crystal structure with modified properties. Earlier studies along these lines have been summarized (45). In this context, it is relevant to inquire into the tightness of packing of PDA crystals.

This structural issue may be addressed by treating a PDA crystal, in this case poly-DCH, as a two component solid. A reasonable estimate of the unit cell volume of poly-DCH, with 2 repeat units per unit cell, is 1030-1040 angstroms^3 (45). Using the reported structure of N-methylcarbazole (46), subtraction of the volume of four alkylcarbazole moieties leaves 40-50 angstroms^3 as the volume occupied by the eight carbons of the backbone. The latter volume is easily less than would be expected (47) for eight unsaturated carbons, and it may be concluded that poly-DCH is tightly packed and that open "channels" are not available in this crystal.

In view of the lack of recorded success of attempts to modify PDA properties by charge transfer chemistry, the observation (48) that treatment of poly-DCH with SbF_5 in pentane resulted in a dark colored solid with a conductivity of 10^{-3} (ohm-cm)$^{-1}$ caught our attention. We decided to study the interaction of poly-DCH with a variety of acceptors and to characterize the chemical, spectral, structural, and electrical properties of the resultant materials on the hypothesis that these could be a fundamental difference between poly-DCH and other PDA.

We found (45) that thermally polymerized DCH undergoes a substantial weight increase on exposure to SbF_5 in perfluorohexane or $SbCl_5$ in CH_2Cl_2 solution at room temperature. The halogens (Br_2 in refluxing CCl_4 and I_2 in refluxing methanol) were found to be markedly less reactive than the antimony halides. From complete elemental analysis, we defined compositions $[(DCH)_2SbF_4]_x$ and $(C_{29}H_{15}N_2)_{0.55}$ $(SbCl_8)_{0.45}$. X-ray diffraction of these solids revealed the pattern of poly-DCH for the SbF_5 material and no diffraction pattern for the $SbCl_5$ material. The solid state spectrum of the SbF_5 material revealed a broad band peaking at

1000-1100 nm which is associated with a partially oxidized PDA backbone. All additional features of the spectrum were previously observed in poly-DCH. The resistivity of these poly-DCH/SbX_5 materials is not less than 10^7 ohm-cm, while pristine poly-DCH is reported (49) to have a resistivity greater than 10^{15} ohm-cm. Materials with resistivity less than 10^7 ohm-cm were not obtained without loss of structural or compositional definition.

DCH polymerized thermally has a fibrous texture and is not single crystalline. When single crystal poly-DCH, prepared by gamma radiation, is exposed to SbX_5 (X = F, Cl), the resultant materials were inhomogeneous by optical microscopy. It was concluded that the Sb species in the thermal poly-DCH reside in defects or vacancies.

It was proposed (17,45) that the partial oxidation of the backbone in poly-DCH is accomplished via initial ionization of the carbazole side chain with subsequent reduction by the backbone. This process is consistent with the reported solid state ionization energies of PDAs (44) and carbazoles (50).

Recently, PDAs were prepared by crystallization of monomer of PTS and DCH in the presence of charge-transfer reagents (51). Conductivities as high as 10^{-4}(ohm-cm)$^{-1}$ were reported and a heterogeneous distribution of the dopant was noted. Spectral and crystallographic properties of these materials were not reported.

The conductivity of poly-DCH was observed (43) to reach 10^{-3}(ohm-cm)$^{-1}$ upon field-driven contact with aqueous NaI/I_2 at a dopant level of 10^{-3} moles per DCH repeat. It was concluded that the doping was probably not homogeneous and that defects might play an important role in the process.

At the present time, there is considerable interest in the use of ion beam techniques for both the preparation and property modification of metals and their alloys, semiconductors, graphite, and polymers (52). Salient results of such investigations include the implantation of slow diffusing species such as In or Sb into lattice sites of silicon (53a), the enhancement of T_c in Si deficient A-15 V-Si superconductors by carbon implantation (53b), and conductivity enhancement in $(CH)_x$ by fluorine implantation (53c) and in poly-p-phenylenesulfide (PPS) by halogen and krypton implantation (53d).

We have recently initiated the study of ion implantation in PDAs using the polymers of PTS, DCH, and TCDU as substrates (54). Our initial studies have used ^{75}As ions at ambient temperature with fluences in the range of $1 \times 10^{13} < \phi < 5 \times 10^{16}$ cm^{-2} at an accelerating voltage of 150 KeV. The implanted materials have been characterized by ESCA, Rutherford Backscattering Spectrometry (RBS), Resonance Raman spectroscopy, transmission spectroscopy, and dc resistivity measurements.

Prior to these studies, 100 keV helium ions were used to introduce traps and recombination centers in poly-PTS (55), and lithium ion implantation in poly-PTS was used to locate a donor level below the conduction band (40).

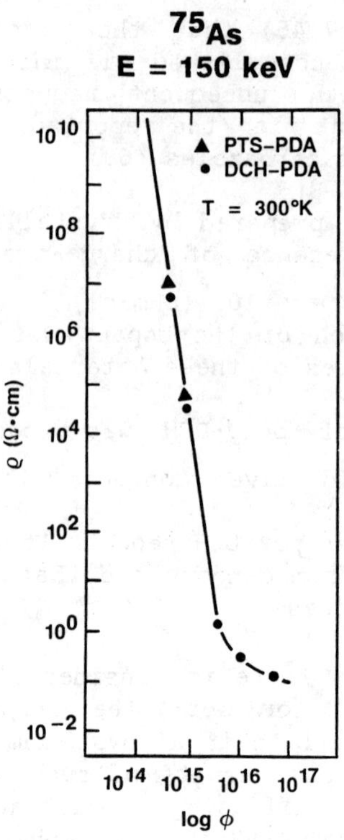

Figure 5. Change in resistivity (ρ) of poly-PTS and -DCH single crystals at 300°K as a function of the fluence (plotted as log ϕ) of ^{75}As ions.

Figure 5 shows the resistivity (as log ρ) of implanted samples of poly-DCH and -PTS as a function of the fluence of ^{75}As ions (as log ϕ). Measurements were made on single crystals with silver paint contacts and the current-voltage characteristics displayed ohmic behavior in the temperature range T=20-290°K. Resistivities of the order of 1 ohm-cm were noted for fluences greater than 5 x 10^{15} cm^{-2} in poly-DCH. Hence, implantation of PDAs with ^{75}As ions provides a viable means for producing materials with markedly reduced resistivities compared to the unimplanted materials. The behavior of the DC resistivity of implanted poly-DCH over the range 20-290°K is consistent with electrical conduction via hopping of carriers between nearest neighbor localized states at elevated temperatures and one dimensional variable range hopping at lower temperatures.

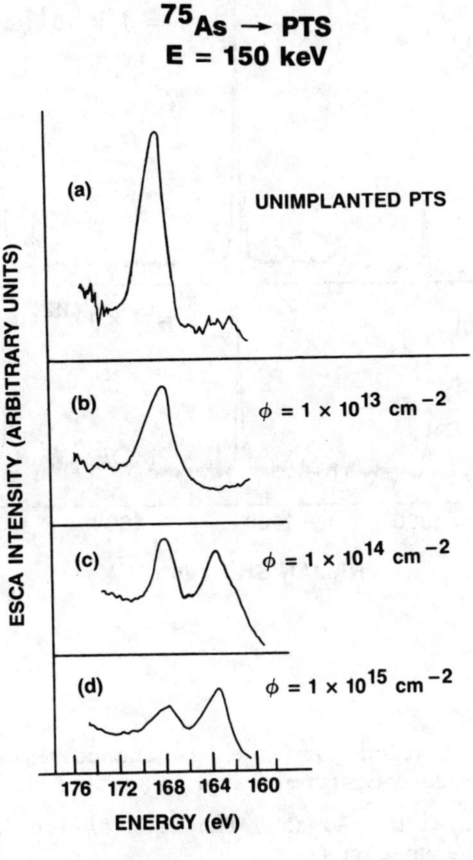

Figure 6. Change in sulfur 2p ESCA spectrum of poly-PTS crystals on exposure to ^{75}As ions as a function of ion fluence (ϕ), which is indicated on the spectrum.

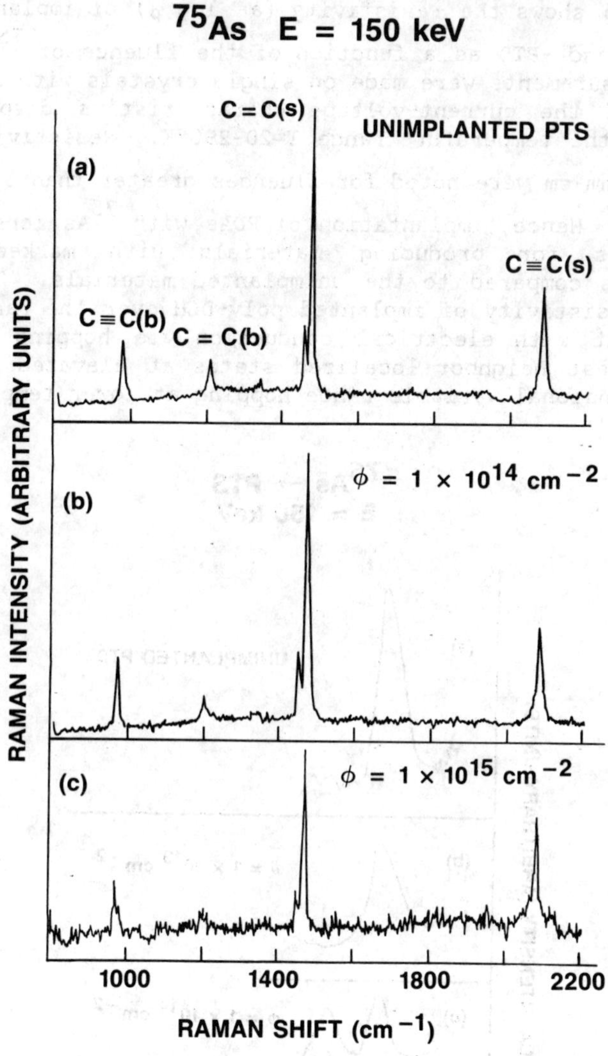

Figure 7. Modification of the resonance Raman spectrum, recorded using 6328 angstrom laser excitation, of poly-PTS crystals implanted with ^{75}As as a function of ion fluence, which is indicated on the spectrum.

The decrease in resistivity brought about by ion bombardment is accompanied by color changes in the implanted material, as well as other spectral changes (54). A green color is noted for poly-PTS implanted at a fluence of 10^{14} cm^{-2} and black is observed for fluences greater than 10^{15} cm^{-2}. The S(2p) ESCA spectrum (Figure 6) shows a new peak at lower binding energy than that of the sulfonate for fluences greater than 10^{14} cm^{-2} in poly-PTS. Figure 7 displays the Raman spectra recorded using laser excitation at 6328 angstroms of unimplanted poly-PTS and samples exposed to fluences of 10^{14} cm^{-2} and 10^{15} cm^{-2} Our spectrum of unimplanted material is in accord with a previous report (56) While the major features of the three spectra in Figure 7 are qualitatively similar, the intensity of the double bond bending mode at 1207 cm^{-1} is gradually decreased at higher fluence. The information we have obtained from RBS, ESCA, and Raman techniques leads to the conclusion that, at low and intermediate fluences of implantation, the side chains of the PDAs are primarily influenced by the ion beam.

ACKNOWLEDGEMENTS

The authors are pleased to acknowledge J. Georger, J. Hryniewicz, S. Meyler, M. Newkirk, and L. Samuelson whose technical assistance facilitated the experimental program summarized above. They also thank Drs. J. Havens and W.F. Manders of the National Bureau of Standards for permission to cite their ^{13}C CP-MAS NMR results prior to publication.

REFERENCES

1. R. Hoffmann, T. Hughbanks, M. Kertesz, and P.H. Bird, J. Am. Chem. Soc, 105, 4831 (1983); M.J. Rice, A.R. Bishop, and D.K. Campbell, Phys. Rev. Lett., 51, 2136 (1983); A.T. Balaban, C.C. Rentia, and E. Ciupitu, Rev. Roumaine de Chim., 13, 231 (1968).
2. A. Karpfen, J. Phys. C., 13, 5673 (1980), and references therein; J.L. Bredas, R.R. Chance, R. Silbey, G. Nicholas, and Ph. Durand, J. Chem. Phys., 75, 255 (1981); J.L. Bredas, R.R. Chance, R.H. Baughman, and R. Silbey, J. Chem. Phys., 76, 3673 (1982).
3. D.J. Sandman, J. Electronic Materials, 10, 173 (1981).
4. R.H. Baughman, J.L. Bredas, R.R. Chance, R.O. Elsenbaumer, and L.W. Shacklette, Chem. Rev, 82, 209 (1982).
5. K.J. Donovan and E.G. Wilson, Phil. Mag., B44, 9 (1981).

6. C. Sauteret, J.P. Hermann, R. Frey, F. Pradiere, J. Ducuing, R.H. Baughman, and R.R. Chance, *Phys. Rev. Lett.* *36*, 959 (1976); *Optics Commun*, *18*, 55 (1976).
7. E. Abraham, C.T. Seaton, and S.D. Smith, *Scientific American*, *248*,(*2*), 85 (1983).
8. P.W. Smith, *The Bell System Technical Journal*, *61*, 1975 (1982).
9. N. Bloembergen, *Science*, *216*, 1057 (1982).
10. "Nonlinear Optical Properties of Organic and Polymeric Materials," D. J. Williams, ed., American Chemical Society Symposium Series 233, 1983.
11. A.F. Garito, C.C. Teng, K.Y. Wong, and O. Zammani' Khamieri, *Mol. Cryst. Liq. Cryst*, *106*, 219 (1984); A.F. Garito and K.D. Singer, *Laser Focus*, *80*, 59 (1982).
12. Y.J. Chen and G.M. Carter, *Appl. Phys. Lett.*, *41*, 307 (1982).
13. J.P. Hermann and P.W. Smith, *Digest of Technical Papers*, XIth International Quantum Electronics Conference, Boston, MA, June 23-26, 1980 (IEEE New York, 1980), paper T6, pp. 656-657.
14. P.W. Smith, private communication to G. M. Carter.
15. M. Lequime and J.P. Hermann, *Chem. Phys.*, *26*, 431 (1977).
16. D. Bloor, D.J. Ando, F.H. Preston, and G.C. Stevens, *Chem. Phys. Lett.*, *24*, 407 (1974).
17. D.J. Sandman, G.M. Carter, Y.J. Chen, S.K. Tripathy, and L.A. Samuelson in Proceedings of the 2nd International Workshop on Molecular Electronic Devices, U.S. Naval Research Laboratory, Washington, D.C., April 13-15, 1983, F. Carter, ed., to be published.
18. G.M. Carter, Y.J. Chen, and S.K. Tripathy in (10), pp. 213-228.
19. G.M. Carter and Y.J. Chen, *Appl. Phys. Lett.*, *42*, 643 (1983).
20. G.M. Carter, Y.J. Chen, and S.K. Tripathy, *Appl. Phys. Lett.*, *43*, 891 (1983).
21. C. Cojan, G.P. Agrawal, and C. Flytzanis, *Phys. Rev. B*, *15*, 909 (1977).
22. G.P. Agrawal, C. Cojan, and C. Flytzanis, *Phys. Rev. B*, *17*, 776 (1978).
23. D.C. Hanna, M.A. Yuratich and D. Cotter, "Nonlinear Optics of Free Atoms and Molecules," D.L. MacAdam, ed., Springer-Verlag (New York, 1979), p.61.
24. B. Reimer and H. Baessler, *Chem. Phys. Lett.*, *55*, 315 (1978).
25. M.L. Shand and R.R. Chance in (10), pp. 187-212.
26. G. M. Carter, Y. J. Chen, J. Georger, J. Hryneiwicz, M. Rooney, M. F. Rubner, L. A. Samuelson, D. J. Sandman, M. Thakur, and S. Tripathy, *Mol. Cryst. Liq. Cryst.*, *106*, 259 (1984).
27. F. Kajzar, J. Messier, J. Zyss, and I. Ledoux, *Optics Commun.*, *45*, 133 (1983).
28. C. Flytzanis in (10), pp. 167-185.
29. G. Wegner, "Molecular Metals," W. E. Hatfield, ed., Plenum Press, New York, 1979, p. 209 ff.
30. M. R. Philpott, *Chem. Phys. Lett.*, *50*, 18 (1977).

31. S. E. Rickert, J. B. Lando, and S. Ching, Mol. Cryst. Liq. Cryst., 93, 307 (1983); S. E. Rickert, J. B. Lando, and S. Ching, in (10), pp. 229-233.
32. R. R. Chance, R. H. Baughman, H. Mueller, and C. J. Eckhardt, J. Chem. Phys, 67, 3616 (1977).
33. V. Enkelmann and J. B. Lando, Acta Crystallogr., B34, 2352 (1978).
34. H. Mueller and C. J. Eckhardt, Mol. Cryst. Liq. Cryst, 45, 313 (1978).
35. J. Havens, W.F. Manders, S. Tripathy, and D.J. Sandman, to be published.
36. J.R. Havens, M. Thakur, J.B. Lando, and J.L. Koenig, Macromolecules, 17, 1071 (1984); G.E. Babbitt and G.N. Patel, Macromolecules, 14, 554 (1981).
37. J.P.C.M. van Dongen, M.J.A. deBie, and R. Steur, Tetrahedron Lett., 1371 (1973).
38. Z. Berkovitch-Yellin and L. Leiserowitz, Acta Crystallogr., B33, 3657 (1977).
39. M. Pope and C. E. Swenberg, "Electronic Processes in Organic Crystals," Oxford University Press, 1982, pp. 673-699.
40. W. Spannring and H. Baessler, Chem. Phys. Lett., 84, 54 (1981).
41. L. Sebastian and G. Weiser, Phys. Rev. Lett., 46, 1156 (1981).
42. A. S. Siddiqui, J. Phys. C: Solid State Phys., 15, L263 (1982).
43. U. Seiferheld and H. Baessler, Solid State Commun., 47, 391 (1983).
44. S. Arnold, J. Chem. Phys., 76, 3842 (1982) A.A. Murashov, E.A. Silinsh, and H. Baessler, Chem. Phys. Lett., 93, 148 (1982).
45. D.J. Sandman, G.P. Hamill, L.A. Samuelson, and B.M. Foxman, Mol. Cryst. Liq. Cryst., 106, 199 (1984).
46. E.G. Popova and L.A. Chetkina, Zh. Strukt, Khim., 20, 665 (1979).
47. A.I. Kitaigorodsky, "Molecular Crystals and Molecules," Academic Press, New York, 1973, Chapter 1, Section A.3.
48. G. Schleier, Doctoral Dissertation, Faculty of Chemistry and Pharmacy, University of Freiburg, 1980.
49. K.C. Yee and R.R. Chance, J. Polym. Sci., Polym. Phys. Edit., 16, 431 (1978).
50. P. Nielsen, A.J. Epstein, and D.J. Sandman, Solid State Commun., 15, 53 (1974).
51. H. Nakanishi, H. Matsuda, and M. Kato, Mol. Cryst. Liq. Cryst., 106, 77 (1984).
52. S.T. Picraux and W.J. Choyke, eds., "Metastable Materials Formation by Ion Implantation," North Holland, New York, 1982.
53. a) J.S. Williams and K.T. Short in (52), pp. 109-115; b) J. R. Gavaler, et al. in (52), pp. 287-293; c) D.C. Weber, et al. in (52), pp. 167-171; d) J.S. Abel, et al., in (52), pp 173-179.

54. B.S. Elman, D.J. Sandman, M.K. Thakur, M. A. Newkirk, E.F. Kennedy, "Ion Beam Modification of Polydiacetylene Crystals," presented at the International Conference on Ion Beam Modification of Materials, Cornell University, Ithaca, NY, July 16-20, 1984; B. S. Elman, M. K. Thakur, D. J. Sandman, M. A. Newkirk, and E. F. Kennedy, J. Appl. Phys., submitted; B.S. Elman, D.J. Sandman, and M.A. Newkirk, Appl. Phys. Lett., in press.
55. U. Seiferheld, H. Baessler, and B. Movaghar, Phys. Rev. Lett., 51, 813 (1983).
56. D.N. Batchelder and D. Bloor, J. Phys. C, Solid State Phys., 15, 3005 (1982).

ELECTRIC FIELD INDUCED OPTICAL SECOND HARMONIC GENERATION AND POLARIZATION EFFECTS IN POLYDIACETYLENE FILMS

P-A. CHOLLET - F. KAJZAR - J. MESSIER

C.E.A. - I.R.D.I.
DEIN/LERA/CEN.SACLAY - 91191 Gif/sur/Yvette Cedex - France

ABSTRACT

Results of electric-field induced second harmonic generation measurements on thin polydiacetylene films performed at 1.064 μm and 1.907 μm are reported. At 1.064 μm one observes large internal polarization fields depending on the polymer thin film morphology. No such polarization effects are observed at 1.907 μm.

1. INTRODUCTION

Polydiacetylenes are known to possess large nonlinear susceptibilities which were measured in cristals (1), in solutions (2) and in thin films by third harmonic generation (3). No second harmonic generation is observed owing to the centrosymmetric character of the polymeric chain. When an external DC-field E_o is applied the symmetry is broken and this allows the second harmonic generation to be observed. Its intensity is proportional to

$$(\chi^{(3)}(-2\omega\,;\,\omega,\omega,0)E_\omega^2\;E_e)^2,$$

where E_e is the effective electric field which is experimented by the polymer and E_ω is the laser field. In this paper we show that the intensity of the second harmonic gives information upon the variation of internal polarization due to the generation and separation of photoinduced charge carriers.

2. EXPERIMENTAL PROCEDURE

2.1 Second harmonic generation measurements

The experimental procedure is depicted in figure 1. Pulses of duration 10 ns were generated by a Nd : YAG laser ($\lambda = 1.064 \mu m$) whose frequency repetition rate f_r was variable from 2.5 Hz to 10 Hz. An hydrogen Raman cell allowed the wavelength to be shifted to 1.907 μm. The laser beam was linearly polarized and its intensity could be continuously varied from 5 μJ to 10 mJ/pulse. The beam ($\phi \sim 8$ mm) was focused on the sample, the spot diameter ranging from 30 to 100 μm.

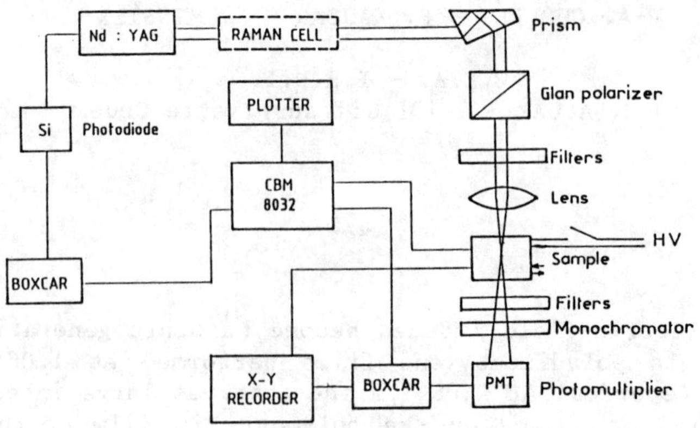

Fig.1 - Experimental set up for second harmonic generation and measurement

The polydiacetylene films were deposited onto silica substrates previously coated with two aluminium electrodes one millimeter apart (much greater than the laser spot diameter). A direct high voltage (0-4000 V) could be switched off or on through a relay. All the experiments were performed under a vacuum better than 10^{-2} Torr. The fundamental beam was rejected by appropriated filters and the second harmonic, selected by a monochromator, was detected by a photomultiplier tube. The pulsed signal delivered by the photomultiplier was integrated by a boxcar whose integration time τ could be varied between 100 ms and 2 s. This allows to select on pulse ($\tau \lesssim \frac{1}{fr}$) or to average over several pulses ($\tau > \frac{1}{fr}$).

2.2 Chemical substances and film morphology

The films were made of polydiacetylenes with different side groups :

2.2.1 Films of 3-BCMU : R-C-C ≡ C-C-R where R is : $-(CH_2)_3OCONHCH_2COOC_4H_9$

This polymer is soluble in several organic solvents and films about 1 μm thick were obtained from solution using several methods :

(i) slow (several hours) and rapid (several minutes) evaporation of a 3g/l solution in chloroform.

(ii) Withdrawal of the substrate (7 cm/mn) from a 15 g/l solution in chloroform.

2.2.2 Langmuir-Blodgett films (100-3000 Å thick) of the cadmium salt : $CH_3-(CH_2)_{11}-C-C \equiv C-C-(CH_2)_8\ COO^-)^2 Cd^{++}$.

2.2.3 Polytoluene sulfonate single cristals whose thickness was about 300 μ m (PTS : $R-C-C \equiv C-C-R$ where R is $CH_3C_6H_4SO_2OCH_2-$).

Unfortunately the samples were damaged by the laser beam at 1.064 μm and experiments could be performed only at 1.907 μm.

3. EXPERIMENTAL RESULTS

3.1 Measurements performed at 1.064 μm

3.1.1 3-BCMU Films

a) Films obtained by the substrate withdrawal technique.

The measured second harmonic intensity $I_{2\omega}$ variations with respect to the external field Eo and the fundamental intensity I_ω are shown in figure 2. $I_{2\omega}$ is found to increase proportionally to I_ω^2 (fig.2a) and E_o^2 (fig.2b) as expected.

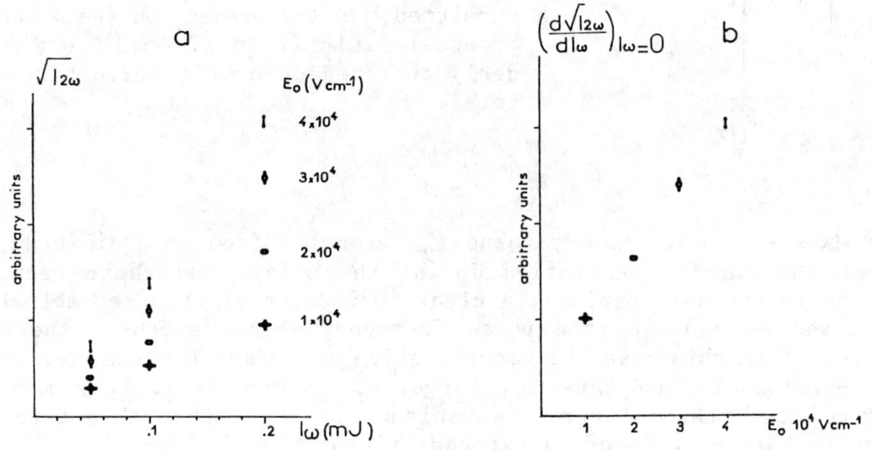

Fig.2 - 3-BCMU films obtained by the substrate withdrawal technique : variation of the second harmonic intensity with respect to the laser beam intensity per pulse (a) and to the DC-field E_o(b). Focal length : 50 cm.

b) Films obtained by evaporation of the solvent.

In this case the second harmonic intensity is found to decrease rapidly at high laser power during the very first pulses from a maximum $I^M_{2\omega}$ to a steady value $I^S_{2\omega}$ (fig.3). This sudden decrease is not due to film damage as can be verified by the constancy of the third harmonic intensity and the reproductibility of the phenomenon. It is very likely due to a rapid polarisation which takes place under the laser beam. The maximum intensity $I^M_{2\omega}$ which is reached during the first pulse increases much less than I^2_ω (fig.4), which proves that a strong polarization takes place during the first 13 ns laser pulse. If the external applied field Eo is switched off suddenly after the steady value $I^S_{2\omega}$ is reached one observes a transient second harmonic peak due the existence of a large remanent polarization field E_p (Fig.3). This polarization is destroyed by the very first laser pulses following the DC-field switching off.

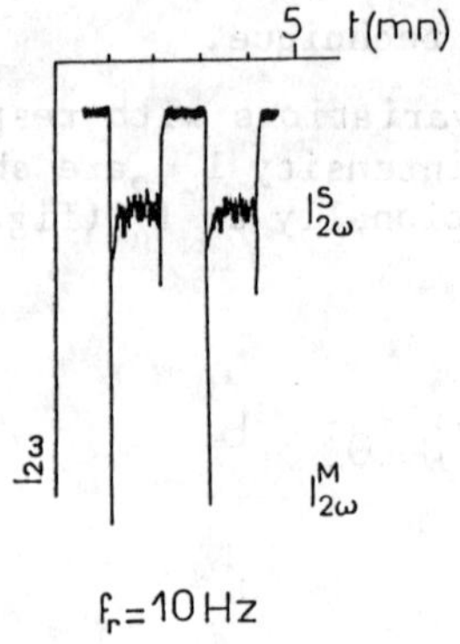

Fig.3 - 3-BCMU films obtained by evaporation of the solvent (thickness 1 μm). Evolution of the second harmonic intensity under the laser pulses (100 μJ/pulse ; focal length 50 cm ; pulse frequency f_r = 10 Hz ; boxcar integration time τ = 100 ms). The film was always submitted to the laser pulses ; the external field (E_o = 10^4 Vcm^{-1}) was suddenly switched on or off through the HV relay.

The existence of a steady value $I^S_{2\omega}$ results from an equilibrum between the carrier generation during the pulses and their recombination in the dark periods. A clear indication of this recombination is given by reducing the pulse frequency which lengthens the dark periods : in this case the recombination increases leading to a smaller polarization and thus to a larger $I^S_{2\omega}$ signal (fig.4). It must be noticed that the polarization vanishes in the dark within a minute even in the presence of an external field Eo = 40 KV/cm.

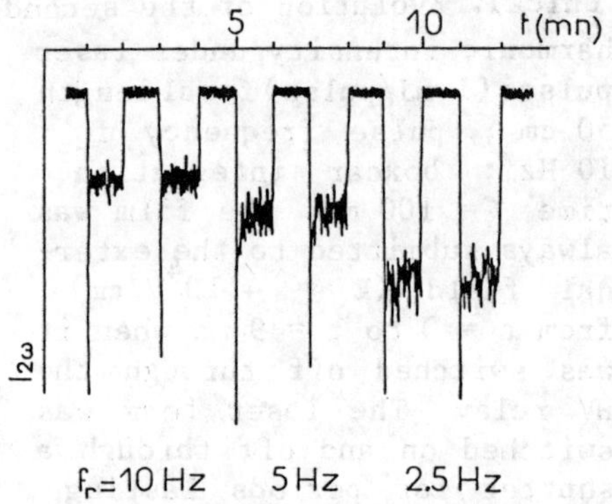

Fig.4 - 3-BCMU film obtained by evaporation of the solvent. Variation of the second harmonic steady value $I^S_{2\omega}$ with respect to the pulse laser frequency. Other experimental conditions are the same as in fig.3 except the film was always submitted to the external field ($E_o = 10^4 Vcm^{-1}$) and the laser beam was suddenly switched on or off through a shutter.

The maximum $I^M_{2\omega}$ increases as I^2_{ω} at low intensity but saturates after, which proves that a strong polarization takes place during the first 13 ns laser pulse (Fig.5a). The initial slope $d(I^M_{2\omega})^{1/2} / dI_{\omega}$ is proportional to E_o^2 (Fig.5b).

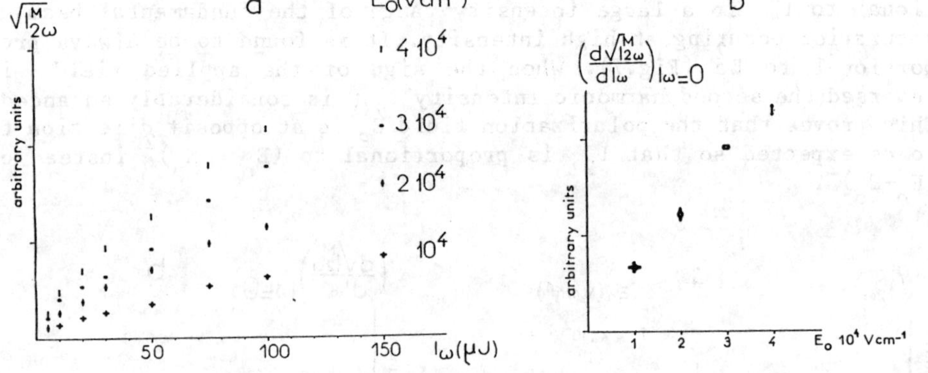

Fig.5 - 3-BCMU film obtained by evaporation of the solvent. Variation the maximum $I^M_{2\omega}$ of the second harmonic intensity with respect to the laser beam intensity per pulse (a) and to the DC-electric field E_o(b) (focal length : 50 cm).

3.1.2 Langmuir-Blodgett multilayers

Similarly as in 3-BCMU films transient polarization phenomena are observed. However the second harmonic intensity $I_{2\omega}$ does not reach a steady value but completely vanishes during the illumination of the LB film by the laser beam. A second harmonic peak is also observed at the DC-field Eo switching off (Fig.6).

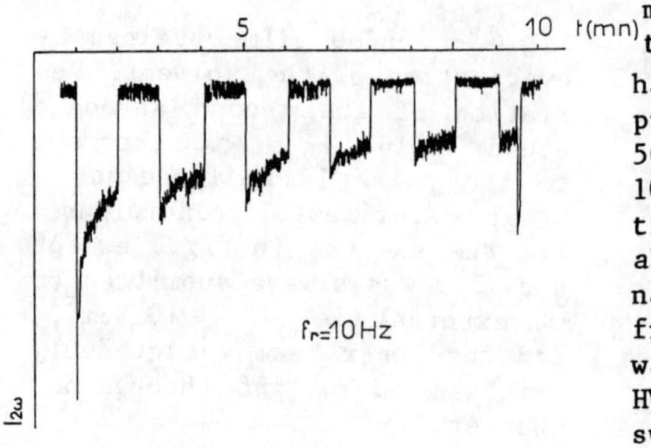

Fig.6 - Langmuir-Blodgett film made up of 100 layers (3000 A thick). Evolution of the second harmonic intensity under laser pulses (1 mJ/pulse) focal length 50 cm ; pulse frequency f_r : 10 Hz ; boxcar integration time τ = 100 ms. The film was always submitted to the external field (E_o = 4.10^4 V/cm) from t = 0 to t = 9 mn when it was switched off through the HV relay. The laser beam was switched on and off through a shutter for periods lasting one minute until t = 9 mn when it continuously illuminated the sample.

In LB films the internal polarization is much durable than in 3-BCMU films and needs several tens of laser pulses to be destroyed. It lasts several hours in the dark at room temperature without noticeable decrease in amplitude. The maximum value I_ω^M is found to be proportional to I_ω^2 in a large intensity range of the fundamental beam, a saturation occuring at high intensity. It is found to be always proportional to Eo^2 (Fig.7). When the sign of the applied field is reversed the second harmonic intensity $I_{2\omega}$ is considerably enhanced. This proves that the polarization field E_p is at opposit direction to Eo as expected so that $I_{2\omega}$ is proportional to $(E_o + E_p)^2$ instead of $(E_o - E_p)^2$.

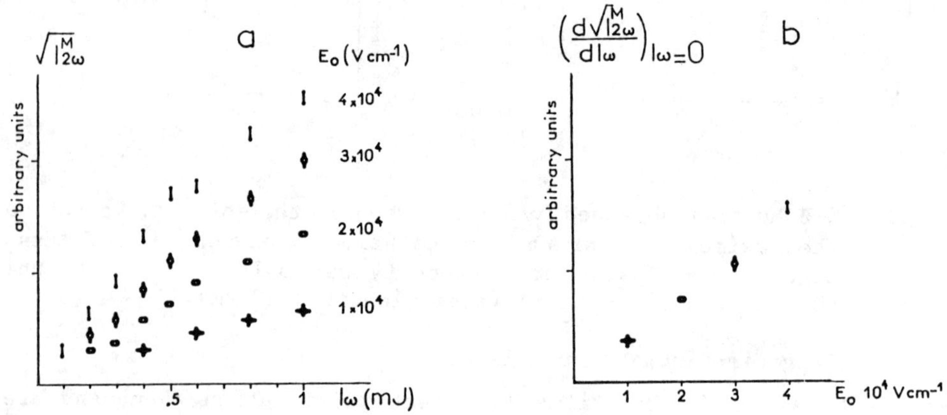

Fig.7 - Langmuir-Blodgett film : variation of the maximum $I_{2\omega}^M$ with respect to the laser beam intensity per pulse (a) and with respect to the DC-electric field E_o (b) (focal length 50 cm).

3.2 Measurements performed at 1.907 μm

No polarization phenomena were observed neither on 3-BCMU films nor on PTS single cristals. Unfortunately the second harmonic intensity generated in LB films was to small to be detected, owing to the smaller sensibiliity of the photomultiplier tube at this wavelength.

4. DISCUSSION AND CONCLUSION

The polarization in PDA films induced by the coupled effects of the external field an the high power laser field shows that charge carriers are generated in the film. This generation is known to occur by a two photon process at 1.064 μm (4). If N is the number of carriers created by unit volum, φ the probability they escape recombination and d their separation distance, the mean polarization by unit volume is given by $P = N\varphi d$.
On the other hand the number of photons absorbed may be evaluated from two photon absorption coefficient β : $dI_\omega / I_\omega = (\beta I_\omega) n dx$, with
$\beta = 5.10^{-48} cm^{-4} s\ (photon)^{-1}\ (repeat\ unit)^{-1}$ (5).
n : $10^{21} cm^{-3}$ is the number of repeat unit per unit volume, dx is the film thickness.
A one millijoule laser pulse focused during 13 ns on a surface of $5.10^{-4} cm^2$ ($I = 150\ MW/cm^2$) leads to a value of $\beta I_\omega n$ equal to 5 cm^{-1}. The number of photons absorbed by unit volume is thus equal to $5.10^{19} cm^{-3}$.
On the other hand the optical second harmonic intensity vanishes completely when the value of the internal field Ep is equal to that of the applied field $Eo = 5.10^4\ V\ cm^{-1}$. This value corresponds to a surfacic charge density $\sigma \sim 10^{+11}\ e/cm^2$ at the boundaries of the laser spot (Fig.8). The spot diameter d being of the order of 100 μm, this surfacic charge leads to a value $N\varphi \sim \frac{\sigma}{d} = 10^{13}\ e/cm^3$ far smaller than the number of photons adsorbed. This shows that the internal polarization may be easily equal to the external field even if the probability of an electron-hole pair to escape geminate recombination is much smaller than unity, or/and separation distance of the pair is much smaller than the laser spot diameter.

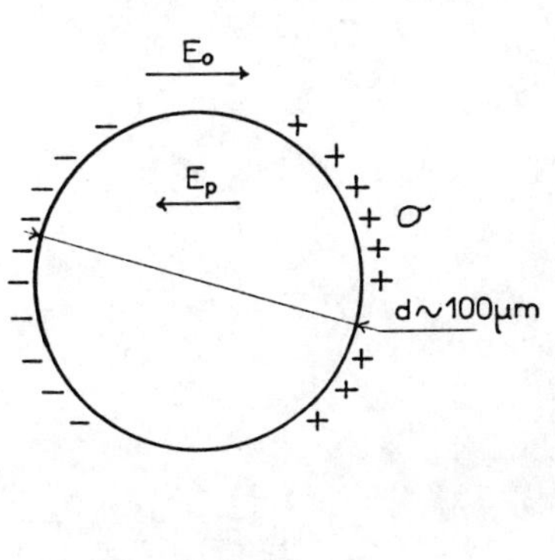

Fig.8 - Setting of the internal polarization E_p through charge drainage at the laser spot boundary.

The polarization intensity and its permanence are drasticaly dependent on the film morphology. The lasting of polarization in Langmuir-Blodgett films is likely to be due to carriers trapping at the boundaries of the crystallites which can be easily seen between crossed polarizers. On the other hand in 3-BCMU films only a transient polarization is observed : This is probably due to the fact that the charge carriers do not pass beyond the places where the polymeric chains are oriented in a perpendicular direction with respect to the external applied field Eo. These places have no special reasons for trapping the charge carriers.

ACKNOWLEDGMENT

We gratefully acknowledge Prof. G. Wegner and Dr. B. Tieke for fruitful discussions and for providing us with diacetylenes.
We also wish to thank J. Banide, R. Gras and A. Lorin for very useful experimental assistance.

REFERENCES

1. Sauteret C. and al. Optical Nonlinearities in One-Dimensional. Conjugated Polymer Crystals. Phys.Rev. Lett. 36 956-959(1976)
2. Kajzar F. and J. Messier. Third Harmonic Generation and two Photon Absorption in a Polydiacetylene Solution. This Issue.
3. Kajzar F., J. Messier, J. Zyss and, I. Ledoux. Nonlinear Interferometry in Langmuir- Blodgett Multilayers of Polydiacetylene. Opt. Commun. 45 133- 137(1983)
4. Reimer B. and H. Bässler. Photoelectric evidence for neutral excited state in fully polymerized PTS single crystals. Chem.Phys. Lett. 55 315-318(1978)
5. Shand M.L. and R.R. Chance. Third-Order Nonlinear mixing in polydiacetylenes solutions. J. Chem. Phys. 69 4482-86(1978).

THIRD HARMONIC GENERATION AND TWO PHOTON ABSORPTION IN A POLYDIACETYLENE SOLUTION

F. KAJZAR and J. MESSIER

CEA - IRDI
DEIN/LERA/CEN.SACLAY
91191 GIF-SUR-YVETTE CEDEX - France

ABSTRACT

Third harmonic generation measurements performed on TS-12 PDA in solution at 1.064 μm and 1.907 μm confirm large values of cubic hyperpolarizabilities of polydiacetylenes. The observed complex value of $\chi^{(3)}$ at 1.064 μm with negative real and imaginary parts is interpreted in terms of two photon resonance and the two photon level is located between 18800 and 28200 cm^{-1}.

1. INTRODUCTION

The nonlinear optical properties of polydiacetylenes are subject of an increasing interest in last years (1-4). This is due to the fact that polydiacetylenes are a one-dimensionnal system of higly polarizable conjugated π electrons. The polarization depends strongly on electron delocalization length. Moreover, the electronic origin of hyperpolarizability implies short response times. On the other hand, the wave dispersed measurements of molecular hyperpolarizabilities yield information about forbidden electronic transitions. In fact the bulk polarization can be developped in external electric field power series :

$$P = \chi^{(1)} : E + \chi^{(2)} : E\,E + \chi^{(3)} : E\,E\,E + \ldots \qquad (1)$$

where $\chi^{(n)}$ are n-th rank three dimensional tensors. For third harmonic generation the corresponding nonlinear susceptibility is given by(5)

$$\chi^{(3)}(-3\omega\ ;\ \omega,\omega,\omega) \propto \sum_{\substack{gnn' \\ n''}} \left\{ \frac{\Omega_{gn''}\ \Omega_{n''n'}\ \Omega_{n'n}\ \Omega_{ng}}{(E_{n''g}-3\omega)(E_{n'g}-2\omega)(E_{ng}-\omega)} + \right.$$

$$+ \frac{\Omega_{gn''}\ \Omega_{n''n'}\ \Omega_{n'n}\ \Omega_{ng}}{(E_{n''g}+\omega)(E_{n'g}-2\omega)(E_{ng}-\omega)} + \frac{\Omega_{gn''}\ \Omega_{n''n'}\ \Omega_{n'n}\ \Omega_{ng}}{(E_{n''g}+\omega)(E_{n'g}+2\omega)(E_{ng}-\omega)} +$$

$$\left. + \frac{\Omega_{gn''}\ \Omega_{n''n'}\ \Omega_{n'n}\ \Omega_{ng}}{(E_{n''g}+\omega)(E_{n'g}+2\omega)(E_{ng}+3\omega)} \right\} \qquad (2)$$

where g, n, n', n" are energy levels of unperturbed system (see Fig.1) and $\Omega_{ij} = \langle i/er/j \rangle$ dipolar transition moments. For the sake of simplicity we have omitted electric field direction vectors. This corresponds to the collinear arrangement as it is case in our experiment.

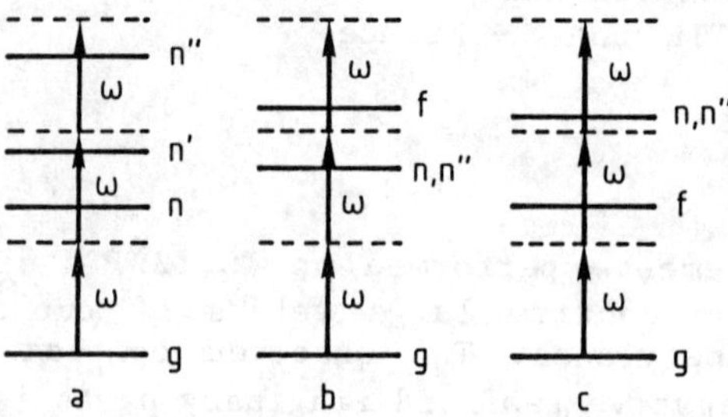

Fig.1 - Schematic representation of unperturbed (——) and virtual (----) levels with two possible positions of two photon level (f).

For photon or multiple photon energies close to the energy difference between fundamental and excited states one should observe a resonance enhancement in nonlinear susceptibilities as well as appearance of an imaginary part in $\chi^{(3)}$. On other hand this imaginary part of $\chi^{(3)}$ will be responsible for two photon absorption processes.

In this paper we present some results of THG measurements in a solution of polydiacetylene at two laser wavelength : 1.064 μm and 1.907 μm. The experimental arrangement, described elsewhere (6), allows not only an exact determination of third order nonlinear susceptibility, but also its phase.

2. EXPERIMENTAL PROCEDURE

THG measurements have been done with a pulsed 10 pps Q-switched Nd :YAG laser at 1.064 μm and at 1.907 μm ; the last wavelength obtained by a Raman shift in high pressure hydrogen cell.

The studied polymer in solution is dodeca 5,7 diyne -1,12 - ylene - bis (p-toluene sulfonate) (TS-12) with side group $R=R'=(CH_2)_4\ OSO_2\ C_6H_4CH_3$ synthetized by Wenz and Wegner (7). The solvent used was DMF.

The optical absorption spectrum of this polymer is shown in Fig.2. The measurements have been done in cells with liquid compartment thickness of 150 μm (because of absorption at harmonic frequency, cf. Fig.2) at 1.064 μm and 1000 μm at 1.907 μm. All measurements were performed in air and corrected for air contribution as described elsewhere (6).

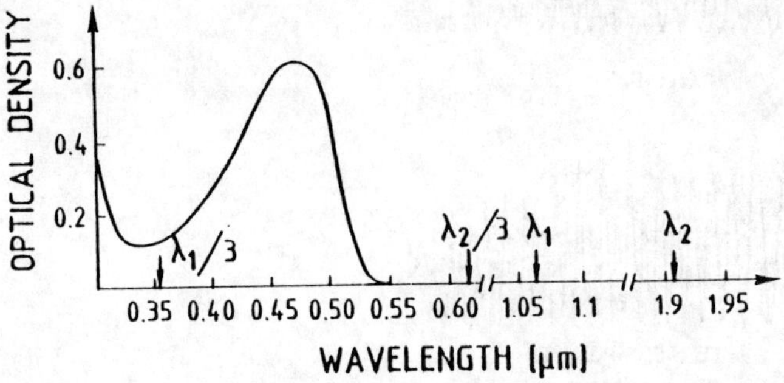

Fig.2 - Optical absorption spectrum of studied polymer. Arrows show fundamental laser wavelengths and those corresponding to harmonic frequencies

The harmonic fields generated in cell windows (G) and in polymer solution (L) (see Fig.3) interfere and the resulting field is given by

$$E^{3\omega} = E^{3\omega}_{G1} + E^{3\omega}_{L} + E^{3\omega}_{G2} \tag{3}$$

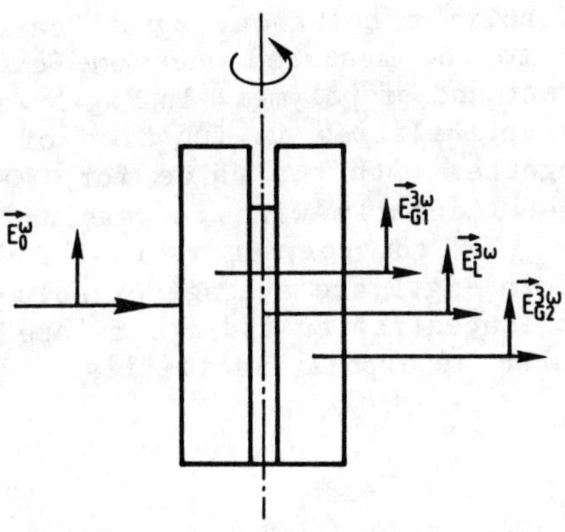

Fig.3 - Experimental arrangement used in the present work. The resulting harmonic field is a sum of harmonic fields created in cell windows and in liquid compartment.

This interference (details in Ref.(6)) allows a determination of absolute value of liquid nonlinear susceptibility and its phase with respect to that of the cell window which is well known. In Fig.4 we show an example of such interference for an empty cell, cell filled

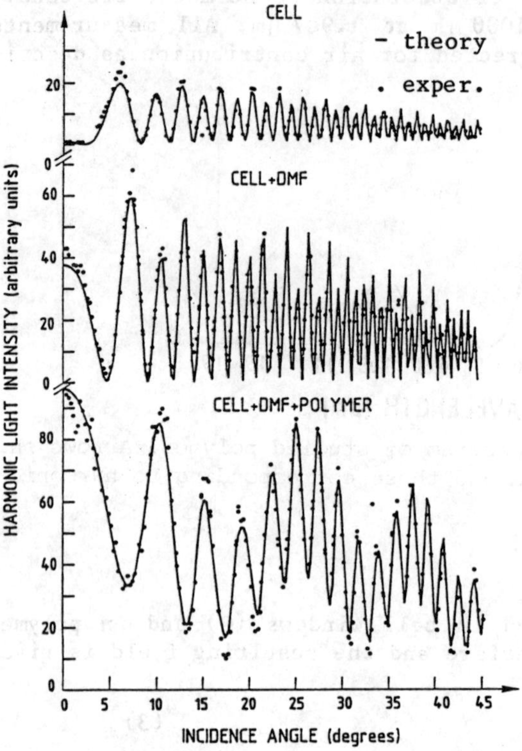

Fig.4 - Fringes patterns for different liquid cell fillings.

with solvent alone and cell with a polymer solution. By a least square fit of calculated intensities to the measured ones one can determine all characteristics of solvent and of polymer. In Fig.5 we have plotted the measured cubic susceptibilities in function of polymer concentration at 1.907 μm together with the value for DMF (6). One can see that $\chi^{(3)}$ increases linearly with increasing polymer concentration. It means that $\chi^{(3)}$ for polymer is real and positive. The values of $\chi^{(3)}$ plotted in Fig.5 are average of solution constituents for a given polymer concentration and are related to the individual microscopic molecular hyperpolarizabilities through following equation :

$$\chi^{(3)} = f \left[N_p \gamma_p + N_s \gamma_s \right] \tag{4}$$

where N_p, N_s is the density of molecules in polymer and solvent, respectively and f is the local field factor, which in the Lorenz approximation is equal to (cf. (8))

$$f = L_\omega^3 L_{3\omega} \tag{5}$$

with $L_{\omega,3\omega} = (n^2_{\omega,3\omega} + 2)/3$ (6)

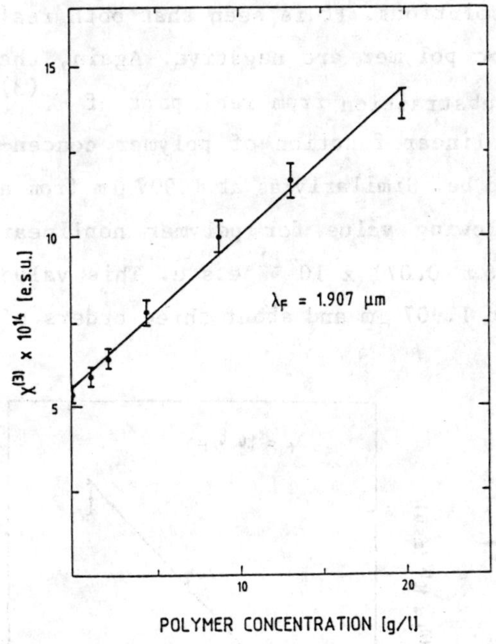

Fig.5 - Concentration variation of cubic susceptibility $\chi^{(3)}$ at 1.907μm.

where $n_{\omega, 3\omega}$ are refractive indices of solution at fundamental and harmonic frequency, respectively.
From an elementary analysis (see e.g. Levine and Bethea (8), Oudar (9), Ledoux and Zyss (10)) one can rewrite Eq. (4) in the following form

$$\chi^{(3)} = x \ \chi_p^{(3)} + (1-x) \ \chi_s^{(3)} \tag{7}$$

where x is the polymer volume fraction in the solution ($x = m_p d_s / (m_p d_s + m_s d_p$ where m_p, m_s are masses and d_p, d_s densities of polymer and solvent respectively : $d_p = 1.4 g/cm^3$ (11) and $d_s = 0.94 \ g/cm^3$) and $\chi_p^{(3)}$, $\chi_s^{(3)}$ corresponding nonlinear susceptibilities. From a least square fit of Eq.(7) to experimental data we have got for the studied polymer at 1.907 m $\chi_{xxxx}^{(3)} = (0.61 \pm 0.04) \times 10^{-11}$ e.s.u. and for DMF $\chi_{xxxx}^{(3)} = (5.6 \pm 0.3) \times 10^{-14}$ e.s.u. compared to the measured value (6) : $\chi_{xxxx}^{(3)} = (5.4 \pm 0.3) \times 10^{-14}$ e.s.u. At 1.064 μm the situation is quite different and in order to explain the shape of interference spectra and the concentration dependence of $\chi^{(3)}$ we had to introduce a phase in cubic susceptibility.

In Fig.6 a we have plotted in complex plane real and imaginary parts of $\chi^{(3)}$ for different polymer solutions. It is seen that both real and imaginary parts of $\chi^{(3)}$ for polymer are negative. Again, the absolute value of $\chi^{(3)}$ (after substraction from real part of $\chi^{(3)}$ for solution that of DMF) is a linear function of polymer concentration (see Fig.6b) as it should be. Similarly as at 1.907 μm from a least-square fit we obtain following value for polymer nonlinear susceptibility : $\chi^{(3)}_{xxxx} = (1.06 \pm 0.07) \times 10^{-11}$ e.s.u. This value is about two times larger than at 1.907 μm and about three orders of magnitude larger than for DMF.

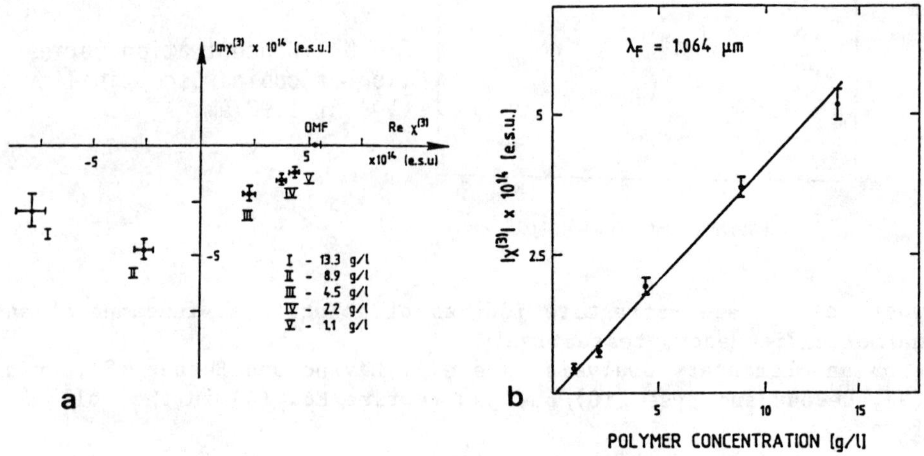

Fig.6 - Real and imaginary parts (a) and modulus (b) of $\chi^{(3)}$ for different polymer concentrations

The measured values of $\chi^{(3)}$ for the studied polymer together with molecular hyperpolarizabilities (per monomer unit)

$$\gamma_p = \chi_p^{(3)} / N_p \tag{8}$$

(where N_p is number of monomer units per unit volume) are compared with corresponding values for several solvents at these two wavelengths in Table I. It is seen that polymer hyperpolarizabilities are from two to three orders of magnitude larger than those of solvents. Moreover, if we take into account the fact that in average a polymer chain contains up to 1.5×10^3 monomer units (7) we obtain the largest observed molecular hyperpolarizabilities (six order of magnitude larger than those of e.g. benzene) naturally for a molecule much larger.

TABLE I

Cubic susceptibilities and molecular hyperpolarizabilities

	$\lambda = 1.064\,\mu m$		$\lambda = 1.907\,\mu m$	
	$\lvert\chi^{(3)}\rvert \times 10^{14}$ e.s.u.	$\lvert\gamma\rvert \times 10^{36}$ e.s.u.	$\chi^{(3)} \times 10^{14}$ e.s.u.	$\gamma \times 10^{36}$ e.s.u.
PDA	1058	419	610	204
$CHCl_3$	6.6[a]	2.6[a]	5.7[a]	2.3[a]
benzene	11.4[a]	4.2[a]	9.9[a]	3.8[a]
CS_2			23[b]	4.4[b]

a - Réf.6
b - Réf.12

3. DISCUSSION

The fact that we observe real and positive value of $\chi^{(3)}$ at 1.907 μm and complex $\chi^{(3)}$ at 1.064 μm means that there exists a resonant two photon state below or above the one photon level. In such case the dominant part in nonlinear susceptibility (Eq.(1)) will be this one containing two photon absorption terms. Neglecting other terms we will have for real and imaginary parts of $\chi^{(3)}$, respectively

$$\mathrm{Re}\,\chi^{(3)}_{xxxx} \propto \frac{E_{fg} - 2\omega}{(E_{fg} - 2\omega)^2 + \Gamma^2} \sum_{n''} \Omega_{gn''}\,\Omega_{n''f} \left[\frac{1}{E_{n''g} - 3\omega} + \frac{1}{E_{n''g} + \omega}\right] \times \sum_{n} \frac{\Omega_{fn}\,\Omega_{ng}}{E_{ng} - \omega} \tag{8}$$

$$\mathrm{Im}\,\chi^{(3)}_{xxxx} \propto \frac{1}{(E_{fg} - 2\omega)^2 + \Gamma^2} \sum_{n''} \Omega_{gn''}\,\Omega_{n''f} \left[\frac{1}{E_{n''g} - 3\omega} + \frac{1}{E_{n''g} + \omega}\right] \times \sum_{n} \frac{\Omega_{fn}\,\Omega_{ng}}{E_{ng} - \omega} \tag{9}$$

where Γ is the dumping term and f denotes two photon level (cf. Fig.1). In Fig.7 we have shown calculated values of real and imaginary parts of $\chi^{(3)}$ as well as its absolute value in function of wavelength for two different cases : two photon level above (Fig.7a) and below (Fig.7b) one photon state, respectively. It is seen that at first case the signs of real and imaginary parts of $\chi^{(3)}$ are in agreement with experimental data at 1.064 μm (both negative). In the second case calculation gives positive real part at 1.064 μm in disagreement with our observation. In fact, if one lookes carefully at formulas (8) and (9) one can conclude that the first situation prevails for $3\omega > E_{fg} > 2\omega$, where ω is the laser frequency (1.064 μm). Thus we can locate two photon level between 0.35 μm and 0.53 μm (28200 and 18800 cm^{-1}, respectively). This result is in agreement with that obtained by Chance et al (2) from four wave mixing experiment and who place the two photon state at 30000 cm^{-1} for yellow solution of 3-BCMU polymer.

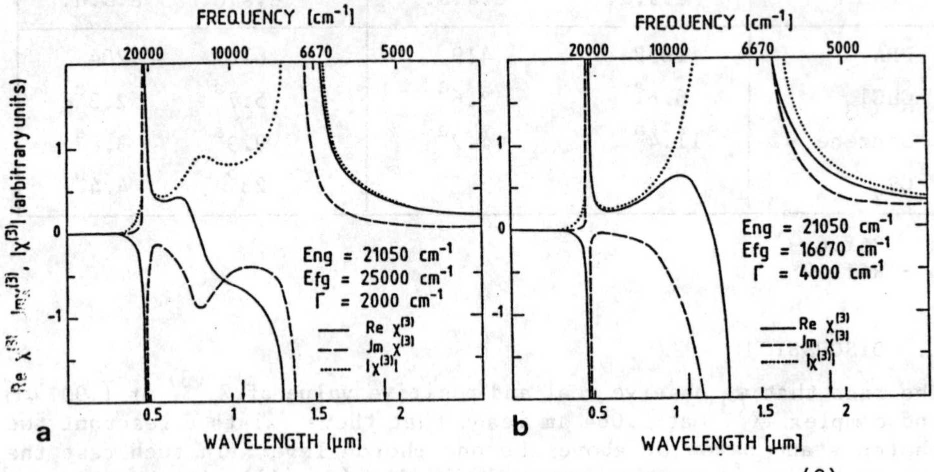

Fig.7 - Calculated modulus, real and imaginary part of $\chi^{(3)}$ for two different positions of two photon level : $E_{fg} > E_g$ (a) and $E_{fg} < 2\omega$ (b) where ω is frequency corresponding to laser wavelength of 1.064μm.

ACKNOWLEDGEMENTS

The authors would like to thank Prof Wegner and Dr. Wenz from Institute für Makromolekulare Chemie from Freiburg for donation of sample used in the present experiment as well as for numerous discussions.

REFERENCES

1. Sauteret C., J.P. Hermann, R. Frey, F. Pradère, J. Ducuing, R.H. Baughman and R.R. Chance. Optical Nonlinearities in One-Dimensional - Conjugated Polymer Crystals, Phys.Rev. Lett.36 (1976) 956-959.
2. Chance R.R., M.L. Shand, C. Hogg, and R. Silbey. Three Wave Mixing in Conjugated Polymer Solutions : Two-photon Absorption in polydiacetylenes. Phys. Rev. B22 (1980)3540-3550.
3. Kajzar F., J. Messier, J. Zyss and J. Ledoux. Nonlinear Interferometry in Langmuir-Blodgett Multilayers of Polydiacetylene. Optics Commun. 45(1983)133-137.
4. Carter G.M., Y.J. Chen and S.K. Tripathy. Intensity-dependent Index of Refraction in Multilayers of Polydiacetylene. Appl. Phys. Lett. 43(1983)891-893, see also D.J. Sandman, G.M. Carter, Y.J. Chen, B.S. Elman, M. Thakur and S.K. Tripathy. Toward Polydiacetylenes as Utile Optical and Electronic Elements. This issue.
5. Hanna D.C., M.A. Yuratich and D. Cotter. Nonlinear Optics of Free Atoms and Molecules, Springer-Verlag, Berlin Heildelberg New York 1979.
6. Kajzar F. and J. Messier. Third Harmonic Generation in Liquids, submitted to J. Chem. Phys.
7. Wenz G. and G. Wegner. Makromol., Molecular Weight Distribution and Solution Properties of a Poly(diacetylene), Chem. Rapid Commun. 3(1982) 231-237.
8. Levine B.F. and C.G. Bethea. Second and Third Order Hyperpolarizabilities of Organic Molecules, J. Chem. Phys. 63 (1975)2666-2682.
9. Oudar J.L. Optical Nonlinearities of Conjugated Molecules Stilbene Derivatives and Nighly Polar Aromatic Compounds, J. Chem. Phys., 67 (1977)446-457.
10. Ledoux I. and J. Zyss. Influence of the Molecular Environment in Solution Measurements of the Second-Order Optical Susceptibility for Urea and Derivatives. Chem. Phys. 73(1982)203-213.
11. Siegel D., H. Sixl, V. Enkelmann and G. Wenz. Polymerization of TS-12 Diacetylene Crystals : Crystal Structures of Monomer and Polymer and Spectroscopy of Reaction Intermediates, Chem Phys. 72 (1982)201-212.
12. Meredith G.R., B. Buchalter and C. Hanzlik. Third-Order Optical Susceptibility Determination by Third Harmonic Generation I., J. Chem. Phys 78(1983)1533-1542.

MECHANICAL AND STRUCTURAL PROPERTIES OF POLYDIACETYLENES

Robert J. Young

Department of Materials, Queen Mary College,
Mile End Road, London, E14NS, UK.

ABSTRACT

The relationship between structure and mechanical properties in polydiacetylene single crystals has been reviewed. It is demonstrated that the ability to produce the crystals with a high degree of perfection has allowed a detailed correlation to be made between their structure and mechanical behaviour.

1. INTRODUCTION

Polydiacetylenes allow a unique opportunity to study the relationship between structure and mechanical properties in polymer crystals. The technique of solid state polymerization [1] enables highly-perfect polydiacetylene single crystals to be produced with macroscopic dimensions. For example single crystal fibres can be grown with lengths in excess of 50 mm [2,3]. Crystalline polymers produced by crystallization from both dilute solution and the molten state are invariably only semi-crystalline [4]. Melt-crystallized polymers are also polycrystalline and have spherulitic structures. Microscopic lamellar single crystals of polymers such as polyethylene can be grown from dilute solution but they are extremely small and can only be studied using electron microscopy. In addition, the molecules fold back and forth within the crystals and the entities are only ~80% crystalline because of the presence of amorphous material in the fold regions [4].

Attempts have been made to determine the relationship between the structure and mechanical properties in semicrystalline polymers. If melt-crystallized polyethylene is subjected to plane-strain deformation such as rolling a sample with a "single-crystal texture" can be produced [5-7]. Although the material is still polycrystalline the crystallites are oriented preferentially in certain directions in the specimens. Deformation studies of such specimens have allowed some information to be obtained concerning the deformation of individual crystals although because of the polycrystalline nature of the specimens certain ambiguities remain [7]. There has also been several reports of studies of the deformation of polymer single crystals by stretching on extensible substrates [8-10]. Because the molecules are invariably oriented perpendicular to the surfaces of the lamellar crystals [7] deformation is limited to directions perpendicular to the chain direction. However, this has allowed important and useful information to be obtained in relation to twinning and martensitic transformations [8-10].

Much of our fundamental understanding of the deformation of metals has been obtained from the study of the behaviour of large single crystals. For many years polymer scientists were frustrated by the lack of polymers with similar morphologies. It was realised that the fundamental problem was the difficulty of arranging the long tangled molecules of a molten polymer into a crystal. Hence there was considerable excitement when polymer chemists started to examine the possibility of producing polymer crystals by solid-state polymerization of single crystal monomers [1]. Some of the first attempts, however, were disappointing. For example, single crystals of trioxane were found to undergo polymerization to polyoxymethylene in the solid state but the polymer produced was found to be polycrystalline with a complex morphology [11]. However, it did allow some limited measurements of the relationship between structure and mechanical properties to be carried out [12, 13]. Wegner [1] has pointed out that the "topotactic" polymerization of trioxane in which there is no lattice control would not be expected to produce good polymer single crystals. He showed that it was necessary for the solid-state reaction to be "topochemical" where there is a direct transition from the monomer molecules into polymer chains without destruction of the crystal lattice. Earlier, Wegner [14] had demonstrated that diacetylenes could be polymerized topochemically to produce highly-perfect single crystals and in this discovery has led onto the development of numerous different diacetylene derivatives, many of which have proved extremely useful for the study of structure/property relations in polymers.

2. STRUCTURE.

A large number of symmetrically and asymmetrically-substituted diacetylenes have now been investigated [1]. However, only relatively few have been used for evaluation of structure/property relations. The main derivatives used are listed in Table I. These

R	Abbreviation
$- CH_2 - O - SO_2 -$ 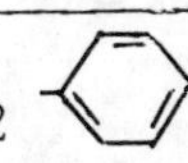 $- CH_3$	TSHD
$- CH_2 - O - CO - NH -$ 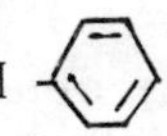	PUHD
$- CH_2 - O - CO - NH - C_2H_5$	EUHD
$- CH_2 - N$ 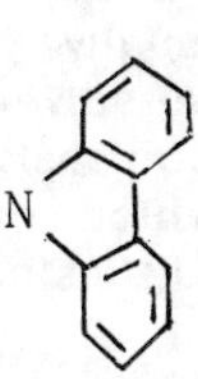	DCHD

TABLE 1. Examples of symmetrical diacetylenes R-C≡C-C≡C-R used in investigations into structure/property relations.

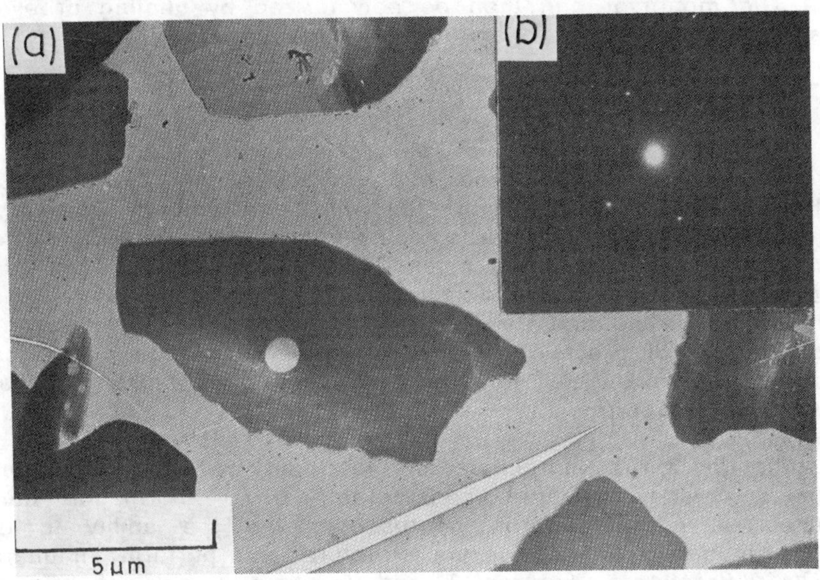

Fig. 1(a) Transmission electron micrograph of lamellar single crystals of polyTSHD grown from dilute solution and thermally polymerized. (b) SADP from double-exposed area in (a).

are all symmetrically substituted diacetylene monomers which can be polymerized topochemically using heat or high-energy radiation to give highly perfect polymer single crystals with a repeat unit of $[RC-C\equiv C-CR]_n$.

2.1 Crystal Morphology.

Polydiacetylene single crystals can be obtained in essentially two crystal forms, either as lozenges or fibres. The morphology is controlled by the conditions under which the monomer is crystallized although the exact reasons why a particular morphology is obtained are not really understood. TSHD is normally found in the form of lozenges from most solvents whereas DCHD is usually obtained as fibres, the aspect ratio of which depends upon the solvent, solution concentration and crystallization temperature. In contrast EUHD can be obtained in three crystal forms, only one of which can undergo solid-state polymerization to give fully-polymerized single crystal fibres [15].

Polydiacetylenes are particularly convenient for studying polymers in the electron microscope. This is because thin monomer crystals can be produced by allowing a droplet of dilute monomer solution to evaporate on a carbon support film on an electron microscope grid [16]. The monomer can then be polymerized by heating or even by exposure to the electron beam in the microscope.

2.2 Crystal structure.

The crystal structures of several polydiacetylenes have been determined to a high degree of accuracy using X-ray diffraction [17-19]. This must be contrasted with conventional polymers for which the crystal structures determined [20] using oriented polycrystalline samples are much more inaccurate. The unique single crystal nature of the polydiacetylenes has enabled the position of the atoms in the unit cells and bond angles to be measured with a high degree of precision. This depth of knowledge of the crystal structures is invaluable for the purpose of relating mechanical behaviour to structure.

Although X-ray diffraction is the most accurate technique to determine crystal structures considerable extra information can be evaluated using an electron microscope with the ability to obtain high magnification images of the structure and perform simultaneous electron diffraction. Figures 1 and 2 show a pair of bright-field images and electron diffraction patterns for lamellar single crystals of polyTSHD and fibres of polyDCHD respectively. In both cases the electron diffraction patterns show clearly that the chain directions of the crystals lie in the plane of the carbon support film [16]. In addition, the diffraction patterns show that the polymers are true

Fig. 2(a) Transmission electron micrograph of polyDCHD crystals showing variety of fibre dimensions. (b) Selected-area diffraction pattern obtained from a polyDCHD crystal.

Fig. 3 (010) lattice image from a thin polyDCHD crystal showing a spacing of approximately 1.2 nm.

single crystals with no amorphous scattering and a high degree of perfection in contrast to single crystals of conventional polymers [4]. Moreover, it means that in polyDCHD there is perfect alignment of the molecules in the fibres. For many years people have been striving with only limited success to attain high degrees of alignment in polymer fibres using techniques such as mechanical orientation [21,22] and solution spinning [23]. In the case of polydiacetylene single crystal fibres this orientation is present as a consequence of their preparation by the topochemical solid-state polymerization reaction. This has tremendous consequences for their mechanical behaviour.

One problem normally encountered with organic materials in the electron microscope is that they are prone to radiation damage whereby the crystal structure is destroyed by exposure to the electron beam [24]. It has been found that most polydiacetylenes are relatively stable but that polyDCHD is particularly outstanding in its ability to resist damage in the electron beam [25] being over 20 times more stable than polyethylene, probably the most widely studied polymer. This high stability has allowed detailed studies to be made of the structure of polyDCHD at high magnification and it has been found that crystal lattice can be imaged directly in the microscope [25,26]. An example of this is shown in Figure 3 where planes parallel to the chain direction are imaged. As well as being an interesting exercise in extending electron microscope studies in polymers imaging the crystal lattice shows the perfect alignment of the polymer molecules in polyDCHD crystals and the relative defect-free nature of polydiacetylene single crystals. Very occasionally chain-end dislocations can be imaged [25] and these are discussed in section 2.4

2.3 Defects in Polymer Crystals.

Crystals of all materials are known to contain defects such as dislocations stacking faults and vacancies but because of problems with obtaining large single crystals [4] and radiation damage in the electron microscope [24], knowledge of such defects in polymers has been somewhat limited. This is unfortunate because there are several reasons why interest should be shown in defects in macromolecular crystals. It is essential from a fundamental viewpoint to know what types of defects exist in polymer crystals and how they compare with defects in other materials. Also it is desirable to know if there is any correlation between the presence of defects and the electrical or mechanical properties of polymer crystals.

The study of defects in metal crystals has been undertaken over the past 30 years using electron microscopy and there is now a high level of understanding of the nature of defects in metal crystals. The chain-folded solution-grown lamellar single crystals of conventional polymers prepared from dilute solution have allowed investigations to be carried out upon certain types of dislocations [27-29]. However, the lamellar morphology of these crystals limit the viewing direction in the microscope to parallel to the chain direction and defects which can only be seen by viewing in other directions cannot be investigated. It is therefore only possible to study a restricted number of defects. Also radiation damage [24] restricts the time available to analyse the defects using different microscope conditions. The advent of polydiacetylene single crystals has allowed detailed studies of several types of defects found in polymer crystals to be carried out because of their better stability in the microscope, their well-defined structure and the ability view crystals at different angles to the chain direction.

2.4 Dislocations in polydiacetylenes.

Detailed examination of polyTSHD crystals, such as the ones seen in Figure 1, by dark-field electron microscopy has shown that in certain areas in the Bragg condition there are small regions which exhibit characteristic black/white contrast patches. An example of this is shown in Figure 4 and it has been demonstrated that these features are edge dislocations with Burgers vectors parallel to the chain direction [30,31]. The density of the dislocations was found to vary from crystal to crystal and the areas of high density were found to have up to $10^{13} m^{-2}$. This is considerably less than typical dislocation densities in metal crystals.

It is of considerable interest to consider the structure of these edge dislocations on a molecular level. The dislocation is sketched schematically in Figure 4b and it can be seen that the extra half-plane of dislocation corresponds to an extra half-plane of chain segments. Such dislocations would be expected to have high line energy as the molecules are stretched along their lengths on one side of the defect and compressed on the other [30]. However, it is thought that these dislocations are present initially in the solution-grown TSHD monomer crystals and "frozen-into" the polymer crystals during solid-state polymerization. Indeed Schermann, Wegner, Williams and Thomas [32] have, by etching TSHD monomer crystals shown from the presence of etch-pits that dislocations are probably present in the monomer and moreover, have shown that such dislocations can help to facilitate thermal polymerization and this has been confirmed by Dudley et al [33] who studied the dislocations using radiation topography. It is thought also that since TSHD monomer has more isotropic elastic properties than the polymer these edge dislocations would be favoured energetically in the monomer [30].

These edge dislocations in polyTSHD have some interesting

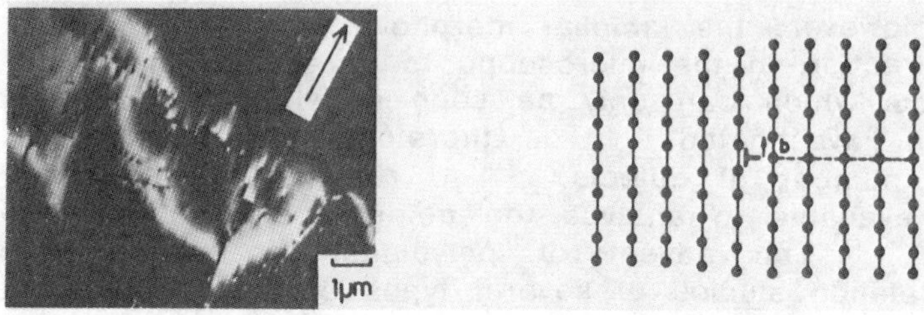

Fig. 4(a) Edge dislocations in pTS single crystal ($\underline{g}$ = [2$\bar{1}\bar{1}$]). (b) Schematic diagram of edge dislocation in pTS.

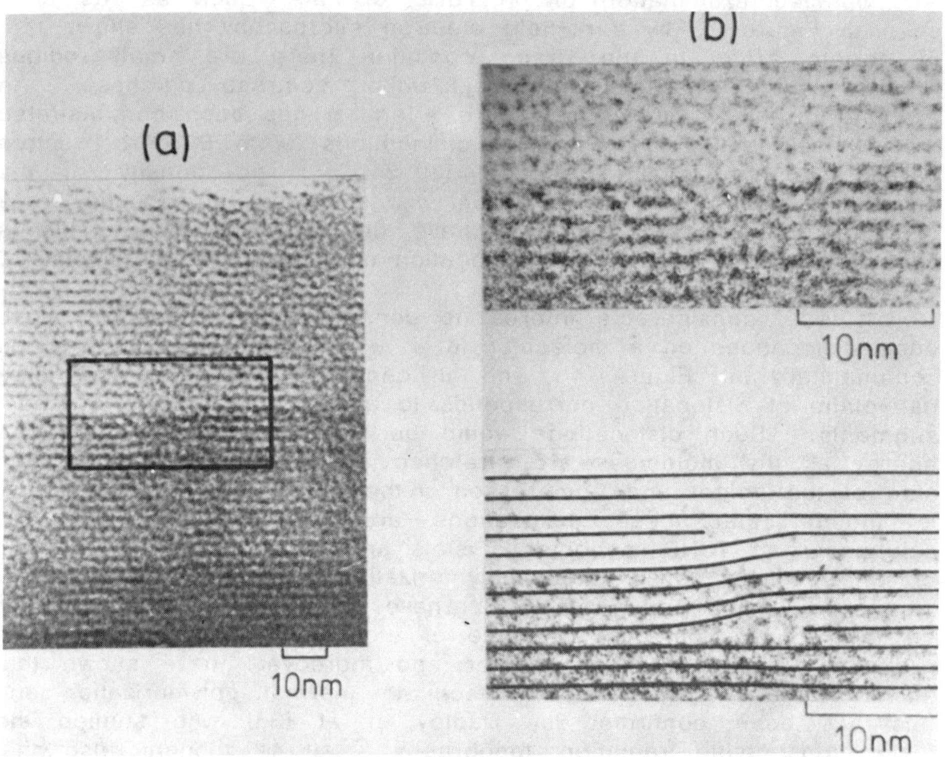

Fig. 5 Lattice image from a polyDCHD crystal showing (010) fringes and containing defects. (a) Lattice image showing planes with 1.22 nm spacing. (b) Enlargements of rectangular area in (a) with the lattice planes sketched showing chain-end dislocation dipole.

properties. They tend to line up in rows parallel to the chain direction (and Burgers vector) and are sometimes found to pack so closely that they can form a small-angle grain boundary [30] which could act as a plane of weakness in the crystal. In addition photoconduction measurements [34] have indicated that polyTSHD contains trap states and the density of traps is found to be approximately the same as that of the dislocations, - implying that these dislocations may be the trap sites for the carriers in the polymer [30].

High resolution electron microscopy has also been used to study defects in polyDCHD [25]. Since the density of dislocations is relatively low most micrographs reveal a defect-free structure as only a relatively small area is viewed at any time. However, occasionally chain-end dislocations such as those shown in Figure 5 can be found [25]. Dislocations of the type shown in Figure 4b would not be seen from images of planes parallel to the chain direction. It is possible that the chain-end dislocation in Figure 5 is again frozen-in from the monomer stage but it would not be expected to be as stable as the edge dislocation of Figure 4 because of a larger Burgers vector. It is therefore more likely that it has been formed as the result of a mismatch during polymerization.

2.5 Stacking Faults.

Virtually nothing is known about the presence or nature of stacking faults in conventional polymer crystals and most of our knowledge in this area now stems from studies in polydiacetylenes [31, 35]. Figure 6 is a bright-held electron micrograph of a relatively thick wedge-shaped polyTSHD crystal. The main features that can be seen are five defects which lie parallel to the chain direction and become narrower as they approach the specimen edge. The defects each display characteristic black/white fringes and it is found that the contrast reverses between bright- and dark-field images [34]. By imaging the defects in different diffraction conditions it has been possible to show that the defects are stacking faults with a displacement vector of $\underline{R}=\frac{1}{2}[\bar{2}11]$, indexing the chain direction in polyTSHD as $\underline{b}$ [35]. It has been shown furthermore that this corresponds to a fault in the stacking of (102) planes and a molecular sketch of the defect is given in Figure 6b. The fault can be accommodated without any disruption of the molecular backbone or the side groups and the presence of these defects in polyTSHD crystals is thought to give rise to easy cleavage on this plane of the polymer [36]. The need to keep the molecular backbone intact means that the only types of stacking faults that can be obtained in polymer crystals are ones due to faults in stacking on planes containing the molecular chains and so one would not expect to be able to image such defects in conventional chain-folded lamellar single crystals which can only be viewed with the electron

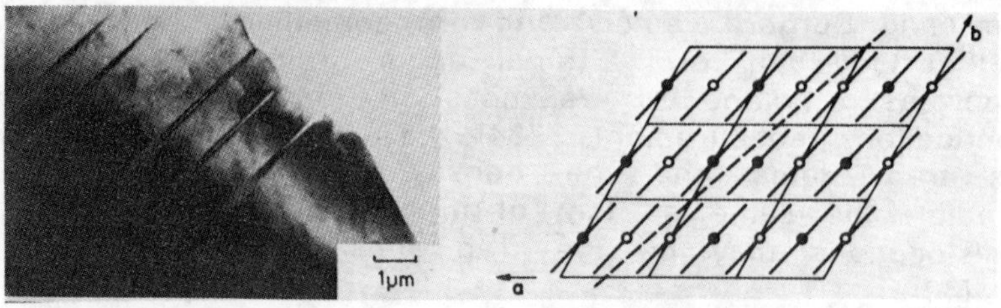

Fig. 6(a) Stacking faults in pTS crystal. (b) Schematic diagram of fault.

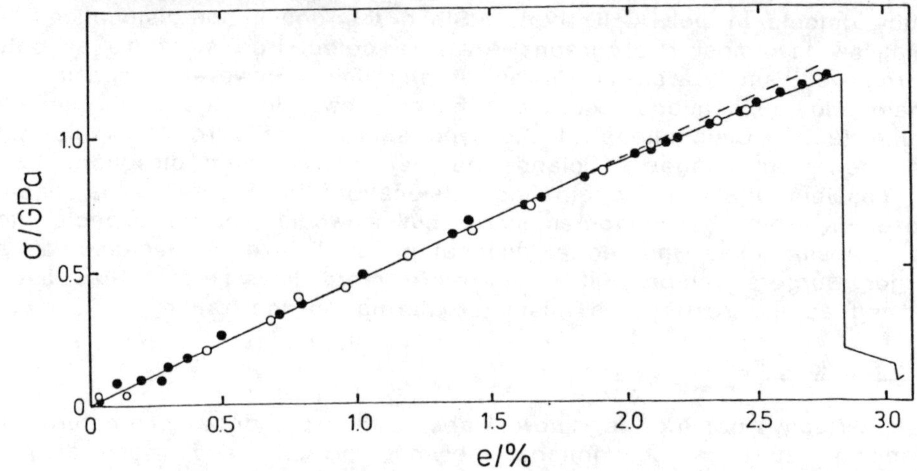

Fig. 7 Stress–strain curve for a polyDCHD single crystal fibre. The closed circles are for loading and the open ones for unloading.

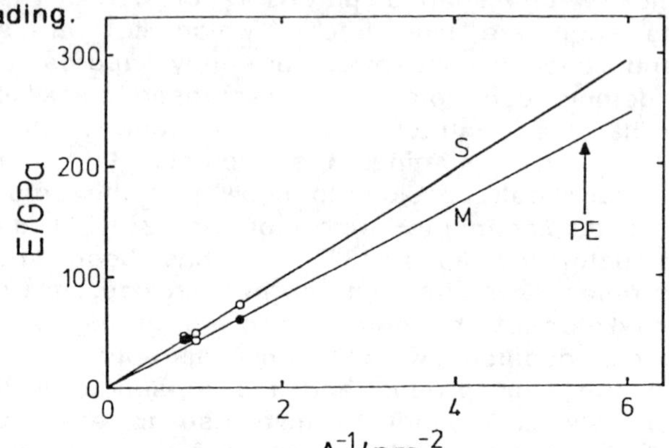

Fig. 8 Dependence of the Young's modulus of polydiacetylene single crystal fibres upon the reciprocal of the area supported by each polymer chain. The line S and the open circles are for values calculated using force-constants determined by Raman Spectroscopy. The line M and the closed circles are for the mechanically-measured ones.

beam parallel to the chain direction [4].

3. ELASTIC DEFORMATION

3.1 Stress/strain Behaviour.

The ability to grow polydiacetylene single crystals to centimeter dimensions has enabled their mechanical behaviour to be evaluated using conventional mechanical testing techniques. In 1974 Baughman, Gleiter and Sendfeld [37] demonstrated that fibre-like single crystals of polyPUHD (Table 1) could be deformed elastically to strains of over 3%. The crystals were found to have high values of Youngs modulus in the chain direction of the order of 45GPa. This work has been extended to other polydiacetylenes with modulus values of 45GPa being determined for polyDCHD [38] and 62GPa for polyEUHD [3].

A typical stress/strain curve for a polyDCHD single crystal fibre is given in Figure 7. The curve is linear up to a strain of about 1.8% and there is a slight decrease in slope above this strain until fracture occurs at a strain of about 2.8%. Loading and unloading takes place along the same path indicating the lack of hysteresis [38]. The fracture strain is found to depend upon the fibre diameter, decreasing as the diameter is increased [3,37,38].

The stress/strain curve in Figure 7 gives an indication of how deformation takes place on the molecular level in polymer crystals. Baughman et al [37] suggested that the slight deviation from Hookes Law above about 2% strain, which is clearly not a yield process, might be due to the anharmonic part of the interaction potential between neighbouring atoms on the polymer chain. Since the polydiacetylene fibres are highly-perfect polymer single crystals the deformation directly involves the stretching and bending of bonds along the polymer backbone. This has been confirmed using Resonance Raman Spectroscopy [39] where it has been shown that the frequencies of the C-C, C=C and C≡C stretching modes in polyEUHD depend upon the deformation of the crystals, decreasing with applied strain. The consequent reduction in the force constants is one of the factors leading to the reduction in the slope of the stress/strain curves at high strains [3,37,38].

3.2 Factors Controlling Young's Modulus.

Baughman and coworkers [37] pointed out that the levels of chain direction modulus displayed by polydiacetylene single crystals were extremely high when account was taken of the relatively high cross-sectional area, A, of the chains due to the relatively large side-groups on the polydiacetylene molecules. This is demonstrated in Figure 8 where the chain direction modulus E of various polydiacetylenes is plotted against the reciprocal of the area supported by each chain in the crystal, 1/A (determined from the knowledge of the crystal structure). It can be seen that the data fall close to a straight line and this shows clearly that the modulus is controlled by the chain cross-sectional area, A. The value of 1/A is also indicated for polyethylene ($A = 0.18\ nm^2$). Assuming the two backbones have about the same stiffness then a modulus of the order of 250 GPa is predicted for polyethylene - in line with current theoretical predictions [40]. This per-chain modulus obtained for polydiacetylene molecules is nearly as high as that of diamond [37] and it indicates the tremendous potential that polymer crystals have as stiff and hence strong materials [41]. Strenuous efforts are currently being made to align molecules in conventional polymers such as polyethylene [21-23]. With polydiacetylene single crystals the alignment is a natural consequence of their topochemical solid-state polymerization. Hence, the way forward with these materials is in the chemical synthesis of monomers with smaller substituent R-groups (Table 1) that will form fibres and polymerize fully to give perfect polymer crystals.

The consequence of incomplete polymerization is demonstrated in Figure 9 where the dependence of E upon the conversion of EUHD to polymer using either heat or γ-rays is shown [3]. It can be seen that there is a linear increase in modulus with the volume fraction of polymer in the crystals for thermal polymerization. The crystals produced by γ-ray polymerization of EUHD are not so perfect and consequently their modulus is lower. Any incomplete polymerization or crystal imperfection will lead to a reduction in fibre modulus.

3.3 Theoretical and Measured Moduli

Although it is the Young's modulus in the chain direction that is of the greatest importance for engineering applications, it must be remembered that a large number of elastic constants are needed to fully describe the elastic behaviour of a crystal [4]. Polydiacetylene single crystals usually possess monoclinic symmetry and therefore have 13 elastic constants. There have been several attempts to calculate the 9 elastic constants for orthorhombic polyethylene [42] but as yet there has been no similar calculation for a polydiacetylene. The calculation ought to be feasible as the crystal structures and atomic positions of several polydiacetylenes are known to a high degree of accuracy [17-19]. However, there are also many different types of covalent bonds in each unit cell and so this will make the calculation extremely complex.

The modulus of polyTSHD crystals deformed in the chain

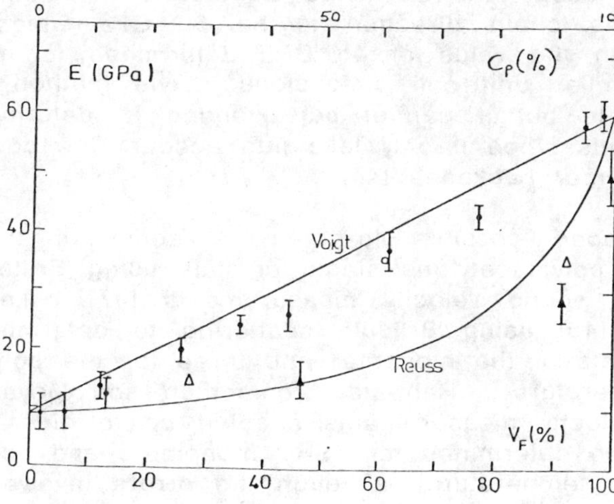

Fig. 9 Variation of Young's modulus with conversion to polymer (C_p) and volume fraction to polymer (V_p). The circles are for thermally polymerized polyEUHD fibres and the triangles for γ-Ray polymerized ones.

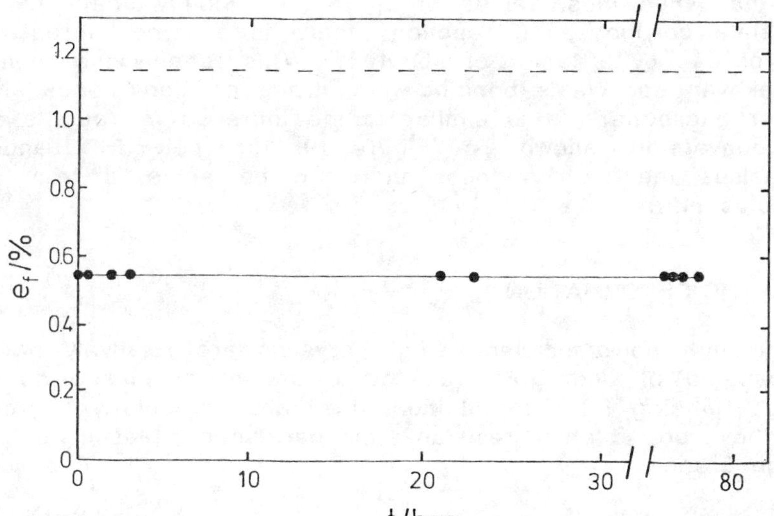

Fig. 10 Variation of fibre strain with time for a polyDCHD fibre held at a constant stress corresponding to 50% of its fracture strain (given by dashed line).

direction has been calculated by Batchelder and Bloor [43] using the method of Treloar [44]. They used force constants measured from Raman spectroscopy and estimated a Young's modulus for polyTSHD crystals in the chain direction to be 50 GPa which is in good agreement with the value of 45 GPa determined for polyDCHD [38] which has similar unit cell dimensions. The method of Treloar is relatively simple but it cannot be extended to determine the other elastic constants since it only take into account the covalent bonding along the polymer backbone [44].

It has been possible to measure some of the 13 elastic constants for polydiacetylene single crystals using Brillouin scattering [45, 46] and sound velocity measurements [47]. Leyrer, Wegner and Wettling [45] using Brilouin scattering to determine six of the elastic constants of the monomer and three for the polymer of TSHD at room temperature. Rehwald, Vonlanthen and Meyer [47] showed that sound velocity measurements enabled up to nine of the elastic constants to be determined for TSHD monomers and polymers over a wide range of temperature, although the errors involved in some of the determinations were rather large. These values of the elastic constants c_{ik} [4] determined for TSHD monomer and polymer are compared in Table 2. It can be seen that on the whole the agreement between the elastic constants determined by the different techniques is relatively good. In particular it can be seen that the values of c_{22} for the polymer which reflects deformation in the chain direction are in close agreement. In addition, it can be seen that while most values of c_{ik} remain mainly unaffected by the solid-state polymerization reaction, there is a large increase in the value of c_{22} by a factor of about 6. This reflects the replacement of weak van der Waals bonding with strong covalent bonds along the polymer backbone. It is similar to the increase in Young's modulus with conversion shown in Figure 9 for polyEUHD and again emphasises the high stiffness that can be achieved from polymer molecules [42].

4. PLASTIC DEFORMATION

Although polydiacetylene single crystals are relatively brittle they are capable of undergoing a limited amount of plastic deformation through twinning [4]. In addition, they are particularly interesting in that they are highly resistant to permanent plastic deformation through creep.

4.1 Creep

One of the most remarkable aspects of the mechanical properties of polydiacetylenes is that it is not possible to measure any time-dependent deformation (or creep) when crystals are deformed in tension parallel to the chain direction [3, 38]. This behaviour is demonstrated in Figure 10 for a polyDCHD single crystal held at constant stress at room temperature and preliminary measurements have indicated that creep could not be detected during deformation at temperatures up to at least 100°C [38].

Creep and time-dependent deformation are normally a serious draw-back in the use of conventional high-modulus polymer fibres such as polyethylene [48] in engineering applications. Such oriented fibres produced by drawing or spinning contain a high density of defects such as chain-ends, loops and entanglements. These allow the translation of molecules parallel to the chain direction during deformation which leads to creep. In contrast since polydiacetylene single crystal fibres contain perfectly-aligned long polymer molecules (cf. Figure 3) there is no mechanism whereby creep can take place even at high temperature. In addition polydiacetylenes tend to degrade rather than melt and this should be contrasted with polyethylene which melts at 140°C and so has only very limited high-temperature applications.

4.2 Twinning

Polydiacetylene single crystals are found to undergo twinning when deformed in compression parallel to the chain direction. This can be best demonstrated using the lozenge-shaped crystals of polyTSHD. When such crystals are compressed parallel to the chain axis they deform by means of a twinning process which involves the molecules kinking over at a well-defined angle [49-52]. Figure 11 shows a twin in a polyTSHD single crystal along with a corresponding diagram of the molecular displacements involved. The process can be differentiated from the formation of kink-bands [53] since the material within the twinned region has the same crystal structure as the undeformed crystal and there is also a mirror image orientation relationship between the deformed and undeformed regions [50].

Until this work upon polydiacetylenes it was thought that twinning in polymer crystals would not be able to take by the kinking of the molecular chains [54]. However, in 1976 Pietralla [55] postulated that this type of twinning might occur in semicrystalline polymers and Bevis [56] subsequently performed a detailed analysis of this 'chain twinning' for the case of polyethylene crystals. Nevertheless, the investigations into deformed crystals of polyTSHD [49-52] provided the first evidence for the chain twinning taking place, although it does appear that a similar twinning process might occur in chain-extended polyethylene lamellae [53].

The ability of polydiacetylenes to undergo chain twinning has

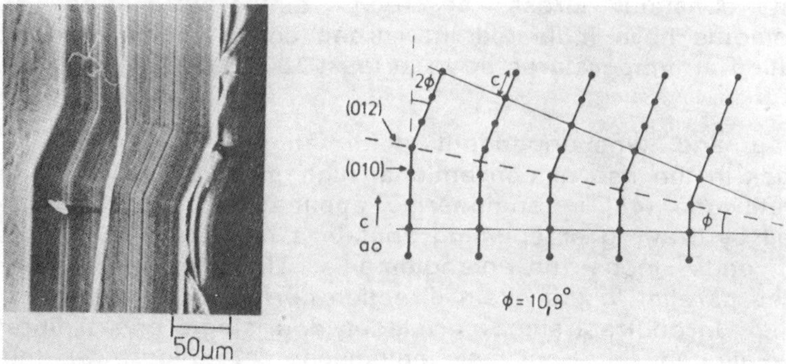

Fig. 11 Twin in a polyTSHD single crystal. Scaning electron micrograph of the twin viewed perpendicularly to the chain direction and corresponding deformation on the molecular level.

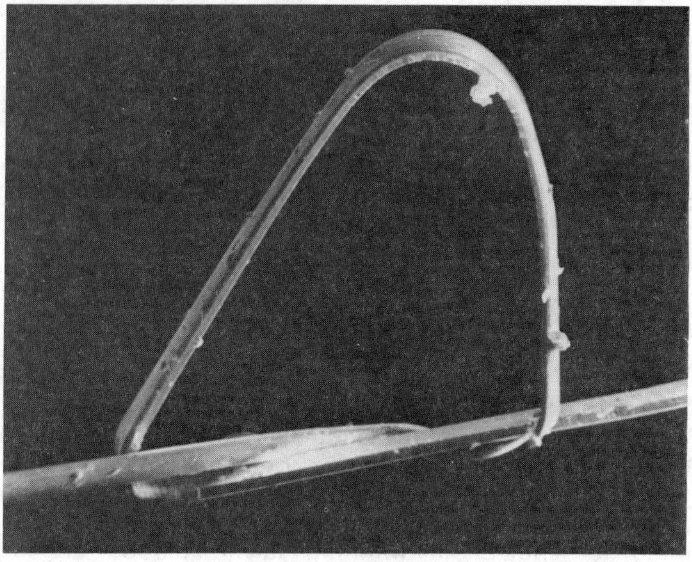

Fig. 12 Scanning electron micrographs of a knotted polyDCHD fibre.

important consequences in the handling of polydiacetylene single crystals. For example, it is possible to knot polyDCHD fibres [38] and an example of a knotted crystal is shown in Figure 12. This process involves a high degree of deformation but polyDCHD is able to cope with these high strains by undergoing chain twinning the inside surface which is subject to compression in the chain direction. There tends to be some cracking parallel to the chain direction, but the crystals remain relatively intact even when pulled into a tight knot [38]. This behaviour has important implications for the use of polydiacetylenes in applications such as composites. For example, the ability to absorb strain by twinning means that the fibres should not break up and undergo 'fibre attrition' as easily as glass fibres during fabrication. Other high-strength fibres such as those of glass and carbon tend to be very fragile and break readily during handling. The ability of polydiacetylenes to undergo limited plastic deformation through twinning means that they tend to be much less fragile than other high-strength fibres.

5. FRACTURE

Polydiacetylene single crystals readily undergo cleavage parallel to the chain direction reflecting the relative strength of covalent bonding compared with the van der Waals bonding between the polymer molecules. The cleavage takes place preferentially on certain crystallographic planes and it was pointed out earlier (Section 2.5) that there is a correlation between the presence of stacking faults and cleavage on (102) planes. However, the most interesting aspect of the fracture behaviour of polydiacetylenes is the high strength that single crystal fibres can exhibit when deformed parallel to the chain direction. This is described in detail below.

5.1 Single Crystal Fibres

Investigations into the fracture behaviour of polydiacetylene single crystal fibres has revealed a strong dependence of the fracture stress, σ_f, upon the fibre diameter, d [3, 37, 38]. This is demonstrated for polyDCHD fibres in Figure 13 and it is similar to the size-dependence reported earlier for inorganic high-strength fibres [57, 58] where the depdndence of σ_f upon d was found to follow a relation of the form

$$\sigma_f \propto 1/d \qquad (1)$$

This behaviour was thought to be due to the presence of surface defects which give rise to a stress concentration when the fibres are deformed. Since the size of the defects was found to scale with the fibre diameter the size dependence was predicted [59]. However, more detailed examination of the data in Figure 13 has shown that equation (1) is not accurately obeyed for polydiacetylene single crystal fibres. In addition,

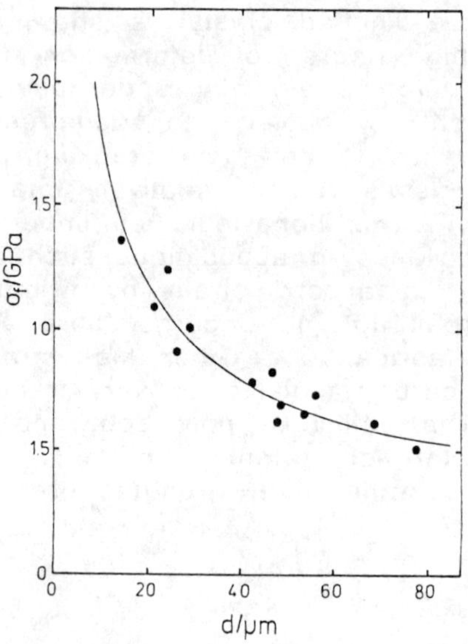

Fig. 13 Dependence of the fracture stress upon effective fibre diameter for polyDCHD fibres.

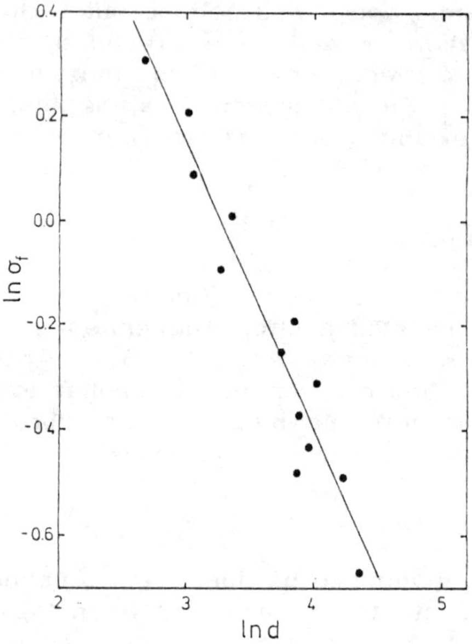

Fig. 14 Log-log plot of fibre strength as a function of effective fibre diameter for the data from Fig. 13. The line is a least squares fit to the points (correlation coefficient = 0.96).

theoretical calculations have shown that a different relationship is expected [3, 38]. The data in Figure 13 have been replotted in Figure 14 in the form of alog-log plot and it can be seen that the dependence of σ_f upon d is given more closely by

$$\sigma_f \alpha 1/d^{0.55} \quad (2)$$

This is precisely the dependence expected from theoretical considerations outlined below [3, 38].

5.2 Theoretical Strength

The magnitude of the stress concentration will depend upon the geometry of surface defects. Examples of defects found on the surfaces of polyDCHD crystals are shown in Figure 15. Although the crystals are of very different sizes it can be seen that the sizes of the steps and notches on the crystals in Figure 15 are about 1/10 of the crystal diameters. It can be shown [58] that in general the stress concentration factor η at the root of a step or notch is given by

$$\eta = \sigma_o/\sigma_a = k_1 + k_2\, a^{1/2}\, \rho \quad (3)$$

where σ_a is the applied stress, σ_o the stress at the root of the defect, a is the notch or step depth, ρ the radius of curvature at the root of the defect and k_1 and k_2 constants which depend upon the geometry of the defects [58, 59].

It can be seen from Figure 15 and it has been shown elsewhere [3] that the parameter a scales with the fibre diameter such that

$$a \sim d/m \quad (4)$$

where m is a constant which is the order of 10. Putting equation (4) into (3), and setting $\sigma_a = \sigma_f$ at fracture and rearranging gives

$$\frac{1}{\sigma_f} = \frac{k1}{\sigma_o} + \frac{k2}{\sigma_o\ (\rho m)^{1/2} d^{1/2}} \quad (5)$$

where σ_o now becomes the strength of a defect-free crystal, i.e. the 'theoretical strength'. Hence it is predicted that a $1/\sigma_f$ versus $d^{1/2}$ should give a straight line of slope $k_2\ \sigma_o\ (\rho m)^{1/2}$ with an intercept of k_1/σ_o. The data from Figure 13 have been replotted according to equation (5) in Figure 16. It can be seen that the data fall on a straight line and give an intercept close to the origin. In addition, knowledge of the detailed geometry of the surface defects allows the value of the theoretical strength, σ_o to be determined from the slope of the line in Figure 16. The theoretical strength of polyDCHD crystals has been determined to be 3 ± 1 GPa [38], the rather large error being due

Fig. 15 Steps on surfaces of polyDCHD crystals (a) SEM micrograph of 15 μm diameter crystal (b) TEM micrograph of 100 nm diameter crystal showing molecular steps.

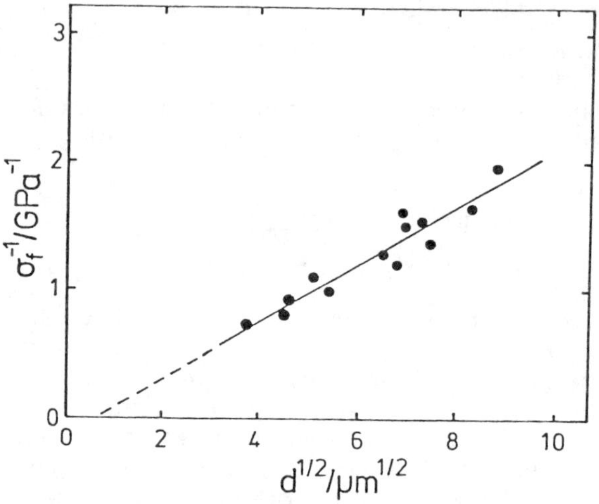

Fig. 16 Variation of the reciprocal of the fibre strength with square root of the fibre diameter for polyDCHD fibres.

to the uncertainty in determining the defect geometry.

5.3 Molecular Fracture

Several years ago Frank [41] pointed out that polymer molecules can have very high values of strength when deformed parallel to their axes. The determination of the theoretical strength of polydiacetylene single crystals has allowed the strength of individual molecules to be estimated. From the knowledge of the crystal structure of polyDCHD [19], it can be shown that each molecule supports a cross-sectional area, A of about 1 nm^2. The theoretical strength therefore corresponds to a force required to break molecules of about 3 nN and a fracture strain of 6-8%. It is of interest to compare this with theoretically calculated values of strengths of covalently -bonded polymer molecules [40, 58, 60]. Kelly [58] has estimated the strength of a polyethylene molecule as 6 nN but this is thought to be rather high. Kausch [60] has shown that a covalently-bonded polymer molecule should be broken by a force of 3nN, which is identical to the value determined experimentally for polyDCHD. A molecular strength of this magnitude corresponds to a fracture stress of the order of 20 GPa for polyethylene single crystals for which the area supported by each molecule is considerably smaller than in polyDCHD. However, it has not yet been possible to make polyethylene single crystals of macroscopic dimensions and even highly-oriented polyethylene fibres are found to have strength values only of the order of 4 GPa [22, 23], well below the theoretical strength.

6. COMPOSITES

It is well-established that the composites produced by incorporating high-modulus fibres in a matrix of epoxy resin or metal can have outstanding mechanical properties. Polydiacetylenes offer considerable promise for reinforcement in polymeric matrices because they have the following properties which have been described in previous sections:

(i) High stiffness.
(ii) High strength.
(iii) Low creep.
(iv) Good thermal stability.
(v) Low density.

Recent investigations [61, 62] into the behaviour of polydiacetylene fibres in epoxy resin matrices have shown that not only such composites have good mechanical properties but that important fundamental details of the mechanisms of fibre reinforcement can also be revealed from their study.

6.1 Reinforcement Mechanisms

It was pointed out in Section 3.1 that the frequencies of certain main-chain Raman active modes were found to change with the level of applied strain [39, 43]. In particular it was found that the C≡C triple bond stretching frequency changes by the order of 20 cm^{-1} for 1% of strain [43] as shown in Figure 17. This property can be used to determine the strain in a polydiacetylene fibre subjected to any general state of stress. The strain can be measured to a high degree of spatial resolution and accuracy as beam diameters of ~ 25 μm can be used and changes in frequency can be determined to ~ 1 cm^{-1}. This idea has been applied [61] to the case of a single fibre in a polymer matrix which is a classical problem in fibre reinforcement [63]. Model specimens have been fabricated and an example is given in Figure 18. The matrix strain can be measured using a strain gauge adhered to the surface of the specimen and the point-to-point variation in strain in the fibre measured using Raman spectroscopy, monitoring the change in C≡C stretching frequency. Typical results are shown in Figure 19 where the fibre strain is given as a function of position along the fibre for different levels of applied matrix strain. The results agree moderately well with theoretical predictions [63]. It can be seen that at higher levels of matrix strain the fibre strain rises from the end to a constant value along the length of the fibre and then falls off at the other end as predicted by the theory. However, there are found to be subtle deviations from the theoretical analysis [63] due to inappropriate assumptions and viscoelastic deformation in the polymer matrix [61]. The technique of measuring fibre strain is being extended to interacting fibres and high volume fraction composites where analytical solutions of the stress distribution in the system do not exist. The measurement of strain using the laser-Raman technique offers considerable scope for both strain measurement in polydiacetylene/epoxy composites and as a novel type of strain gauge needing no physical connections to the monitoring equipment.

6.2 High Volume Fraction Composites

Some preliminary measurements have been made of the mechanical properties of composites consisting of aligned polyDCHD fibres (~10 mm long) in an epoxy resin matrix [62]. The modulus is as might be expected found to increase with the volume fraction of fibres. Values of over 20 GPa have been measured for volume fractions in excess of 50%. The composites have been fractured in tension and tensile strengths of up to 150 MPa have been obtained [62]. A typical fracture surface is shown in Figure 20 and it can be seen that there has been a combination fibre fracture and pull-out.

This work upon the mechanical properties of polydiacetylene/epoxy composites has proved to be extremely

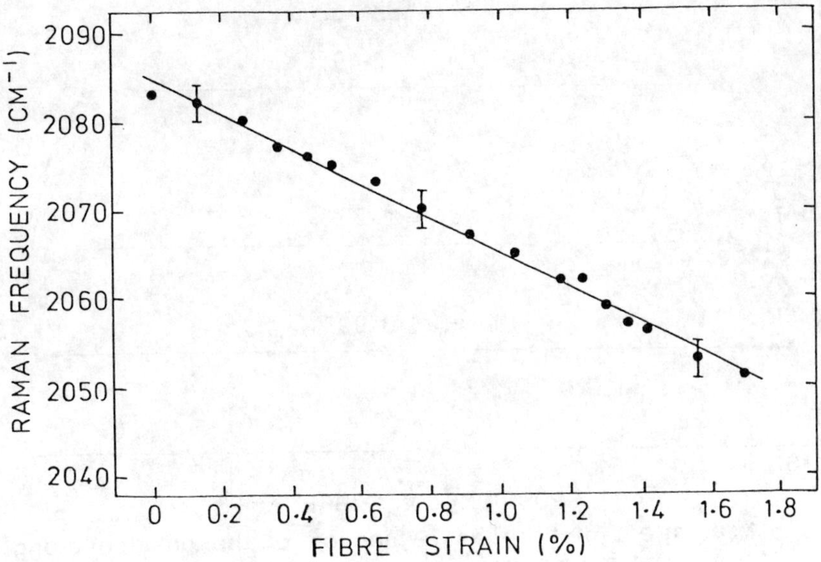

Fig. 17 Strain dependence of the 2085 cm^{-1} Raman line of a free-standing polyDCHD fibre.

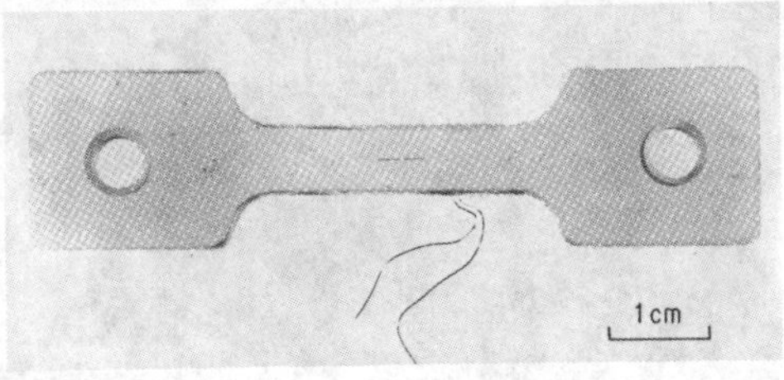

Fig. 18 Single fibre composite specimen. The polyDCHD fibre is located at the centre of the specimen while the resistive strain gauge is attached to the outer surface.

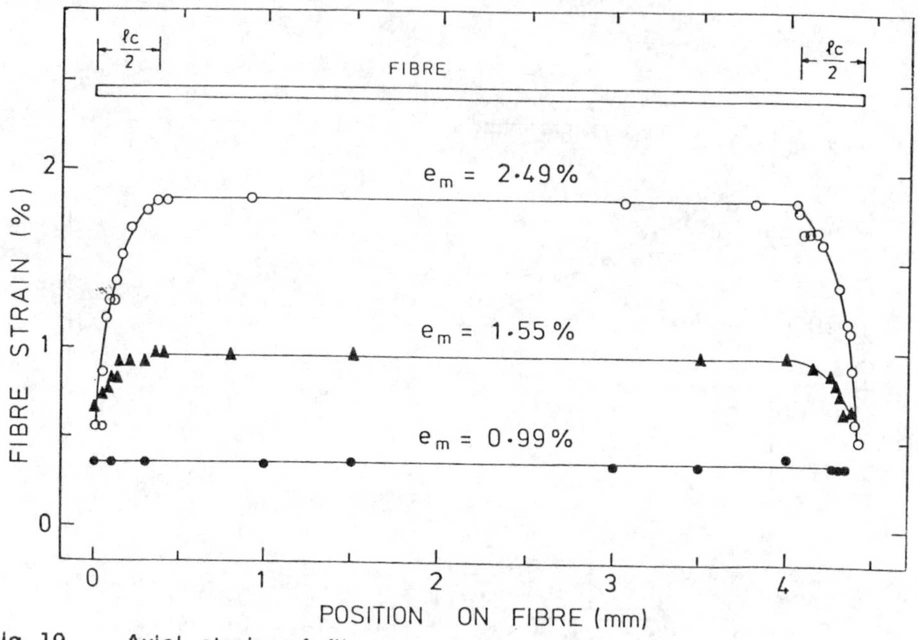

Fig. 19 Axial strain of fibre as a function of the position along the fibre. The matrix strain is given by e_m.

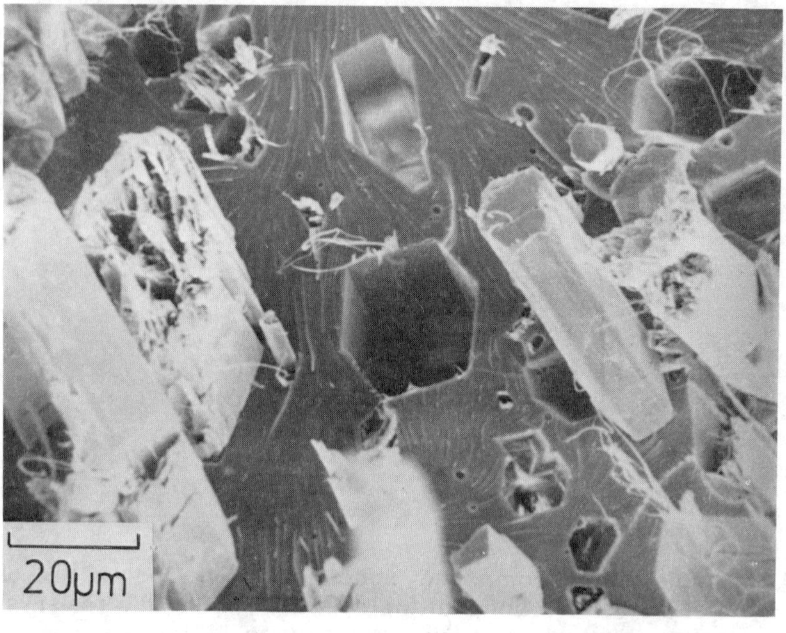

Fig. 20 Scaning electron micrographs of a polyDCHD reinforced epoxy resin composite.

interesting from both a theoretical and practical viewpoint. It is obviously an area where more experiments need to be done. It is important to know how the properties of the composites are affected by fibre diameter, orientation and surface pretreatment. In addition other properties such as fatigue and impact are important in practical situations and so need to be evaluated.

7. CONCLUSIONS

Investigations into the relationship between structure and mechanical properties in polydiacetylenes has enabled considerable new insight to be obtained concerning structure/property relationships for polymers in general. Electron microscope studies have allowed defects such as dislocations and stacking faults to be studied in detail for the first time. Because of their highly-perfect structure polydiacetylene crystals are found to have their theoretical Young's moduli and this has enabled predictions to be made concerning the levels of stiffness that could be achieved from other polymer molecules. The extended-chain nature of the crystals has allowed a new twinning mechanism to be discovered which takes place by molecules kinking and is activated by compression parallel to the chain direction. The crystals are found to exhibit high levels of mechanical strength and do not creep when deformed in tension parallel to the chain direction, the level of strength being governed by the presence of defects. Analysis of this behaviour has enabled the strength of the individual polymer molecules to be determined as about 3 nN, a value of considerable interest to the understanding of the deformation of all types of polymer fibres. Finally, it has been shown that polydiacetylene fibres offer considerable scope as reinforcing fibres in composites. Strain measurement using Raman spectroscopy has allowed a detailed study of the mechanisms of stress transfer between a fibre and the matrix to be made and composites containing high volume fractions of aligned fibres have been found to have promising mechanical properties.

It is suggested that future studies should be directed towards the preparation of fibres with higher stiffness and strength and evaluation of the properties of such fibres in composites.

ACKNOWLEDGEMENTS

The author is grateful to Professor D. Bloor and Dr. D.N. Batchelder for introducing him to polydiacetylenes and for their continued encouragement and help. He would also like to thank Mr. D. Ando and Mr. I.F. Chalmers for help with preparation of monomers. Finally he must extend his gratitude to Dr. C. Galiotis, Dr. R.T. Read and Mr. P.H.J. Yeung who are the research workers who have performed the bulk of the experimental work.

REFERENCES

1. G. Wegner, Pure and Appl. Chem., 49 (1977) 443.
2. K.C. Yee and R.R. Chance, J. Poly. Sci., Poly. Phys. Ed., 16 (1978) 431.
3. C. Galiotis and R.J. Young, Polymer, 24 (1983) 1023.
4. R.J. Young, 'Introduction to Polymers', Chapman and Hall, London, 1981.
5. I.L. Hay and A. Keller, J. Mater. Sci., 1 (1966) 41.
6. R.J. Young and P.B. Bowden, J. Mater. Sci., 8 (1973) 1177.
7. P.B. Bowden and R.J. Young, J. Mater. Sci., 9 (1974) 2034.
8. P.H. Geil, 'Polymer Single Crystals', Interscience, New York, 1963.
9. P. Allan and M. Bevis, Proc. Roy. Soc., A341 (1974) 75.
10. P. Allan, E.B. Crellin and M. Bevis, Phil. Mag. 27 (1973) 127.
11. E.H. Andrews and G.E. Martin, J. Mater. Sci., 8 (1973) 1315.
12. E.H. Andrews and G.E. Martin, J. Mater. Sci., 9 (1974) 1507.
13. Y.R. Patell and J.M. Schultz, J. Macromol. Sci. Phys., B7 (1973) 445.
14. G. Wegner, Z. Naturforsch., 24b (1969) 84.
15. C. Galiotis, R.J. Young, D.J. Ando and D. Bloor, Makromol. Chem. 184 (1983) 1083.
16. R.T. Read and R.J. Young, J. Mater. Sci., 14 (1979) 1968.
17. D. Kobelt and E.F. Paulus, Acta Cryst., B30 (1974) 232.
18. E. Hädicke, H.C. Mez, C.H. Krauch, G. Wegner, J. Kaiser, Angew. Chem., 83 (1971) 253.
19. P.A. Apgar and K.C. Yee, Acta. Cryst. B34 (1978) 957.
20. B. Wunderlich, 'Macromolecular Physics, Vol. 1', Academic Press, London, 1973.
21. G. Capaccio and I.M. Ward, Poly. Eng. Sci., 15 (1975) 219.
22. W.T. Mead, C.R. Desper and R.S. Porter, J. Poly. Sci., Poly. Phys. Ed., 17 (1979) 859.
23. A.J. Pennings and K.E. Meihuizen in 'Ultra-High Modulus Polymers', Ed. A. Ciferri and I.M. Ward, Applied Science, London, 1979, p. 117.
24. D.T. Grubb, J. Mater. Sci., 9 (1974) 1715.
25. R.T. Read and R.J. Young, J. Mater. Sci., 19 (1984) 327.

26. R.T. Read and R.J. Young, J. Mater. Sci. Lett., 16 (1981) 2922.
27. P.H. Lindenmeyer, J. Poly. Sci., C15 (1966) 109.
28. V.F. Holland and P.H. Lindenmeyer, J. Appl. Phys., 36 (1965) 3049.
29. J. Petermann and H. Gleiter, Phil. Mag., 25 (1972) 813.
30. R.J. Young and J. Petermann, J. Poly. Sci., Poly. Phys. Ed., 20 (1982) 961.
31. R.J. Young, R.T. Read and J. Petermann, Inst. Phys. Conf. Ser. No. 61: (1981) Ch. 10, 475.
32. W. Schermann, G. Wegner, J.O. Williams, J.M. Thomas, J. Poly. Sci., Poly. Phys. Ed., 13 (1975) 753.
33. M. Dudley, J.N. Sherwood, D. Bloor and D. Ando, J. Mater. Sci. Lett., 1 (1982) 479.
34. K.J. Donovan and E.G. Wilson, Phil. Mag., B44 (1981) 9.
35. R.J. Young, R.T. Read, J. Petermann, J. Mater. Sci., 16 (1981) 1835.
36. D.N. Batchelder, Q.M.C. Physics Dept., Private Communication.
37. R.H. Baughman, H. Gleiter and N. Sendfeld, J. Poly. Sci., Poly. Phys. Ed., 13 (1975) 1871.
38. C. Galiotis, R.T. Read, P.H.J. Yeung, R.J. Young, I.F. Chalmers and D. Bloor, J. Poly. Sci., Poly. Phys. Ed., to be published.
39. C. Galiotis, R.J. Young and D.N. Batchelder, J. Poly. Sci., 21 (1983) 2483.
40. A.J. Kinloch and R.J. Young, 'Fracture Behaviour of Polymers', Applied Science, London, 1983.
41. F.C. Frank, Proc. Roy. Soc., A319 (1970) 127.
42. A. Odejima and T. Maeda, J. Poly. Sci., C15 (1966) 55.
43. D.N. Batchelder and D. Bloor, J. Poly. Sci., Poly. Phys. Ed., 17 (1979) 569.
44. L.R.G. Treloar, Polymer 1 (1960) 95.
45. R.J. Leyrer, G. Wegner and W. Wettling, Ber. Bunsenges. Phys. Chem., 82 (1978) 697.
46. V. Enkelmann, R.J. Leyrer, G. Schleier and G. Wegner, J. Mater. Sci., 15 (1980) 168.
47. W. Rehwald, A. Vonlanthen and W. Meyer, Phys. Stat. Sol. (a) 75 (1983) 219.
48. G. Capaccio, G. Gibson and I.M. Ward in 'Ultra-High Modulus Polymers', Ed. A. Ciferri and I.M. Ward, Applied Science, London, 1979, p.1.
49. R.J. Young, D. Bloor, D.N. Batchelder and C.L. Hubble, J. Mater. Sci., 13 (1978) 62.
50. R.J. Young, R. Dulniak, D.N. Batchelder and D. Bloor, J. Poly. Sci., Poly. Phys. Ed. 17 (1979) 1325.
51. R.J. Young and J. Petermann, Makromol. Chem., 182 (1981) 621.

52. R.T. Read and R.J. Young, Phil. Mag. A42 (1980) 629.
53. R.J. Young in 'Developments in Polymer Fracture', Applied Science, London, 1979.
54. A. Keller, private communication.
55. M. Pietralla, Coll. Polym. Sci., 254 (1976) 249.
56. M. Bevis, Coll. Polym. Sci., 256 (1978) 234.
57. S.S. Brenner, J. Appl. Phys., 33 (1962) 33.
58. A. Kelly, 'Strong Solids', Clarendon Press, Oxford, 1966.
59. D.M. Marsh in 'Fracture in Solids'. Eds. D.C. Drucker and J.J. Gilman, Interscience, New York, 1963.
60. H.H. Kausch, 'Polymer Fracture', Springer-Verlag, Berlin, 1978.
61. C. Galiotis, P.H.J. Yeung, R.J. Young and D.N. Batchelder, J. Mater. Sci., 19 (1984) 3640.
62. C. Galiotis, P.H.J. Yeung and D.N. Batchelder, to be published.
63. H.L. Cox, Brit. J. Appl. Phy., 3 (1952) 72.

Elastic constants c_{ik}/GPa	Monomer [45]	Monomer [47]	Polymer [45]	Polymer [47]
c_{11}	5.5	8.7	∿ 7	12.9
c_{22}	6.7	9.7	43.3	43.7
c_{33}	3.5	8.2	<9.2	-
c_{44}	1.0	-	1.07	-
c_{55}	3	3.4	∿ 3	6.9
c_{66}	0.9	1.5	1.03	-
$\|c_{15}\|$	∿ 1	-	<3.1	-
$\|c_{35}\|$	1.0	-	-	-
$\|c_{46}\|$	0.05	-	∿0.03	-

TABLE 2. Elastic constants c_{ik} determined at room temperature for TSHD monomer and polymer using Brillouin scattering [45] and sound velocity measurements [47].

POLYMERIZED DIACETYLENE MONOMERS IN MONOLAYER AND MULTILAYER FILMS, CHARACTERIZATION and POSSIBLE APPLICATIONS

Jerome B. Lando

Case Western Reserve University
Department of Macromolecular Science
Cleveland, Ohio 44106

The first publication of work on polydiacetylenes, polymerized in monolayers and multilayers, was published jointly by groups at the University of Mainz and the University of Freiburg in West Germany and at Case Western Reserve University in the United States. It is interesting to note that the idea for this work was developed during a six month period in 1974, when Professors Ringsdorf, Wegner and Lando were together at the University of Mainz. Their subsequent work was supported by a Nato grant which made possible further mutual interaction.

The experimental basis for this subject is that certain diacetylene monomers consisting of a polar hydrophilic head group at one end of the diacetylene moiety and a sufficiently long hydrophobic hydrocarbon tail at the other end can be spread at the gas-water interface. Such a thin film is unimolecular in thickness and can be compressed and subsequently transferred onto a suitable substrate under an appropriate lateral surface pressure. By repeating this technique several layers (called multilayers) can be built up on a substrate surface. The technique was originally devised by Langmuir and Blodgett using salts of saturated fatty acids. In most cases the monolayers can be polymerized either at the gas-water interface or after transfer onto a suitable substrate. Polymerization takes place between monomer molecules within the same layer. The resultant oriented polymeric films consist of conjugated double and triple bonds along the chain axes.

The scope of this review is limited to an elucidation of the salient features of the polymerization, morphology and structure of diacetylene monolayers at the gas-water interface and in multilayers. Subsequently, some possible applications of these systems will be discussed.

Polymerization and Characterization

Diacetylene monomers of the type $H-(CH_2)_n-C{\equiv}C-C{\equiv}C-(CH_2)_m-X$ (where X is CH_2OH, COOH, or monovalent or divalent salts of the carboxylic acid) have been polymerized as monolayers at the gas-water interface, as multilayers and as bulk crystals. The compounds will be denoted as $C_nC_m - CH_2OH$, etc. Much of the data that has been collected using the series $C_{(10-16)}C_8$ [(1,5-11)] although a detailed study of multilayer and bulk polymerization has been performed on shorter side chain materials $C_{10}C_2$. [(12,13)]

If one considers the polymerization of the compounds in the series $C_{(10-16)}$ C_8-COOX and $C_{10}C_2$-COOH the optical data, obtained during and after polymerization, seems somewhat confusing. In all cases the monomer is uncolored. However polymerized bulk crystals are blue,[9,14] whereas the polymerization of monolayers at the gas-water interface or in multilayers having no register between layers yields red material at full conversion, although a blue color develops initially[(6,11)]. Polymerization in multilayers having crystallographic register between layers gives the same results as bulk crystals.[(12-13)] Although the kinetics of polymerization vary depending on whether the acid, or a monvalent or divalent salt is utilized, the resulting colors appear not to be effected.

The picture that arises from these data is that the stable form of the polymer is the red form. A phase transformation from a solid solution of polymer in the monomer structure (blue form) to the polymer structure (red form) can only take place if enough mobility is present (either in the polymerization of monolayers or misregistered multilayers). This conclusion is supported by the fact that swelling in a number of nonsolvents or heating induces a change in the polymer from the blue to red form that is not reversible.[(9)]

The morphology and structure of monolayers of $C_{16}C_8$ - COOLi polymerized at the nitrogen water interface by UV radiation has been investigated in detail.[(7,8)] That the typical phase transition takes place from the blue to the red form during polymerization is demonstrated in Figure 1.[(6)] The morphology of these layers is dependent on the spreading conditions used to obtain the monomer monolayer.[(7)] Morphological studies were conducted using an optical microscope with crossed polaroids after picking up the monolayer on a glass slide. Spreading the monomer in a relatively poor solvent benzene-chloroform yields small monomer and thus small polymer monolayer crystallites blue green or yellow green in color in a patchwork. Spreading the monomer in chloroform results in spherulitic crystallization (on the order of 1mm in diameter) of the side chains. The locus of the polymer chain axis in crystallites that make up the spherulites is either a righthanded or lefthanded spiral depending on the sequence of the spherulite sectors yellow-green, black, blue-green or blue-green, black, yellow-green in a clockwise sense. The blue-green sectors represent one approximate chain orientation, the yellow green another and in the dark regions the chains must be either parallel or perpendicular to the polarizers. If the monomer is spread in chloroform into a chloroform-nitrogen atmosphere blue and yellow single crystals 35Å thick and on the order of millimeters in lateral dimensions can be observed. Again the color depends on polydiacetylene chain orientation with respect to the plarizers. Note that all these polymerized monolayers are red in color and absorb strongly in the green.

Studies of the crystal structure of the polymerized monolayers were undertaken using electron diffraction data.[8] The kinematic approximation could bemade, since a monolayer crystal contained only 1/2 the a-axis repeat of the normal structure of both the monomer and the polymer. The monomer structure is monoclinic with a 33° tilt of the hydrocarbon side chains (a-axis) from the bc plane towards the b-axis. Thus c is the unique axis and the monolayer is 31Å thick (a-axis 62Å). A complete structure of the polymer has been performed. The unit cell was orthorhombic with a = 70Å (monolayer thickness 35Å) b = 8.11Å and c = 4.89Å (chain axis). The monolayer contains 2 monomer repeats in 2 chains (unit cell contains 4 monomer repeats in 4 chains) related by a glide along the b-axis. The final residual was 9% indicating an excellent fit between calculated and observed intensities.

Figure 2[8] is a typical electron diffraction pattern. A distinct smearing of the intensity was observed in the b* direction and increased with the ℓ index. A possible origin of this may be due to a small disordering in the packing. The type of disorder can be derived from the Fourier transform as being due to a slip in the c direction between polymer chains. In some diffraction micrographs where little beam destruction occurred, the smeared reflections are seen to really consist of many distinct spots.

To determine the nature of the disorder experimental lattices were drawn with various kinds of disorder and then placed in a laser optical diffraction apparatus. Through trial and error, optical diffraction patterns very similar to the monolayer diffraction patterns were obtained. The lattice which produced these patterns had a shift in the c direction at systematic intervals of about 24 Å along b, or three unit cells. This gave the effect of a superstructure and results in the closely spaced spots on the upper layer lines. The shift in the c direction averaged about 0.8 Å to simulate the correct positions of the low-intensity "superstructure" spots. Although the average shift is 0.8 Å, this had to vary from 0.0 to 1.6 Å to produce the observed smearing.

Monolayers of these polymerized monolayers ($C_{16}C_8$ - COOLi) were collapsed at the gas-water interface to self-supporting fibers.[15] The monolayer fibers were found to have an unusual morphology in that while the polymeric backbones have a random orientation, the hydrocarbon side chains are all perpendicular to the fiber axis. The fiber d spacings determined from x-ray diffraction agree well with electron diffraction data from unperturbed polymeric monolayers. Tensile tests of the monolayer fibers indicate average strength and high modulus. It is somewhat unusual that fibers with larger crystallite size had a longer extension to break along with the expected higher modulus. This could be related to the systematic disorder in the polymer crystals, discussed above, which would allow a slip mechanism in these essentially completely crystalline fibers.

The morphology and structure of two other monolayer forming diacetylene monomers has also been investigated in detail,[12,13] $C_{10}C_2$ - COOH and $C_{10}C_2$ - CH_2OH. The conclusions that can be reached from both infrared spectra and diffraction experiments on monomer and polymerized bulk crystals and monomer and polymerized Langmuir-Blodgett multilayers are as follows.

Although special conditions were necessary to produce good multilayers from these relatively short hydrophobic chain molecules there was crystallographic register between layers. The crystal structures of both monomer and polymerized multilayers were identical to those of the bulk crystals. A phase change from the polymer in the monomer orthorhombic structures to their stable monoclinic polymer structures occures upon Co^{60} γ irradiation continued well after complete polymerization. It is interesting that in reverse of the $C_{16}C_8$ case these shorter chains undergo an orthorhombic to monoclinic phase transition after polymerization.

On polymerization the $C_{10}C_2$-COOH crystals crack parallel to the c-axis both in the bulk crystalline form and in multilayers, while the alcohol does not. (See Figure 3) This was attributed to the larger volume change upon phase transition. It also may well relate to the fact that hydrogen bonding is not three dimensional in the acid but is of the dimer type along the a-axis.[13]

Possible Applications of Polydiacetylene Monolayers and Multilayers

The following is a brief description of a by no means exhaustive list of possible applications of polydiacetylene monolayers.

A number of studies[16-20] of the nonlinear optical properties of polydiacetylene multilayers have been undertaken. It is obvious from these investigations that there is a strong possibility that these materials have applications in integrated optics as wave guide materials. However this will be dependent on the ability to control morphology and structure in these layers, thus indicating the importance of the basic studies reviewed above.

Considerable interest in polydiacetylenes multilayers as possible semiconductors and photoconductors has been expressed. A preliminary study of C C COOH and C C COOLi bilayers deposited on lock and key electrodes has been undertaken.[21] Samples exhibited initial conductivities on the order of $10^{-7}\ \Omega^{-1}$ cm which decayed at varying rates depending on sample composition to $10^{-11}\ \Omega^{-1}\ cm^{-1}$. This initial space charge current was found to be strongly dependent on the presence of small amounts of water. A current increase was observed of over three orders of magnitude in the acid upon doping with iodine. Current flow decreased, however, back to original values after removal of the iodine. Polydiacetylene bilayers were also found to exhibit photo conductive current increases of over an order of magnitude.

Subsequent to an investigation of a Josephson junction (cryogenic superconducting switching device) using a poly(vinyl stearate) monolayer as the thin insulating layer[22] a $C_{16}C_8$ - COOH polymerized monolayer was utilized.[23] Critical currents at zero voltage of approximately 240 amperes/cm^2 were observed at 4.2K using lead-indium alloy electrodes. These results are interesting in that the monolayer was 35Å thick. Since there is an exponential dropoff of I_c with insulator thickness the results are quite promising.

Recently a number of studies[24-28] have demonstrated that molecules with polar end groups with two hydrophobic chains each containing diacetylene groups form spherical vesicals upon sonication. After polymerization these vesicals are good barriers to the flow of dye molecules. Thus they could be used as a drug delivery system in a controlled release mode if the diffusion rate is modified by incorporation of unpolymerizable material.

CONCLUSION

The major conclusion of this review is that the investigation and control of mode of polymerization, structure and morphology is not only of scientific interest but will be of great importance in developing any future applications, such as those described in this paper.

ACKNOWLEDGEMENT

The financial support of this work by the National Science Foundation under Grant DMR 81-1141 is gratefully acknowledged.

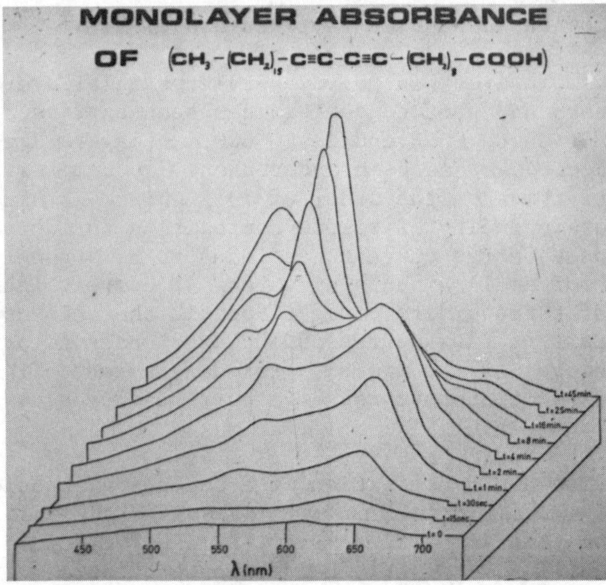

Fig. 1. Absorbance of a monolayer of $C_{16}C_8$ - COOH (1) vs. polymerization time (constant area: 23 $Å^2$/molecule, N_2 atmosphere) (2) vs. wavelength

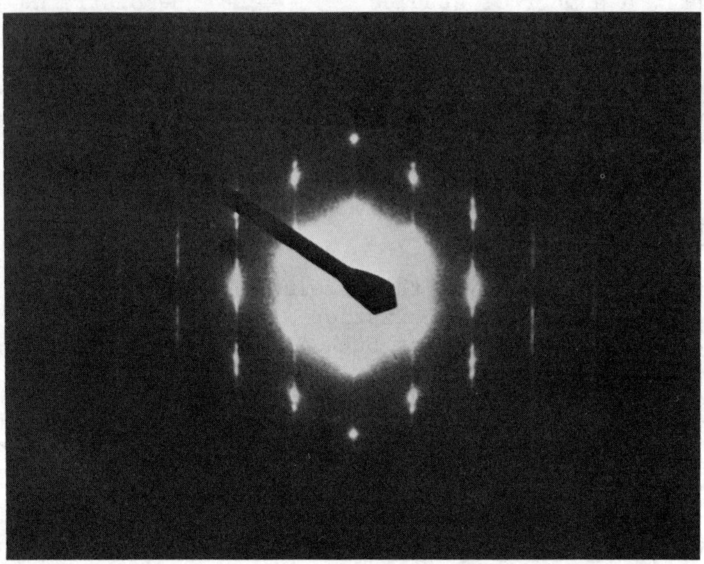

Fig. 2 Electron diffraction of accidently registered bilayer.

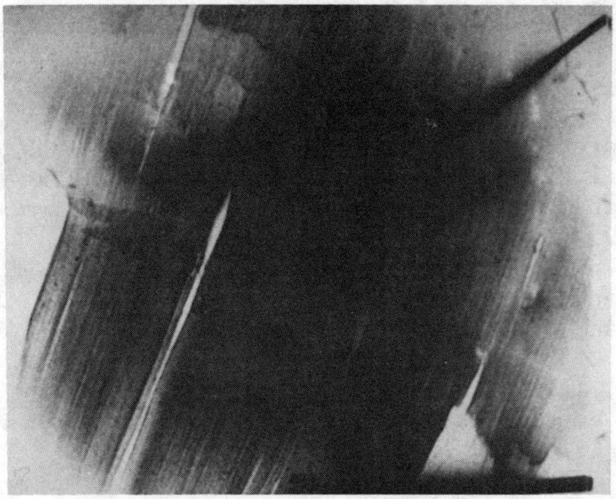

Fig. 3. Electron micrograph of a multilayer film of $C_{10}C_2$ COOH

1. B. Tieke, H.J. Graf, G. Wegner, B. Naegele, H. Ringsdorf, A. Banerjie, D.Day and J.B.Lando, Kolloid-Zeitschrift, 255, 521 (1977).
2. I. Langmuir, Trans. Faraday Soc., 15, 62 (1920).
3. K. B. Blodgett, J. Am. Chem. Soc., 57, 1007 (1935).
4. K.B. Blodgett and I. Langmuir, Phys. Rev., 51, 964 (1937).
5. D.R. Day and H. Ringsdorf, J. Polymer Sci., Polymer Letters Ed., 16 205 (1978).
6. D.R. Day and H. Ringsdorf, Makromol. Chem., 180, 1059 (1979).
7. D.R. Day and J.B. Lando, Macromolecules, 13 1478 (1980).
8. D.R. Day and J.B. Lando, Macromolecules 13, 1483 (1980).
9. B. Tieke, G. Lieser and G. Wegner, J. Polymer Sci., Polymer Chem. Ed., 17, 1631 (1979).
10. B. Tieke, V. Enkelmann, H. Kapp, G. Lieser and G. Wegner, J. Macromol. Sci-Chem., A15, 1045 (1981).
11. B. Tieke, G. Lieser and K. Weiss, Thin Solid Films, 99, 95 (1983).
12. M. Sarkar and J.B. Lando, Thin Solid Films, 99, 119 (1983).
13. M. Sarkar, M. Thakur and J.B. Lando, Macromolecules submitted.
14. D.R. Day and J.B. Lando, unpublished work.
15. D.R. Day and J.B. Lando, J or Polymer Sci., Polymer Physics Edition, 19, 165 (1981)
16. C.W. Pitt and L.N. Walpita, Thin Solid Films, 68, 101 (1980).
17. F. Grunfeld and C.W. Pitt, Thin Solid Films, 99, 249 (1983).
18. F. Kajzar and J. Messier, Thin Solid Films, 99, 117 (1983).
19. G.M. Carter, J.J. Chen and S.K. Tripathy, ACS Symposium Series, 283, 213 (1983).
20. G.M. Carter, Y.J. Chen and S.K. Tripathy, Appl. Phys. Lett., 43, 891 (1983).
21. D.R. Day and J.B. Lando, J. Appl. Polymer Science, 26, 1605 (1981).
22. G.L. Larkins, Jr., E.D. Thompson, E.Ortiz, C.W. Burkhart and J.B. Lando, Thin Solid Films, 99, 277 (1983).
23. J.B. Lando, C.W. Burkhart, F.L. Larkins, Jr. and E.D. Thompson, unpublished work.
24. H.H. Hub, B. Hupfer, H. Koch and H. Ringsdorf, J. Macromol Sci.,-Chem., A15, 701 (1981).
25. H. Koch and H. Ringsdorf, Makromol. Chem. 182, 255 (1981).
26. A. Akimoto, K. Dorn, L. Gras, H. Ringsdorf and H. Schupp, Agnew. Chem., Int. Ed. Engl., 20, 90 (1981).
27. J.H. Fendler, Acc. Chem. Res., 13, 7 (1980).
28. S.L. Regen, B. Czech and A. Sing, J.Am. Chem. Soc., 102, 6638 (1980).

DIACETYLENIC PHOSPHOLIPID POLYMERS.

David Samuel Johnston and Dennis Chapman

Department of Biochemistry and Chemistry, Royal Free Hospital School of Medicine (University of London) Rowland Hill Street, London NW3 2PF, U.K.

Phospholipids, phosphatidylcholines, have been synthesized which contain the diacetylene group in their acyl chains. These lipids may be dispersed in water or deposited in Langmuir-Blodgett multilayers and their polymerisation initiated with ultraviolet light. The polymers are optically active and thermochromic. A similar polymerisation can be initiated in the membranes of micro-organisms which have been grown in the presence of diacetylenic fatty acids.

A wide variety of polymerisable groups have been incorporated into amphiphatic compounds and it has been shown that these compounds will polymerise when they are dispersed in water. Many contain the photolabile diacetylene group, their synthesis prompted by the initial observation of Wegner,[1] that molecules with conjugated triple bonds form crystalline polymers upon irradiation with ultraviolet light. Our interest is in the lipid molecules which make up cell membranes. We have produced cell membrane lipids containing the diacetylene group by direct synthesis[2] and by dispersing diacetylenic fatty acids in the growth medium of micro-organisms that are unable to synthesize fatty acids.[3] These mutant cells utilise the added diacetylenic fatty acid to make membrane phospho- and glycolipids.

We had a number of objectives in mind when we set out to make these molecules:- biomembranes have potential therapeutic value. When dispersed in water the most ubiquitous lipid component of biomembranes, the diacyl phosphatidylcholines, form closed spherical structures known as liposomes. Liposomes have aroused considerable interest as containers with which to introduce drugs to the body.

However the instability of liposomes, which may be composed of a single bilayer of lipid molecules and their degradation in the body have so far restricted their application. Covalently linking the lipid molecules of the membrane might be expected to increase its stability and render it less susceptible to attack by enzymes. Phosphatidylcholines usually lie in the outer lamella of biomembranes where they confer an inertness on the cell exterior. For example phosphatidylcholines, in contrast to phosphatidylethanolamines, do not trigger blood clotting.[4] Using the Langmuir-Blodgett technique, uniform oriented layers of lipid can be transferred to solid substrates. In principle layers with the polar group outermost should mimic the surface of biomembranes and with appropriate lipids, phosphatidylcholines say, make the substrate biocompatible. However multilayers are fragile and subject to rearrangement. As in the case of liposomes there seemed to be a possibility that their stability could be improved by polymerisation. Finally we wished to investigate the effect that linking lipids had on the protein component of biomembranes. By linking the lipid molecules the fluidity of the membrane would be increased isothermally.

PROPERTIES OF DIACETYLENIC PHOSPHATIDYLCHOLINES

Two types of diacetylenic phosphatidylcholine have been synthesized, one with a diacetylene group in the acyl chain attached to the 2-position of glycerophosphorylcholine (mixed chain) and one with diacetylene groups in both acyl chains. (same chain) In each case the triple bonds occupy the 10 and 12 positions in the acyl chain although lipids with chains containing from 20 to 31 carbon atoms have been made. It was discovered that like phosphatidylcholines which have fully saturated chains the temperature (Tc) at which the gel to liquid-crystalline transition takes place increases as the length of the diacetylenic acyl chain increases.[2] However at equivalent chain lengths the diacetylenic phospholipids have much lower Tcs. This is presumably a result of the diacetylene group hindering chain packing. This is also reflected in pressure-area isotherms of monolayers of these lipids, the monolayers occupying greater areas under set conditions than monolayers of the equivalent phosphatidlycholine with fully saturated hydrocarbon chains.

POLYMERISATION

In common with other surfactants containing the diacetylene group, diacetylene containing phosphatidylcholines only polymerise in the solid state, in this case when they are cooled below Tc.[2] Polymerisation occurs either when the lipid is dispersed in water in the form of liposomes or compressed in KBr discs. That diacetylene containing phosphatidylcholines dispersed in water do form closed lamellar structures has been confirmed by electron microscopy and the observation that the spherical bilayers revealed by microscopy are osmotically active.[5] Upon irradiation diacetylenic phosphatidylcholines turn red, the intensity of the colour depending on the length of the acyl chain and the number of diacetylenic chains in the molecule. Lipids with only one diacetylenic acyl chain yield more intensely coloured polymers than lipids with two diacetylenic acyl chains. The longer the acyl chain, the more intense the colour.[5] However the amount of monomer remaining in each case is similar and it seems that differences in the intensity of colour are not caused by differences in polymer yield. It has been suggested that the rate of termination in mixed chain lipid polymerisations is lower, higher molecular weight polymers are formed and these higher molecular weight polymers have a greater extinction coefficient.[6] The size of liposomes influences conversion. The comments above apply to large multilamellar types. If dispersions are vigourously probe sonicated until only small unilamellar liposomes remain, no polymerisation occurs on irradiation.[2] Most diacetylene polymers are highly crystalline an insoluble in all solvents. However, with one exception, phospholipi polymers will dissolve in chloroform or chloroform-alcohol mixture: The exception is the polymer formed when same chain lipid is subjected to prolonged irradiation. These polymers are probably cross-linked. A polymer chain cross-linking reaction is only possible when the monomer contains more than one diacetylene group.[5]

SPECTRA

In solution polydiacetylene phospholipids are thermochromic. i.e. their spectra change with temperature. Their spectra also depend c solvent type. Thus in pure chloroform or at 60°C the visible absorption maximum lies at approximately 460nm. and it shifts to 520nm. when the solution is cooled to 0°C or chloroform is replac by methanol.[2,7] The shifts in absorption spectra are related to the flexibility of the polydiacetylene chain, a reduction in flexibility caused by a reduction in temperature or transfer to a poorer solvent causing a shift in the spectrum to longer waveleng The visible absorption band is optically active and the polymers have circular dichroic spectra. (G factor 10^{-3}) The sign and the strength of the dichroism varies considerably and changes in circular dichroism are related to the changes which temperature a solvent cause in the visible spectrum. The optical activity of th polymers has been explained in terms of the chiral 2-position of glycero-phosphorylcholine head group favouring a packing of one particular screw sense.[8] Some of the characteristics of both visible and CD spectra suggest that bands in the spectra of poly aggregates may arise from exciton (energy transfer) interactions between the chains.

POLYMERISED LIPOSOMES

Electron microscopy shows that polymerisation of diacetylene lipids which have been dispersed in water to form liposomes does not alter the structure of the liposomes. They also remain osmotically active. Polymerisation reduces the permeability of the liposomes to glycerol. This may be a consequence of triple bonds being converted into less polar double bonds. Polymerisation also reduces the propensity of liposomes to precipitate, the more so the lower the temperature. E.s.r. measurements with labelled fatty acid probes did not reveal any changes in the fluidity of the bilayer after its polymerisation. However this may be because polymer and remaining monomer separate into domains and the spin labels lie in the monomer domain.[5]

LANGMUIR-BLODGETT MULTILAYERS

Phosphatidylcholines with two diacetylenic acyl chains form stable monolayers. If these monolayers are compressed to a film pressure of 40 dynes-cm. multilayers can be built up on a variety of substrates.[9,10,11] We have developed two techniques for polymerising these layers and leaving the outer layer oriented so that the polar groups are exposed. In the first procedure the layers are polymerised under water. In the second the monolayer of diacetylenic lipid is replaced by a non-polymerisable one before the substrate is raised for the last time. After the diacetylene layers have been polymerised the single non-polymerised surface layer can be washed away leaving the ionic head groups of the glycerophosphorylcholine group exposed.

POLYMERISATION OF MEMBRANES CONTAINING PROTEINS

There are two ways in which membranes of diacetylenic lipids containing intrinsic membrane proteins can be obtained; either proteins extracted from natural membranes with detergent can be reconstituted into synthetic diacetylenic phosphatidylcholines or the growth medium of micro-organisms incapable of synthesizing their own fatty acids can be enriched with diacetylenic fatty acid. In this laboratory, Ca^{2+}-ATPase from sarcoplasmic reticulum and bacteriorhodopsin from the purple membrane of Halobacterium halobium have been reconstituted into diacetylenic phosphatidylcholines.[6] Provided the more reactive mixed-chain lipids are used polymerisation can be achieved before the protein is denatured by the UV irradiation. Both proteins remain active within polymeric bilayers.

Incorporation of diacetylenic fatty acids into biomembrane phospho- and glycolipids has been accomplished with the bacterium,

Acholeplasma laidlawii.[12] When A. laidlawii cells were grown in the presence of the C_{20} diacetylenic fatty acid up to 90% of the membrane acyl chains were derived from the medium. Similarities between the diacetylenic acid and oleic acid in regard to uptake, capacity to support growth and distribution among the lipid classes suggest that the diacetylenic lipids are arranged in the lipid bilayer in a manner similar to oleate containing lipids. When irradiated at $0^{o}C$, up to 97% of the lipids in the membrane polymerised. In intact cells the glycolipids appeared to polmerise more readily than the phospholipids although the physical properti of these polymers and those formed by the synthetic phosphatidylcholines were very similar. Polymerisation of the membrane was accompanied by a loss in activity of the intrinsic membrane protein, NADH oxidase. By contrast, the activity of an extrinsic membrane protein, ribonuclease, was unaffected.

1. Wegner, G. Topochemical Reactions of Monomers with Conjugated Triple Bonds. Makromol. Chem. 154 (1972) 35-46.
2. Johnston, D.S., S. Sanghera, M. Pons and D. Chapman. Phospholipid Polymers - Synthesis and Spectral Characteristics. Biochim. Biophys. Acta. 602 (1980) 57-69
3. Alonso, A., J. Leaver, D. Johnston, S. Sanghera, C. Villaverde and D. Chapman. Polymerisation of Diacetylenic Fatty Acids in Cultures of Bacillus cereus. Biochim. Biophys. Acta. 712 (1982) 292-298.
4. Zwaal, R.F.A. and H.C. Hemker. Blood Cell Membranes and Haemostasis. Haemostasis 11 (1982) 12-39.
5. Johnston, D.S., L.R. McLean, M.A. Whittam, A.D. Clark and D. Chapman. Spectra and Physical Properties of Liposomes and Monolayers of Polymerisable Phospholipids Containing Diacetylene Groups in One or Both Acyl Chains. Biochemistry 22 (1983) 3194-3202.
6. Hayward, J.A., J.R. Cooper, F. Castelli, D.C. Lee, D.S. Johnston, Q.F. Ahkong and D. Chapman. manuscript in preparation
7. Pons, M., D.S. Johnston and D. Chapman. A Study of the Spectra of Diacetylenic Phospholipid Polymers in Solvents and Dispersions. J. Polym. Sci., Polym. Sci. Ed. 20 (1982) 513-520
8. Pons, M., D.S. Johnston and D. Chapman. The Optical Activity and Circular Dichroic Spectra of Diacetylenic Phospholipid Polymers. Biochim. Biophys. Acta. 693 (1982) 461-465
9. Johnston, D.S., S. Sanghera, A. Manjon-Rubio and D. Chapman The Formation of Polymeric Model Membranes From Diacetylenic Fatty Acids and Phospholipids. Biochim. Biophys. Acta 602 (1980) 213-216.
10. Albrecht, O., D.S. Johnston, C. Villaverde and D. Chapman. Stable Biomembrane Surfaces formed by Phospholipid Polymers. Biochim. Biophys. Acta. 687 (1982) 165-169.
11. McLean, L.R., A.A. Durrani, M.A. Whittam, D.S. Johnston and D. Chapman. Preparation of Stable Polar Surfaces using Polymerisable Long-Chain Diacetylene Molecules. Thin Solid Films. 99 (1983) 127-131.
12. Leaver, J., A. Alonso, A.A. Durrani and D. Chapman. The Biosynthetic Incorporation of Diacetylenic Fatty Acids into the Biomembranes of Acholeplasma laidlawii A Cells and Polymerisation of the Biomembranes by Irradiation with Ultraviolet Light. Biochim. Biophys. Acta. 727 (1983) 327-335.

HOMOEPITAXY AND IN-PLANE ANISOTROPY IN LANGMUIR-BLODGETT FILMS

I R Peterson

GEC Research Laboratories
East Lane
Wembley UK

ABSTRACT

Several possible applications of Langmuir-Blodgett (LB) films, are ruled out at the moment by their poor crystallinity. Homoepitaxy and in-plane anisotropy are two phenomena which could be used to achieve better order, but while they have been observed on many occasions there is little agreement on the conditions required to reproduce them. This paper discusses the present state of the author's research. Homoepitaxy can now be achieved routinely with a number of materials, while in-plane anisotropy has been shown to produce significant ordering, although in a more more complicated manner than previously thought. Possible future directions for polydiacetylene LB films are discussed.

INTRODUCTION

It has been known for a long time that Langmuir-Blodgett (LB) films of the long-chain fatty acids and their salts have a regular layer structure (1,2,3,4) which is almost free of defects (5,6) and whose thickness can be controlled to high precision. These organic thin films can thus be compared to the superlattices of inorganic semiconductors produced by the molecular beam epitaxy (MBE) technique (7).

The great interest at present being shown in MBE can be ascribed to the control of composition and crystalline perfection which it permits, together with the excellent carrier mobilities in semiconductors amenable to the technique. The LB technique, too, can be used to fabricate semiconductors: Lieser, Tieke, Wegner,

Day and Lando (8,9,10) have shown that it is possible to fabricate LB films of polydiacetylenes (PDAs), and Donovan, Wilson, Spannring and Bässler (11,12,13) have shown that PDAs have unusual and outstanding charge transport properties. However, the conduction in PDA LB films so far reported (14) is disappointing, and this can be ascribed to their very poor crystalline order.

Charge transport is not the only option: there is growing interest in the possibilities for optical processing using organic materials (15,16). LB films of PDA and other materials have been proposed as passive (17,18) and active (19) elements in an integrated optical technology. Here again, excessive scattering due to poor crystallinity (20) has been a major obstacle to the implementation of these ideas.

A third area in which film crystallinity is proving to be important is that of insulating and tunnelling layers (21,22,23). Recent work at Lancaster University (24,25) has demonstrated that conduction through LB films is normally dominated by grain boundary effects. In all three areas, then, practical exploitation of the LB technique is waiting on substantial improvements in film crystallinity.

The present paper reports on two phenomena, homoepitaxy and in-plane anisotropy, which can be used to control the crystalline structure of an LB film. Both of these have a venerable history, and in fact were observed by Germer and Storks (26) in 1938, only four years after the LB technique was first reported (27). However, both phenomena have since been surrounded by controversy, with no general agreement on the conditions required to reproduce them.

HOMOEPITAXY

The phenomenon of LB film homoepitaxy has elsewhere been called interlayer registration. It has been observed with a number of materials (26,28,29), including diacetylenes (30,31), using the techniques of infrared spectroscopy (28), electron diffraction (29,30,31) and polarised microscopy (32), which provide the following information:

(a) Figure 1 is the TED pattern of 21 layers of 22-tricosenoic acid on a thin alumina support. The sharpness of the diffraction spots shows that all 21 layers are in register.

(b) Figure 2 shows a typical curve of the δCH_2 bending-mode absorption versus direction of polarisation. The constancy of shape of this curve with different numbers of layers dipped in different directions is evidence for homoepitaxy.

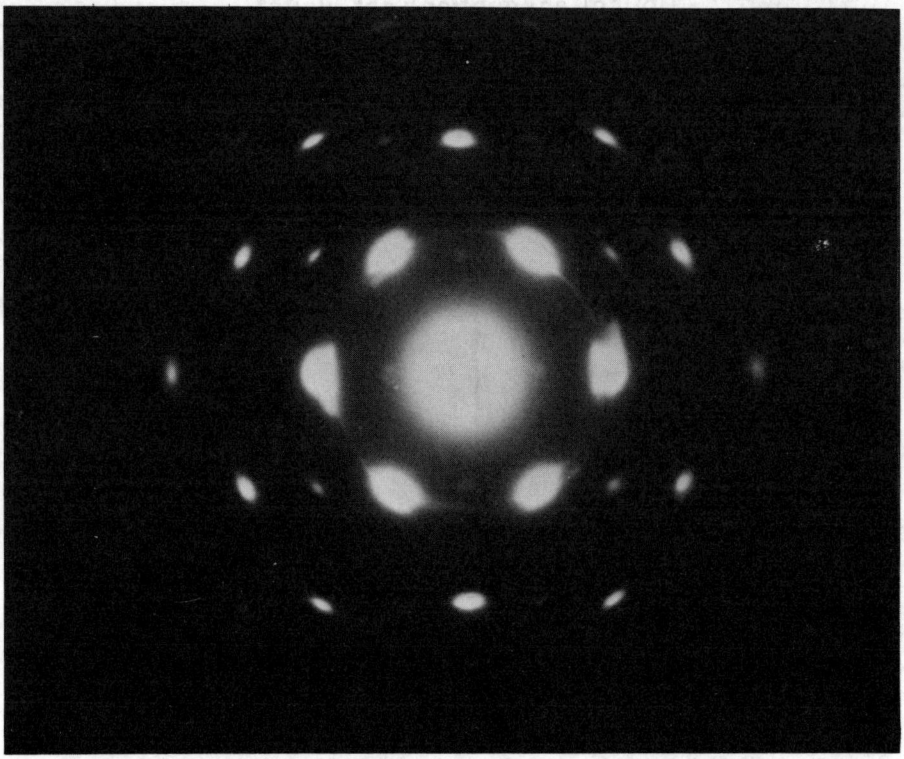

Figure 1: The TED pattern of 21 layers of 22-tricosenoic acid on a thin alumina support

(c) Figure 3 gives the reflection micrograph between crossed polarisers of a liquid crystal layer in contact with a 170-layer film of 22-tricosenoic acid on a hydrophobic silicon substrate partially covered with a stearic acid step structure. The change of crystallite size at the stearic acid boundary, and the visibility of the 2- to 4-layer stearic acid discontinuity, is evidence for epitaxial deposition.

However, all three techniques are similar in relying on order being present in the initial monolayer. The evidence for homoepitaxy is the preservation of this order with continued deposition. It follows that the sensitivity of each technique depends on the degree of order present, and equally on its spatial resolution, as a sufficiently small area of film approximates to a single crystal. In the latter respect, polarised microscopy is at least two orders of magnitude better than the other two.

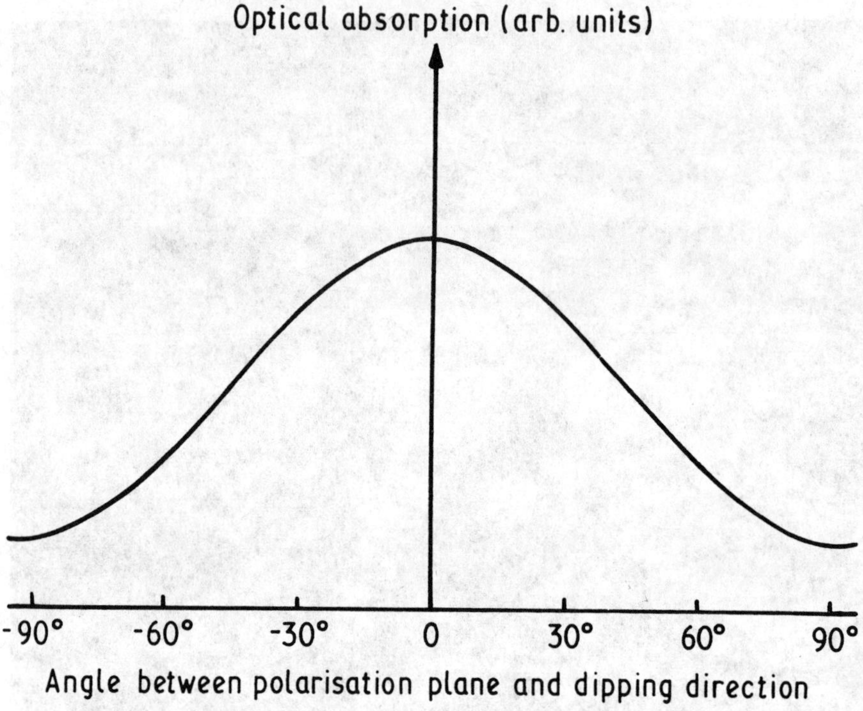

Figure 2: Typical curve of the δCH_2 bending-moment absorption versus plane of polarisation at normal incidence

Using polarised microscopy, Veale et al (33,34) have shown that monolayers of a number of saturated fatty acids with chain lengths of up to 23 are deposited epitaxially from a pure water subphase. Epitaxy is inhibited (perhaps only partially) with longer chain lengths and in the presence of cadmium.

When LB films are deposited epitaxially in this way, their diffraction patterns are sharp enough for unambiguous structural determinations. The only such study (35), and most others not using epitaxy (8,36,37,38) with a few exceptions (26,39) have identified structures fitting into Kitaigorodski's classification for the packings of aliphatic chain molecules. Many materials are polymorphic (8,26,35) displaying several of the allowed packings, depending on deposition conditions. This is reasonable in view of the preponderance of aliphatic chain in the molecular structures studied, and suggests that heteroepitaxy between LB monolayers of different materials should be a common phenomenon. Such heteroepitaxy has in fact been observed with fatty acids (34), and is apparent in Figure 3.

Figure 3: Reflection micrograph between crossed polarisers of the following structure: (1) silanised silicon substrate (2) step structure of 0,2 and 4 layers of stearic acid with increasing immersion (3) 170 layers of 22-tricosenoic acid and (4) several μm thickness of liquid crystal E7

Since the polymerisation of diacetylenes is topochemical, it is to be expected that the polymerisation kinetics will depend strongly on the molecular packing. Unfortunately, there is as yet insufficient data available to make any correlations.

IN-PLANE ANISOTROPY

Although homoepitaxy results in a film with increased crystalline order, it is not immediately helpful for any of the three applications discussed in the introduction. It would not be expected to reduce charge carrier scattering in PDA films, it is likely to increase optical scattering because the coherence length of crystallite boundaries now approaches the wavelength of light along the film normal as well as along the other two directions, and it produces films with grain boundaries aligned through the film, maximising the probability of metal penetration. Improvements in all these respects call for increased crystallinity of the initial monolayer.

A second phenomenon which can possibly be harnessed for this is in-plane anisotropy. As with homoepitaxy, the phenomenon has been observed in fatty acids (26,32,40) and diacetylenes (31), using the same three techniques, IR spectroscopy (29,40,41), electron diffraction (26,35) and polarised microscopy (32), plus a fourth, electron spin resonance (42).

Once again, polarised microscopy appears to be the best all-round technique because of its high resolution coupled with its imaging facility over a wide field of view. There are a number of variants, involving liquid crystal alignment (43,44), vacuum evaporation of a metal layer (45,46) and homoepitaxial deposition of a thick LB film (32). Of these, the last is least damaging to the monolayer, least subject to artefacts and least difficult to interpret, and has produced evidence for several types of anisotropic behaviour.

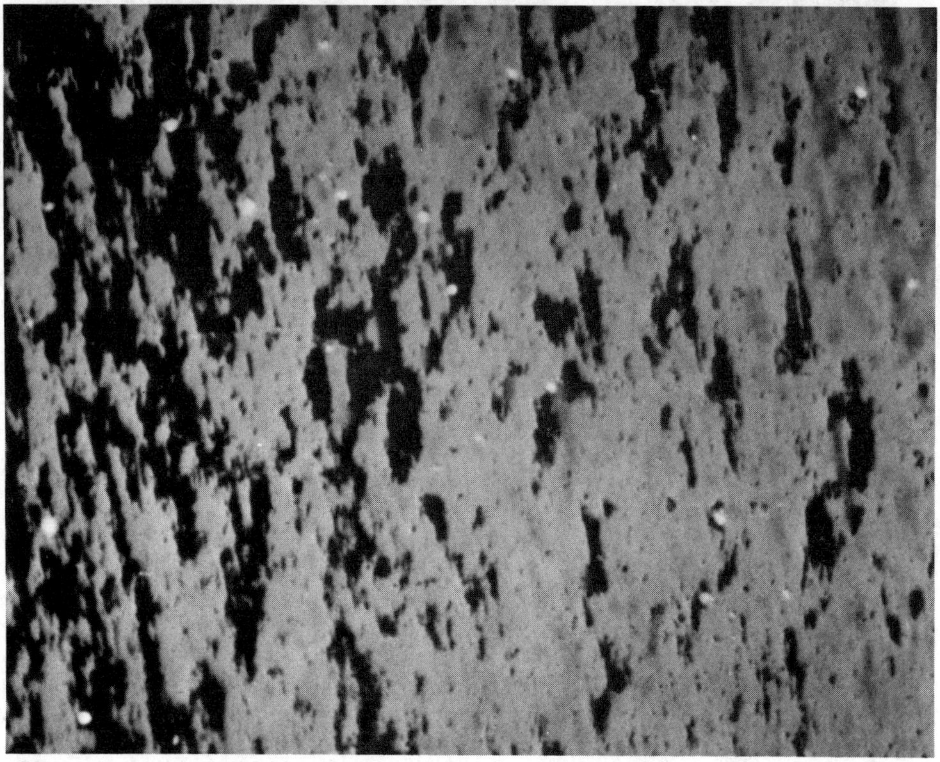

Figure 4: Reflection micrograph between crossed polarisers of 170 layers of 22-tricosenoic acid on silanised silicon. Polarisation direction at +27° to dipping direction

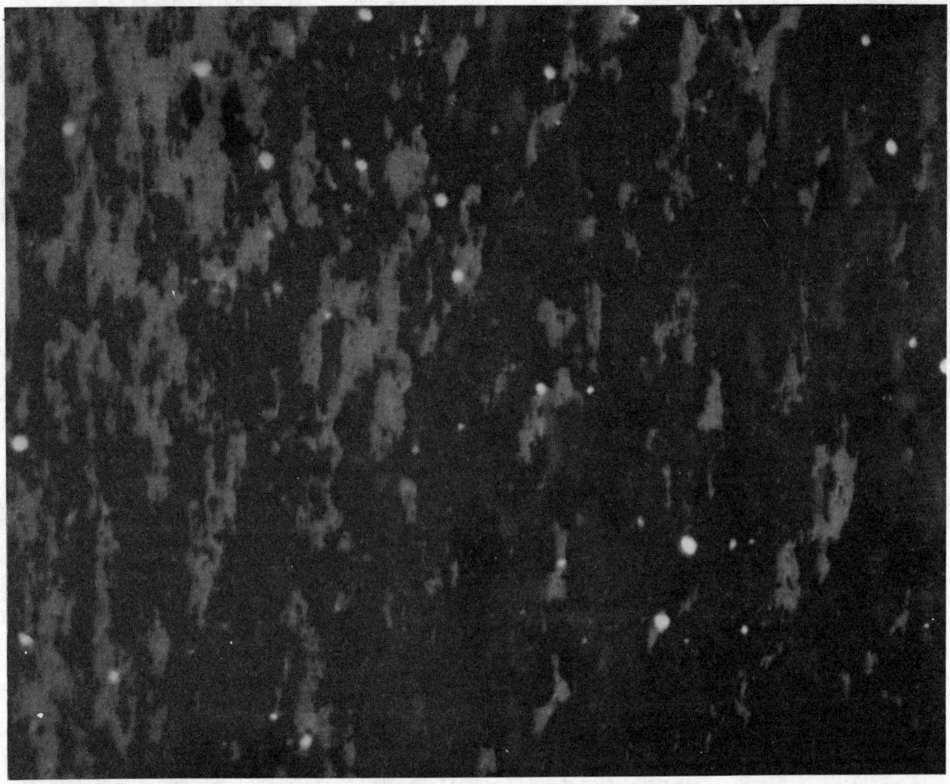

Figure 5: As for Figure 4, but with polarisation direction at -17° to dipping direction

A commonly observed type of anisotropy is illustrated in Figures 4 and 5, here encountered with 22-tricosenoic acid deposited on to a silanised silicon wafer. The two figures represent the same area of film, but with the polarisation direction at +27° and -27°, respectively, to the direction of dipping. Crystallites not extinguished in the first are extinguished in the second, and vice versa, indicating a very pronounced ordering of the film, although perhaps of limited technological use. As the probability distribution function (PDF) of crystallite orientation has two maxima, this type of anisotropy will be called bimodal.

A more interesting type of anisotropy, trimodal, is illustrated in Figure 6, here encountered with an aged monolayer of 22-tricosenoic acid deposited on to an evaporated aluminium layer. The three peaks of the PDF are at ±27° and 0° approximately. The low-deviation crystallites are seen to form a continuous matrix, with very good long range order.

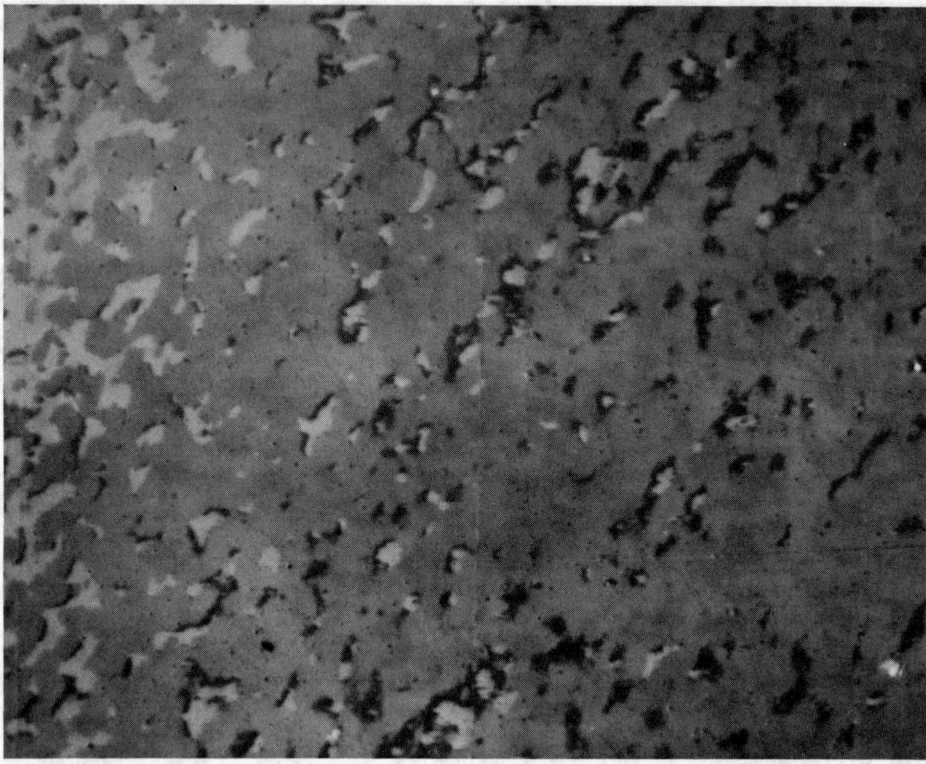

Figure 6: Reflection micrograph between crossed polarisers of 171 layers of 22-tricosenoic acid on vacuum-deposited aluminium. Polarisation direction at ~25° to dipping direction

A sample prepared under these conditions was responsible for the excellent diffraction pattern shown in Figure 1, which has been identified as due to the R(011) packing of the film molecules. It is reasonable to identify this with the low deviation phase. The other two crystallite types have not been unambiguously identified.

While clearly of major importance, further studies of trimodal anisotropy have been plagued by lack of reproducibility. It is thought that the essential conditions for the effect are related to extremely low levels of cations or other impurities, present in the subphase and the LB film material.

Of course, the effects of water turbulence and convection currents would be expected to cause deviations from perfect crystallinity even if the ultimate goal of monomodal deposition were realised. An alternative, capable in principle of perfect order, is to look for a lattice-matched inorganic substrate on to which the monolayer deposits epitaxially. Although this has been claimed (47), it has not been substantiated.

CONCLUSION

The above findings are directly relevant to the study of diacetylene films. Since homoepitaxy does not occur in the presence of multivalent metal ions, the emphasis should be shifted to materials which can be deposited on a pure water subphase - some have already been demonstrated. It is predicted that these materials will also display heteroepitaxy with fatty acid monolayers.

Much work remains to be done to improve the large-scale order of the initial monolayer, and this could conceivably be achieved either by discovering the conditions for reproducible monomodal anisotropic deposition or by finding a lattice-matched inorganic substrate. It will also be necessary to determine which of the possible packing schemes leads to the best polymerisation of diacetylenes, and how best to achieve it.

In spite of the uncertainty which still surrounds them, homoepitaxy and in-plane anisotropy seem likely to become increasingly important phenomena in LB film technology.

BIBLIOGRAPHY

1 Matsuda, A., M. Sugi, T. Fukui, S. Iizima, M. Miyahara and Y. Otsubo. J. Appl. Phys., 48, (1977) 771-774

2 Henke, B.L. Advan. X-ray Anal. 7 (1964) 460-488

3 Srivastava V.K. and A.R. Verma. Sol. St. Commun., 4 (1966) 367-371

4 Nicklow, R.M., M. Pomerantz and A. Segmüller. Phys. Rev. B. 23 (1981) 1081-7

5 Larkins Jr., G.L., E.D. Thompson, E. Ortiz, C.W. Burkhart and J.B. Lando. Thin Solid Films, 99 (1983) 277-282

6 Mann B. and H. Kuhn. J. Appl. Phys., 42 (1971) 4398-4405

7 Cho, A.Y. Thin Solid Films, 100 (1983), 291-317

8 Lieser, G., B. Tieke and G. Wegner. Thin Solid Films, 68 (1980) 77-90

9 Day D. and J.B. Lando. Macromolecules, 13 (1980) 1483-1487

10 Tieke B. and G. Lieser. J. Coll. Interface Sci., 88 (1982) 471-486

11 Donovan K.J. and E.G. Wilson. Phil. Mag. B., 44 (1981) 9-29

12 Wilson, E.G. Chem. Phys. Lett., 90 (1982) 221-224

13 Spannring W. and H. Bässler. Chem. Phys. Lett., 84 (1981) 54-58

14 Day D.R. and J.B. Lando. J. Appl. Polymer Science, 26 (1981) 1605

15 Williams, D.J. ed, "Nonlinear Optical Properties of Organic and Polymeric Materials", American Chemical Society Symposium Series 233 (1983)

16 Garito A.F. and K.D. Singer. Laser Focus, 80 (1982) 59

17 Pitt C.W. and L.M. Walpita. Electron Lett., 12 (1976) 479-481

18 Pitt C.W. and L.M. Walpita. Thin Solid Films, 68 (1980) 101-127

19 Carter, G.M., Y.J. Chen and S.K. Tripathy. Appl. Phys. Lett., 43 (1983) 891

20 Grunfeld F. and C.W. Pitt. Thin Solid Films, 99 (1983) 249-255

21 Winter C.S. and R.H. Tredgold. J. Phys. D., 17 (1984) L123-L126

22 Dharmadasa, I.M., G.G. Roberts and M.C. Petty. Electron Lett., 16 (1980) 201-202

23 Vincett P.S. and G.G. Roberts. Thin Solid Films, 68 (1980) 135-172

24 Tredgold R.H. and C.S. Winter. J. Phys. D., 14 (1981) L185-8

25 Tredgold, R.H., A.J. Vickers and R.A. Allen. J. Phys. D., 17 (1984) L5-L8

26 Germer L.H. and K.H. Storks. J. Chem. Phys., 6 (1938) 280-293

27 Blodgett, K.B. J. Am. Chem. Soc., 56 (1934) 495

28 Chollet P.-A. and J. Messier. Thin Solid Films, 99 (1983) 197-204

29 Peterson, I.R., G.J. Russell and G.G. Roberts. Thin Solid Films, 109 (1983) 371-378

30 Sarkar M. and J.B. Lando. Thin Solid Films, 99 (1983) 119-126

31 Lieser, G., B. Tieke and G. Wegner, Thin Solid Films, 68 (1980) 77-90

32 Peterson, I.R. Thin Solid Films, 116 (1984) 357-365

33 Veale G. and I.R. Peterson. J. Colloid & Interface Science, To be published

34 Veale, G., I.R. Girling and I.R. Peterson. Submitted to Thin Solid Films

35 Peterson I.R. and G.J. Russell. Phil. Mag. A., 49 (1984) 463-473

36 Rabolt, J.F., F.C. Burns, N.E. Schlotter and J.D. Swalen. J. Chem. Phys., 78 (1983) 946-952

37 Stephens J.F. and C. Tuck-Lee. J. Appl. Cryst. 2 (1969) 1-10

38 Russell, G.J., M.C. Petty, I.R. Peterson, G.G. Roberts, J.P. Lloyd and K.K. Kan. J. Mat. Sci. Lett., 3 (1984) 25-28

39 Day D.R. and J.B. Lando. Macromolecules, 13 (1980) 1483

40 Koyama, Y., M. Yanagishita, S. Toda and T. Matsuo. J. Colloid & Interface Sci, 61 (1977) 438-445

41 Takenaka T. and K. Nogami. J. Colloid & Interface Sci, 40 (1972) 409-416

42 Messier J. and G. Marc. J. Physique, (Paris), 32 (1971) 799-804

43 Staromlynska, J., F.C. Saunders and G.W. Smith. Mol. Cryst. Liq. Cryst., 109 (1984) 233-243

44 Hiltrop K. and H. Stegemeyer. Ber. Bunsenges Phys. Chem. 85 (1981) 582-588

45 Ackermann, J. Ph.D. Thesis, Georg-August-Universität Göttingen (1981)

46 Steele, S.C., M.N. Wybourne and D. Möbius. Thin Solid Films, 99 (1983) 117-118

47 Fariss, G., J. Lando and S. Rickert. Thin Solid Films, 99 (1983) 305-315

WORKING GROUP REPORTS

The current status of research on Polydiacetylenes and outstanding problems were considered by five working groups. The members of these groups were:

1) Polymerization Reaction

Professor H. Sixl (Chairman), Dr. H. Eckhardt, Dr. W. Guttler, Dr. N. Balcioglu, Dr. E. Prusik, Dr. Y. Yardley, Dr. M. Dudley, Dr. M. Goodyear, Dr. P.C. Morgan.

2) Electronic Properties of PDA Crystals

Professor H. Bassler (Chairman), Dr. D.N. Batchelder, Prof. G. Weiser, Dr. H. Kiess, Dr. I. Ledoux, Dr. J. Orenstein, Dr. K. Donovan, Dr. D. Sandman, Dr. S. Allen, Dr.J-P. Aime, Miss M. Drury, Dr. S. Mann.

3) Structural Problems

Professor M. Schott (Chairman), Dr. R.J. Young, Dr. R.M. Baughman, Dr. V. Enkelman, Dr. P.A. Albouy.

4) Theory

Professor R. Silbey (Chairman), Dr. A. Karpfen, Dr. E.G. Wilson, Dr. V. Bodart, Dr. S. Suhai, Dr. B. Movaghar, Professor. E. Mulazzi.

5) PDA Solutions and Films

Professor G. Wegner (Chairman), Professor F. Wudl, Dr. J. Messier, Professor J.B. Lando, Dr. I. Petersen, Dr. R. Priest, Dr. F. Kajzar, Dr. S. Tripathy, Professor D. Chapman, Dr. R. Sethi, Dr. R.R. Chance.

1) Polymerization Reaction

Under favourable packing conditions, diacetylene single crystals exposed to heat, radiation, or pressure react to form high quality polydiacetylene single crystals. Structural aspects of monomer reactivity are fairly well understood (at least at a qualitative level) via models based on least motion principles. However, there remain many unanswered questions regarding the detailed chemistry and physics of reaction initiation, propagation, and termination.

The overall activation energy for the diacetylene reaction is typically 1.0 eV. The major contributor to this energy is undoubtedly the initiation process, which is generally agreed to be the formation of a biradical dimer. There are no measurements to date which shed any light on the question of other contributors to E_a, namely activation barriers for propagation and termination. The temperature dependence of the photopolymerization process (E_a^p is generally observed to be the range 0.1 to 0.3 eV, but see below) could originate from several sources: the dynamics of the excited state reaction leading to the biradical dimer, chain propagation, and/or chain termination. A key experiment identified by the working group is the determination of the dependence of chain propagation length on reaction temperature in a photopolymerization experiment; 4BCMU is thought to be an ideal system here since the polymer is soluble and therefore amenable to conventional molecular weight measurements. Such an experiment would be just a first step in sorting out the energetics; additional experiments are needed specifically addressing questions related to chain propagation and termination. With regard to the latter, the only specific idea that has been put forward as a termination mechanism is a hydrogen insertion reaction to form cyclopropene structures at the chain end; experiments are needed to examine this possibility.

There is some disagreement in the literature regarding the temperature dependence of the photoreaction in PTS. There are published reports of both positive and negative activation energies. This problem needs to be resolved. In doing so, it should be kept in mind that the photoreaction is bimolecular at low temperatures (two excited states reacting to form the dimer) and unimolecular at high temperatures (one excited state plus one ground state); the intermediate temperature region should be probed.

In general, there is a relatively high level of understanding of the photopolymerization intermediates: short chain biradicals, bicarbenes, etc.. Most of these low temperature isolation experiments have been conducted on PTS; extension to other monomers is needed. A question has been raised as to whether the shortest oligomer observed spectral is a dimer (as currently assigned) or a trimer.

2) Electronic Properties of PDA Crystals

The linear optical properties (UV-visible and Raman) of PDA crystals have been thoroughly characterized and are reasonably well-understood. The lowest energy optical transition is typically located at about 2.0 eV and is excitonic in origin. Distortion of the backbone due to deliberately induced disorder or strain caused by side group interactions shifts this transition to higher energies. Crystal strain (e.g. polymer chains in the monomer lattice) can shift the transition either way. More work, theory and experiment, is needed to sort out and understand these effects.

The nonlinear optical properties, though of much interest, are not well-understood. It is generally agreed that PDA crystals have unusually large nonlinear indices. However, the role of two-photon absorption in these processes needs to be clarified. Work on PDA solutions suggests that a PDA crystal should have a two-photon state near in energy to the one photon state. If this is true, applications in non-linear devices are limited. For detailed nonlinear characterization of PDA crystals, higher quality materials are required. The role of photo-induced defects and the origin of observed photo-induced absorption requires additional study as well.

Though it is generally accepted that the quantum yield for fluorescence in PDA crystals is quite low, more work needs to be done to set limits on the yield and to determine whether or not observed low-level emissions are defect-related. At present, there is no good explanation of the low fluorescence yields. A related question is the location of the triplet state of a PDA crystal; the approach to this question should probably start with PDA solutions.

Electric field induced absorption is observed at about 2.4 eV in PDA crystals and has been assigned as a band-to-band transition which correlated well with photoconduction onsets. Could this forbidden transition also be correlated with the two photon state suggested by solution measurements?

There are many unanswered questions regarding photocarrier generation and transport in PDA crystals. It is generally agreed that carrier transport is dispersive at long times, i.e., at times long compared with that required for the carriers to encounter a trap. After pulse excitation photocurrents, and drift velocities, are observed to have dependences on applied field and time of $E^{1-\alpha}$

and $t^{-\alpha}$. Because of this a clear carrier transit cannot be observed but there is an urgent need for experiments to study separately electron and hole motion. The temperature dependence of transport is not properly understood but may result from a temperature dependence of the exponent α. The defects responsible for trapping have not been positively identified, though chain ends are an obvious possibility. Similarly the trap time has not been measured and studies in the picosecond time domain are required to do this. Short time measurements are also necessary to observe transport before trapping occurs. The question of the field dependence of the drift velocity before trapping should be addressed. The gross defects introduced by Donovan and Wilson (see p. 165) should be investigated further. Do they act as recombination centres or can the carriers be recovered by field reversal? These measurements and those using very short light pulses are essential in the verification, or rebuttal, of the very high initial mobility claimed for carriers in PDAs.

Theoretical work into both acoustic- and optic-polarons should be extended to provide models for interpretation of the experimental data.

Doping of PDA crystals with strong oxidizing or reducing agents has been carried out in a number of laboratories. The largest conductivities achieved thus far are about 0.001 S/cm compared to 1000 S/cm obtained with other doped conjugated polymers. Ion implantation can, however, be utilised to obtain higher conductivities but still only of the order 1 S/cm. More work is required to chose among several possible explanations for the much lower conductivities found for PDA crystals, e.g., reduced diffusion, reaction with sidegroups, etc..

3) Structural Problems

Structural studies are especially important with regard to monomer reactivity and have already contributed significantly to understanding in this area. Further work is encouraged along the following lines: 1) establish the role of side-group structure, as a "guide" in determining reactivity, 2) carry out detailed structural characterizations during evolution of monomer-polymer solid solution, and 3) determine the role and nature of defects which interrupt polymerization. The scope of these studies should extend beyond the well-studied PTS case.

The chromism observed in certain PDA systems (especially the urethane substituted polymers) requires structural study if the origin of these effects are to be established. Correlation of optical properties with direct determinations of backbone geometry and electron densities should yield a better understanding of chromic transitions. Possible origins are: fundamental change of state in PDA electronic structure, structural (and therefore electronic) changes induced by side group interactions, and electronic effects of side groups. Colour changes in disordered systems should be correlated to direct measurements of the disordered structures.

In the solution area, key questions are: 1) What constitutes a good microscopic description of the PDA chains in good solvents (yellow solutions) and in poor solvents (red or blue solutions)? 2) Are the red or blue solutions composed of aggregates? 3) What factors control the absorption lineshapes in the yellow solutions? 4) In PDA solutions which have no capability of hydrogen bonding, what causes the ordering of the chains to produce the red colour (e.g. side group crystalization)? 5) Is the role of hydrogen bonding in urethane substituted PDAs really established? These questions are discussed in more detail in the report of the Solution and Film Working Group.

Structural studies are also encouraged in the search for new materials in the PDA area - materials with new reaction chemistry, or materials with promise of new electronic phenomena.

4) Theory

Theoretical work in the PDA area has concentrated on the optical properties of PDA oligomers and chains. For the most part this work has been semi-empirical (e.g. Huckel theory) but has had semi-quantitative success in explaining experimental observations. Ab initio level calculations have firmly established that the acetylenic ground state is lowest in energy, but are generally too costly to extend to optical properties. Little attention has been paid to the development of a quantum mechanical description of the reaction process. An important step in this direction would be consideration of the various intermediates that have been identified: biradicals, bicarbenes, etc.. More attention also should be paid to the effects of defects and disorder on electronic properties, as well as to the calculation of force constants for the PDA chain. With regard to carrier transport, charged defect state calculations are relevant for intrachain and interchain motion of carriers. Finally, virtually all theoretical work in the PDA area has concentrated on single chain properties; consideration of the effect of the condensed phase is needed. In all areas, close coordination of theoretical work with experiment is encouraged.

5) PDA - SOLUTIONS, FILMS AND MONOLAYERS

Report of the Working Group prepared by G. Wegner

G. Wegner

Max-Planck-Institut fur Polymerforschung,
Jakob-Welder-Weg 11, D-6500 Mainz, FRG

1 SOLUTIONS OF PDA

Although the solubility of certain PDA and the surprising colour shifts taking place on dissolution and reprecipitation have been noted since the beginning of modern research on PDA chemistry, the subject has only recently developed into an area of research and controversy in its own right (1). The reasons for this are manyfold but among others it is realized that soluble PDA may serve as model objects for the general questions related to the conformational properties of polyconjugated macromolecules. In addition a number of interesting temperature and solvent dependent colour shifts have been observed in solutions of selected PDA, seemingly indicating the occurrence of intramolecular phase transitions. The nature of these transitions and consequences in terms of electronic and hydrodynamical properties are presently discussed controversially (1-3). Thirdly, a rather large number of soluble PDA has now become available, so that substituent effects such as H-bonding may be separated from the properties of the "pure" PDA backbone; in addition substituent effects as such can now be easily studied. It is convenient to discuss the problems of present and possible future research following an experimental classification due to the lack of theoretical understanding.

The origin of the problem is demonstrated by Fig. 1 showing typical UV-Vis-spectra observed for a solid single-crystal PDA "as polymerized", the same PDA dissolved to a dilute monodisperse solution and reprecipitated to a thin, transparent polycrystalline film (1). The two peaks observed for a partially polymerized single crystal of PTS-12 (1) with λ_{max} = 570 nm and 535 nm belong

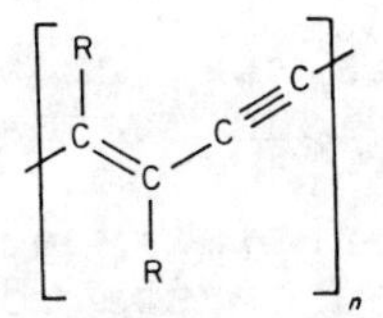

	1	$(CH_2)_4OSO_2C_6H_4CH_3$
R—	2	$(CH_2)_3OCONHCH_2CO_2$-n-C_4H_9

to the extended chains of the PDA molecule in two different modifications of the surrounding lattice. The peak at 570 nm will disappear at higher conversions due to a change in modification.

As a generalization it may be said that PDA in single cyrstal surrounding (extended chains) exhibit absorption spectra with maximum around 600 nm which shifts to approximately 450 nm on dissolution. Reprecipitation with concurrent recrystallization does not reconstitute the original spectrum but rather produces a spectrum with at least two, in some cases three peaks, one of which is always somewhat red-shifted from the peak of the dissolved PDA, the other somewhat blue-shifted from the single crystal spectrum. Only in exceptional cases do films of polymer cast from solution show spectra which are similar to that of the crystal, e.g. Poly-3BCMU (4).

a) Nature of the "yellow" solutions

The large spectral shift observed when dissolving PDA molecules has been treated by Chance and co-workers and others invoking the concept of effective conjugation length (4-6). The PDA molecule in solution is thought to consist of planar segments of the back-bone extending over several π-bonds which are electronically decoupled from adjacent segments by bond rotations. A distribution of segment lengths is invoked in order to treat the observed increase in half-width of the absorption peak of the dissolved vs. the single-crystal sample. This model is sketched in Fig. 2b. The same model is used to interpret the shift in Raman-frequencies generally observed comparing single-crystal and dissolved PDA samples in the region of double- and triple-bond vibrations (6). Finally, ^{13}C-nmr data concerning the signal of the sp-hybridized C-atom may be interpreted similarly - all data pointing to the existence of an "effective conjugation length" of 5-7 repeat units for the dissolved PDAs in their "yellow" solutions (1). It is important to note that such an interpretation relies strictly on the model shown by Fig. 2b in which adjacent planar segments are totally decoupled from each other electronically; it is thus assumed that rules dserived from the investigation of low molecular weight homologuous series of unsaturated oligomers (7,8) do apply to the present case as well. The chain conformation of the yellow solution of PTS-12 and other PDAs was investigated by light scattering and viscosity studies. It is generally agreed

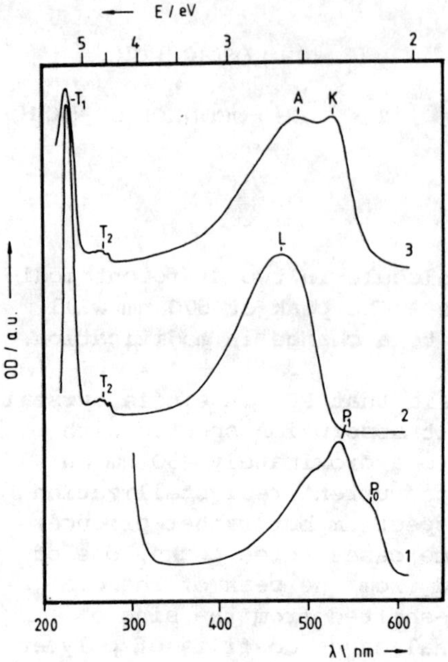

Fig. 1

UV-vis spectra of PTS-12 invarious states of aggregation:
(1) partially polymerized single crystal of thickness d=0.1 mm at 5 % conversion;
(2) solution of the polymer in 1,2-dichloroethane at a concentration $c=3x10^{-5}g\ cm^{-3}$;
(3) recrystallized film of thickness d $<10\ \mu$m.

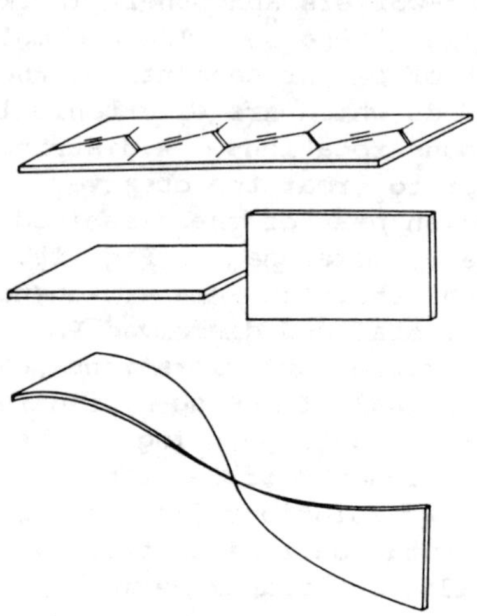

Fig. 2

Shape of the PDA chain:
(a) planar structure in the crystal;
(b) two segments in solution in terms of the Kuhn chain (note that adjacent segments are electronically decoupled due to bond rotation and that the bond connecting two segments has the character of a local defect);
(c) part of wormlike chain with continuous curvature of the chain skeleton.

that PDA molecules exhibit the conformation of a random coil in their yellow solutions, earlier claims of a rod like nature (9) were not substantiated. Wegner and co-workers (1,3) analyzed their data on light scattering and hydrodynamic properties of PTS-12 in dilute solutions in terms of a Porod-Kratky-chain (work-like chain) thus creating a model of a chain with continuous deformation of the backbone as depicted by Figure 2c. This model explains the observed solution properties very well and seems to be in conflict with the model used to interpret the spectral data. In terms of the worm-like chain the meaning of "effective conjugation length" is obviously quite different (and has yet to be found from the definition used to develop the "broken-plane" model shown in Fig. 2b).

b) Nature of the "blue" solutions and phase transition

Some PDAs show dramatic colour changes, if the solvent quality of the yellow solutions is decreased by either adding a non-solvent or by lowering the temperature below a critical threshold value. In the case of Poly-3BCMU (2) a change from yellow-orange to deep blue is observed which was first described by Patel et al. (10) and interpreted as an intramolecular broken-plane to rod transition. This idea was taken up by A.J. Heeger and co-workers (2) who added further experiments concerning the change in hydrodynamic behaviour as observed by dynamic light scattering in the course of the phase transition. A strictly one-dimensional and intramolecular treatment of the data was used to explain the features as a random coil-to-rod transition.

Wegner and co-workers questioned these results and on the basis of their light scattering results believe that the observed yellow to blue transition is intimately linked to the precipitation of the PDA (3). According to their data isolated rod-like molecules of Poly-3BCMU do not exist in solution. The spectral shift under debate is shown by Figure 3.

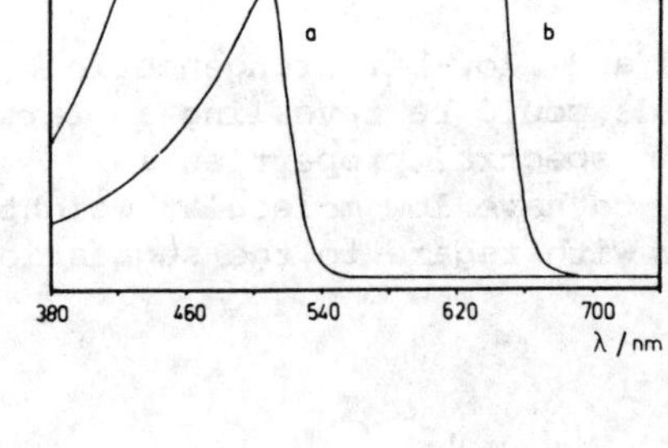

Fig. 3

UV-vis spectra of solutions of P3BCMU: (a) "yellow" solution in chloroform, (b) "blue" solution obtained after addition of n-hexane (6/4 $CHCl_3$/hexane).

A preliminary phase diagram with regard to this transition has been published (10), but it is agreed that this is not a phase diagram in terms of thermodynamics; in other words, it still needs to be demonstrated, if there exists a true equilibrium between the "yellow" and the "blue" state; in addition, the order of the phase transition is unclear since all effects are obscured by hysteresis. However, the phase diagram and subsequent kinetic measurements show that the solvent ratio required for the colour transition does not depend on polymer concentration; this observation argues for (but perhaps does not prove) a single-chain, intramolecular origin for the transition. (See Chance et al paper in these proceedings.) Such measurements do not, however, address the question of the equilibrium configuration of the chains in the blue phase, i.e., the question of individual chains versus aggregates.

c) Suggested work

It is strongly suggested to further investigate the solution properties of PDA both in their "yellow" and their "blue" state. The molecular weight dependence of the hydrodynamic properties should be able to differentiate between possible chain conformations (e.g. $\eta \sim M^{\alpha}$, $0.5 < \alpha < 1$ for random coils, $\alpha = 2$ for rods); polydispersity effects should, however, be properly taken into account. Similarly orientation in an electric field (Kerr-effect) could be used.

For rod-like molecules theory predicts the formation of liquid-crystalline textures which have not been found so far. The gelation of the "blue" Poly-3-BCMU observed under certain conditions is also of relevance and investigation of the nature of the gel could throw light onto the chain conformation as well.

It is strongly recommended that third parties take part in the measurements in order to avoid experimental errors. Furthermore, other experimental methods should be applied to resolve the present ambiguities. Such methods could be nmr- and dielectric-relaxation as well as time resolved fluorescence studies in order to study the dynamics of the local change in conformation or, in terms of the "broken-plane" model the nature and life-time of the defects linking two adjacent planar segments.

Measurements of spectral changes and flow-birefringence in capillary flow under high shear fields could be revealing as far as chain elongation and its effect on spectral properties is concerned. It would also be helpful to have low molecular weight model compounds available especially with regard to the suggestion that cis-trans isomerism may play a role in the observed solution properties (11)

Equally important is theoretical work, necessary to predict changes in the electronic structure of polyconjugated molecules on chain deformation (bending, folding, torsion).

Finally, it is worth mentioning that polymers such as PTS-12 (1) and 3-BCMU (2) can be melted without decomposition; the melts show liquid crystalline behaviour which disappears upon further increase in temperature. A sharp transition to the isotropic state has not been observed. It is recommended to do further research on this subject which could be relevant to the general area of polymeric thermotropic liquid crystals as well as to the further discussion of the temperature dependence of chain stiffness in PDA molecules.

2. MONOLAYERS, LANGMUIR-BLODGETT-(LB)-MULTILAYERS AND OTHER LAYERED STRUCTURES

Molecules of amphiphilic nature which contain the diacetylene moiety and which give rise to stable solid-analogue films at the air-water interphase can generally be polymerized by UV-irradiation. The polymerization proceeds according to the topochemical rules. A solid monomolecular film of the corresponding polydiacetylene is thus formed (12).

The solid-analogue film of the monomer at the air-water interphase may be transferred to suitable substrates (glass, metal, plastic) by the LB-technique; multilayer structures are built-up by consecutive transfer of individual layers. The multilayer structures are polymerized by UV or high-energy radiation with complete retention of the layer structure. Polydiacetylene multilayer films are thus formed (12-15).

If the diacetylene moiety is introduced into phospholipids or similar double-layered membrane and/or micelle/vesicle-forming molecular structures, these special forms of organisation may be polymerized as well providing a solid-like arrangement of the reactive groups is attained. Vesicles and micelles of polydiacetylenes have been prepared in a great variety following this principle (12,16).

The current interest in the polymerization of layered structures with molecular control of thickness and chemical structure is considerable. Among others the following reasons are given:

LB- and monomolecular films:

They are used to practice the handling of molecules for build-up of higher organized structures; methods to attain new properties or create molecular devices through special layer sequences with or with out inclusion of dye-molecules as antenna-, transmitter- and receiver-molecules are investigated. The intra-layer structure is optimized to either obtain a two-dimensional single-crystal-like character or an istropic distribution of very small domains depending on the desired use as, e.g., non-linear optic device or waveguide for integrated optics or as a micro-photoresist for the highest resolution of line elements. In addition, the use of LB layers or polydiacetylenes has been discussed as dielectric medium in microcapacitor tunnel-devices (17).

Vesicles, Liposomes, Micelles:

The polymerization serves to stabilize micellar structuraes permanently without affecting the other properties (transport, surface, interaction) in a severe manner. Consequently the properties of micelles, membranes and vesicles can be studied at length, once they have been stabilized (12). Models for membrane transport, cell adhesion, cell-cell-recognition, protein-lipid interaction have been envisaged and verified. Furthermore, enzymes or drugs can be immobilized inside the micelles of defined shape and surface properties thus creating biocompatible delivery systems. Research in this area is strongly encouraged both for practical and theoretical reasons. Several problems have to be solved in order to be able to fully control the build-up of layered structures and their properties. Among others the reproducible formation of single crystal-like textures in the LB-films is an unsolved problem. This is also valid for the attainment of orientation of chain axes within the layers with respect to a given frame of reference. Similarly, the domain size and structure of domain boundaries is unclear in polycrystalline LB-layers. Progress in the fabrication of optical devices is unlikely, unless these problems are solved. Another problem is related to the ageing of the monomer- and polymer layers due to loss of initially bound water in the head-head region of the layers. This has prevented exact structure determination of the packing of the hydrophilic head-groups in the LB-layers.

In the polymerized membranes, vesicles and micelles very little is known on the details of chain arrangement including chain-ends and defects. Thus, details of the transport properties cannot be explained without further work.

REFERENCES

1. Early work as well as the present status of experimental finding are reviewed in G. Wenz, M.A. Muller, M. Schmidt, G. Wegner. Macromolecules 17 (1984) 837.

2a. K.C. Lim, L.A. Fincher, A.J. Heeger, Phys.Rev.Lett. 50 (1983) 1934.

2b. K.C. Lim, M. Sinclair, S.A. Casalnuovo, C.R. Fincher, F. Wudl, A.J. Heeger, Mol.Cryst.Liq.Cryst. 105 (1984) 329.

3. M.A. Muller, M. Schmidt, G. Wegner. Makromol.Chem.Rapid Commun. 5 (1984) 83.

4. R.R. Chance, G.N. Patel and J.D. Witt, J.Chem.Phys. 71 (1979) 206.

5. R.R. Chance. Macromolecules 13 (1980) 396.

6. M.L. Shand, R.R. Chance, M. Le Postollec, M.J. Schott. J.Polym.Sci.Polym.Lett. 19 (1981) 529 and Phys.Rev.B 25 (1982) 4431.

7. H. Kuhn, Fortschr.Chem.Org.Naturst. 16 (1958) 169.

8. H. Kuzmany, Phys.Status Solidi B 97 (1980) 521.

9. G.N. Patel, E.K. Walsh, J.Polym.Sci.Polym.Letters Ed. 17 (1979) 203.

10. G.N. Patel, R.R. Chance and J.D. Witt. J.Polym.Sci.Polym. Letters Ed. 16 (1978) 607 and J.Chem.Phys. 70 (1979) 4387.

11. A.J. Berlinksy, F. Wudl, K.C. Lim, C.R. Fincher, A.J. Heeger, J.Polym.Sci.Polym.Phys.Ed. 28 (1984) 847.

12. L. Gros, H. Ringsdorf, H. Schupp. Angew.Chem. 93 (1981) 311. H. Koch, H. Ringsdorf. Macromol.Chem. 182 (1981) 255.

13. B. Tieke, G. Wegner, D. Naegele, H. Ringsdorf. Angew.Chem.Int.Ed.Engl. 15 (1976) 764.

14. B. Tieke, H.J. Graf, G. Wegner, B. Naegele, H. Ringsdorf, S. Banerjie, D. Day, J.B. Llando. Colloid Polym.Sci. 225 (1977) 521.

15. D.S. Johnston, S. Sanghera, A. Manjon-Rubio, D. Chapman. Biochimica Biophys.Acta 602 (1980) 213.

16. Many references will be found in the special volumes on LB-layers of Thin Solid Films 68 (1980) and 99 (1983) as well as in J.Mol.Cryst.Liq.Cryst. 106 (1984). (Special issue on the Symposium on Order in Polymers.)

SUBJECT INDEX

Schwind